SIMULATION
FOR APPLIED GRAPH THEORY
USING VISUAL C++

SIMULATION
FOR APPLIED GRAPH THEORY
USING VISUAL C++

SHAHARUDDIN SALLEH

ZURAIDA ABAL ABAS

CRC Press
Taylor & Francis Group
Boca Raton London New York

CRC Press is an imprint of the
Taylor & Francis Group, an **informa** business

A CHAPMAN & HALL BOOK

CRC Press
Taylor & Francis Group
6000 Broken Sound Parkway NW, Suite 300
Boca Raton, FL 33487-2742

© 2016 by Taylor & Francis Group, LLC
CRC Press is an imprint of Taylor & Francis Group, an Informa business

No claim to original U.S. Government works

Printed on acid-free paper
Version Date: 20160513

International Standard Book Number-13: 978-1-4987-2101-1 (Hardback)

Library of Congress Cataloging-in-Publication Data

Names: Salleh Shaharuddin, 1956- | Abas, Zuraida Abal.
Title: Simulation for applied graph theory using Visual C++ / Shaharuddin
Salleh and Zuraida Abal Abas.
Description: Boca Raton : Taylor & Francis, 2017. | "A CRC title." | Includes
bibliographical references and index.
Identifiers: LCCN 2016008211 | ISBN 9781498721011 (hard back)
Subjects: LCSH: Graph theory--Data processing. | Microsoft Visual C++
Classification: LCC QA166 .S24 2017 | DDC 511/.502855133--dc23
LC record available at http://lccn.loc.gov/2016008211

Visit the Taylor & Francis Web site at
http://www.taylorandfrancis.com

and the CRC Press Web site at
http://www.crcpress.com

Printed and bound in the United States of America by Publishers Graphics,
LLC on sustainably sourced paper.

To Ruby, Ila, Liya, and Azri.

To Mas Hariadi, Ahmad Aqhil, and Nuranna.

Thank you for all the support.

Contents

Preface

Simulation for Applied Graph Theory Using Visual C++ has been written to promote the use of Visual C++ in scientific computing. C++ is a beautiful language that has been responsible for shaping the modern world today. The language has contributed to many device drivers in electronic equipment and serves as the main engine in much of today's great software, and as a tool for research, teaching, and learning.

Graph theory is a diverse area of mathematics that deals with the abstraction of problems into mathematical structures called graphs that consist of objects and the pairwise interaction between the objects. A typical engineering problem, for example, involves thousands of interacting objects and finds its solution by first modeling its form into a graph. For example, there are many interacting variables to consider in designing the traffic flow of the streets in a big city, and reducing the problem into a graph simplifies the whole solution. Most real-world problems are nonlinear in nature and they have many interacting variables. One step in its solution is the reduction of the problem into a graph which helps in producing the solution by applying the properties of the graph.

Many problems involving graph theory are nondeterministic polynomial (NP) time-complete where the solutions have exponential complexities as the sizes of the problems grow. The solutions often involve tedious and massive calculations that may not be possible without the use of a computer. Simulations on a computer are essential and, therefore, good techniques in programming contribute to good solutions. C++ is an excellent choice to do the simulations due to its support for object-oriented programming and high-performance numeric support.

This book will help graduate and undergraduate students, who are interested, to work on simulation problems on applied graph theory. The book has been written with the main objective of teaching simulation using the C++ programming language and applying it to some common problems in graph theory. It is our aim to promote C++ as a language for numerical simulation and modeling. C++ has all the necessary ingredients for numerical computing due to its flexible language format, its object-oriented methodology, and its support for high numerical precisions. Due to stiff competition, in the past C++ popularity has suffered from the emergence of several new languages such as Java and C#. These new languages have been developed with the main objective to handle web and network programming requirements. However, due to its flexibility, C++ is still dominant and widely practiced.

This book will not duplicate other good books in the market today that mostly touch on the fundamental concepts of graph theory. Many such books have algorithmic approaches for the topic, while some even include C++ programming in its simulation approach. In this book, we integrate simulation with visualization through user-friendly interfaces using Visual C++ for some of the most common graph theoretical problems. Simulation is an important component of any research and it is performed either through a primary language like C++, C#, and Java, or through a secondary language like MATLAB® and Mathematica. A good applied researcher is somebody who is strong in both the development of new findings and simulation for verifying the findings. This book works on these requirements to help in the simulation for project development involving graph theory.

Graph theory and C++ are two separate and diverse areas. It will not be possible to discuss all topics in graph theory in a single book. We selected some of the most common: introductory concepts in Chapter 1, graph creation using C++ in Chapter 2, graph coloring in Chapter 3, shortest path problem in Chapter 4, minimum spanning tree in Chapter 5, maximum clique problem in Chapter 6, and graph triangulation in Chapter 7. To benefit researchers who are doing work on applied areas we also include some selected applications of graph theory in Chapters 8, 9, and 10. The chapters should provide the reader with good simulation skills and working examples for producing good simulation work on applied graph theory areas such as optimization and network design.

In completing the work on this book, the authors would like to thank several people whose involvement have directly or indirectly contributed in its preparation. We thank Prof. Datuk Ir. Dr. Wahid Omar, vice chancellor of Universiti Teknologi Malaysia and Prof. Datuk Dr. Shahrin Sahib, vice chancellor of Universiti Teknikal Malaysia Melaka for their strong policy of encouraging publication through book writing. We thank our colleagues at Universiti Teknologi Malaysia: Norsarahaida Mohd. Amin, Zainal Abdul Aziz, Rohanin Ahmad, Ali Selamat, Ali Hassan, Zaitul Marlizawati, Arifah Bahar, Norzieha Mustapha, and K. Visvanathan. Special thanks are due to our colleagues at Universiti Teknikal Malaysia Melaka who provided support and encouragement: Burairah Hussin, Zaheera Zainal Abidin, Abdul Samad Shibghatullah, Ahmad Fadzli Nizam Abdul Rahman and Mohamad Raziff Ramli. We also extend our thanks to our research friends, Prof. Dr. Albert Zomaya of University of Sydney, Australia, Prof. Dr. K.L. Teo of Curtin University, Australia, Prof. Dr. Stephan Olariu of Old Dominion University, the United States, and Dr. Ismail Khalil of Johannes Kepler University, Austria.

Additional material is available from the CRC website: http://www.crcpress.com/product/isbn/9781498721011.

Shaharuddin Salleh
ss@utm.my

Zuraida Abal Abas
zuraidaa@utem.edu.my

MATLAB® is a registered trademark of The MathWorks, Inc. For product information, please contact:

The MathWorks, Inc.
3 Apple Hill Drive
Natick, MA 01760-2098 USA
Tel: 508 647 7000
Fax: 508-647-7001
E-mail: info@mathworks.com
Web: www.mathworks.com

Authors

Shaharuddin Salleh is currently a professor of computational mathematics at the Center for Industrial and Applied Mathematics, and the Department of Mathematical Sciences, Universiti Teknologi Malaysia. He obtained his PhD at the same university, his MS at Portland State University, and BS at California State University Chico. Prof. Salleh is the author of six books, including two for Wiley-Interscience, Hoboken, New Jersey, in 2005 and 2008; and one for Kluwer Academic Publishers (now Springer) in 1999. His research interests are numerical computing, parallel computing, applied graph theory, and wireless sensor networks. Prof. Salleh has published over 150 technical papers in journals and conferences.

Zuraida Abal Abas is currently a senior lecturer at the Department of Industrial Computing, Faculty of Information and Communication Technology, Universiti Teknikal Malaysia Melaka. She obtained her PhD in mathematics at Universiti Teknologi Malaysia; her MSc in operational research at London School of Economics, United Kingdom; and her BSc in industrial mathematics at Universiti Teknologi Malaysia. Her research interests are operational research, applied graph theory, modeling, and simulation.

1

Graph Theory

1.1 Introductory Concepts

Graph theory is a branch of mathematics that studies the properties of graphs for applications to problems in everyday life, particularly in mathematics, science, and engineering. A graph provides a useful visualization of the problem that relates its components effectively. Some good discussions of graph theory and its applications can be found in [1–3].

One good example of the usefulness of graph theory is in scheduling the flights for an airline traveling to a set of cities in a way that will minimize the traveling costs involved. This is an optimization problem where the initial solution is obtained by reducing the problem to a graph. Here, the nodes are the cities where the planes travel while the edges are the traveling costs involved. The traveling costs between the cities vary according to factors such as time, duration, and other operational costs. In order to minimize the total traveling costs, the planes need to be scheduled according to priority order by considering the costs involved. A reduction of this priority order into the form of a graph is the first step.

A *graph*, denoted as $G(V, E)$ or simply G, is a mathematical structure consisting of two components: the set $V = \{v_i\}$ of n *nodes* or *vertices* for $i = 1, 2, ..., n$, and the set $E = \{e_k\}$ of m *edges* or *arcs* connecting the nodes, for $k = 1, 2, ..., m$. The graph is finite if m and n are fixed numbers. The *cardinality of the nodes* $n = |V|$ is the number of nodes of the graph, also called the *order* of the graph. The *cardinality of the edges* $m = |E|$ is its number of edges, also called the *size* of the graph. In a graph, the nodes and edges may represent entities such as cities and roads linking the cities, respectively, in a geographical region.

Figure 1.1 shows a graph $G(V, E)$ on the left with $n = |V| = 8$ and $m = |E| = 12$ in its generic form (Figure 1.1a), and the same graph with labeled nodes (Figure 1.1b). The nodes are labeled as v_i for $i = 1, 2, ..., n$ while e_k represents the edges for $k = 1, 2, ..., m$ as shown in Figure 1.1b. The label for edge e_k is more commonly represented as e_{ij}, which denotes the edge between v_i and v_j. An edge with an arrow showing its direction is called a *directed edge*. A graph with directed edges is called a *directed graph*.

An edge between two nodes in the graph denotes the relationship between them. If an edge between two nodes exists, then the two nodes are said to be *adjacent*. Otherwise, the two nodes are not adjacent. An edge between a node with itself is shown as a loop in the graph. Otherwise, if no loop exists, then the node is said to have no edge with itself. Multiple edges may exist between two nodes to illustrate multiple relationships between them. For example, three edges between two nodes may represent three different channels for communication between the two nodes. A graph is called a *simple graph* if it has no loops or multiple edges.

Figure 1.2 shows two graphs with multiple edges; one is undirected (Figure 1.2a) while the other is directed (Figure 1.2b). Multiple edges may, for example, denote the availability

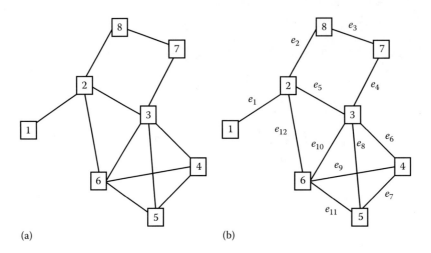

FIGURE 1.1
Graph $G(V, E)$. (a) Unweighted graph and (b) weighted with the weights on the edges as shown.

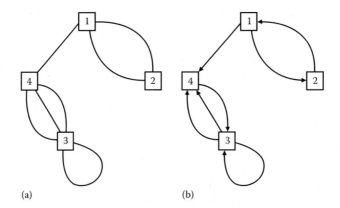

FIGURE 1.2
Two graphs with multiple edges. Two graphs with multiple edges in (a) undirected and (b) directed edges.

of several channels for communication between the two edges. In Figure 1.2a, a loop exists on v_3 while the pair (v_1, v_2) has two edges and (v_3, v_4) has three edges. This implies (v_1, v_2) can communicate with each other using two edges while (v_3, v_4) communicate in three ways, each with no specific directions mentioned. The directed edges in Figure 1.2b denote the partial orders or flows between the nodes. For example, the single directed edge from v_1 to v_4 suggests the unidirectional flow between the two edges. A bidirectional flow exists between v_1 and v_2, while v_3 and v_4 can communicate through three different edges for a tridirectional flow.

The adjacency relationship between pairs of nodes in a graph gives rise to the adjacency matrix. An *adjacency matrix* $A = [a_{ij}]$ for $i, j = 1, 2, ..., n$ is defined as a square 0-1 matrix of size $n \times n$ consisting of binary values of 0 and 1, where

$$a_{ij} = \begin{cases} 1 & \text{if nodes } v_i \text{ and } v_j \text{ are adjacent} \\ 0 & \text{otherwise} \end{cases} \qquad (1.1)$$

The adjacency matrix is related to the *incidence matrix*, which is defined as matrix $B = [b_{ij}]$ of size $m \times n$ for $i = 1, 2, \ldots, n$ and $j = 1, 2, \ldots, m$, where

$$b_{ij} = \begin{cases} 1 & \text{if edge } e_j \text{ is incident on } v_i \\ 0 & \text{otherwise} \end{cases} \tag{1.2}$$

Referring to the graph in Figure 1.1, the incidence matrix is constructed with the rows as the node numbers and the columns as the edges. The adjacency and incidence matrices for the graph in Figure 1.1 are

$$A = \begin{bmatrix} 0 & 1 & 0 & 0 & 0 & 0 & 0 & 0 \\ 1 & 0 & 1 & 0 & 0 & 1 & 0 & 1 \\ 0 & 1 & 0 & 1 & 1 & 1 & 1 & 0 \\ 0 & 0 & 1 & 0 & 1 & 1 & 0 & 0 \\ 0 & 0 & 1 & 1 & 0 & 1 & 0 & 0 \\ 0 & 1 & 1 & 1 & 1 & 0 & 0 & 0 \\ 0 & 0 & 1 & 0 & 0 & 0 & 0 & 1 \\ 0 & 1 & 0 & 0 & 0 & 0 & 1 & 0 \end{bmatrix}$$

$$B = \begin{bmatrix} 1 & 0 & 0 & 0 & 0 & 0 & 0 & 0 & 0 & 0 & 0 & 0 \\ 1 & 1 & 0 & 0 & 1 & 0 & 0 & 0 & 0 & 0 & 0 & 1 \\ 0 & 0 & 0 & 1 & 1 & 1 & 0 & 1 & 1 & 1 & 0 & 0 \\ 0 & 0 & 0 & 0 & 0 & 1 & 1 & 1 & 0 & 0 & 0 & 0 \\ 0 & 0 & 0 & 0 & 0 & 0 & 1 & 1 & 1 & 0 & 1 & 0 \\ 0 & 0 & 0 & 0 & 0 & 0 & 0 & 0 & 1 & 1 & 1 & 1 \\ 0 & 0 & 1 & 1 & 0 & 0 & 0 & 0 & 0 & 0 & 0 & 0 \\ 0 & 1 & 1 & 0 & 0 & 0 & 0 & 0 & 0 & 0 & 0 & 0 \end{bmatrix}$$

The *degree* or *valency* of a node is the number of edges incident on the node. A node with degree 0 is an *isolated node* because it is not adjacent to any node. A node with degree 1 is a *leaf*, while the edge incident on a leaf is called a *pendant*. A node is called the *dominating node* of the graph if its degree is $n - 1$, which is the maximum degree in the graph. The degree sequence of a graph is the nonincreasing sequence of the nodes. For example, the degree sequence of the graph in Figure 1.1 is {5, 4, 4, 3, 3, 3, 2, 1}, with v_3 having a maximum degree of 5 and v_1 having a minimum degree of 1. It can be seen that if E is finite, the total sum of the degrees is twice the number of edges, or

$$\sum_{i=1}^{n} \deg(v_i) = 2|E| \tag{1.3}$$

Associated with the degree is the degree matrix $D = [d_{ij}]$, which is a diagonal matrix with

$$d_{ii} = \sum_{j=1}^{n} a_{ij} \qquad\qquad (1.4)$$

From Figure 1.1, the degree matrix of the graph is

$$D = \begin{bmatrix} 1 & 0 & 0 & 0 & 0 & 0 & 0 & 0 \\ 0 & 4 & 0 & 0 & 0 & 0 & 0 & 0 \\ 0 & 0 & 5 & 0 & 0 & 0 & 0 & 0 \\ 0 & 0 & 0 & 3 & 0 & 0 & 0 & 0 \\ 0 & 0 & 0 & 0 & 3 & 0 & 0 & 0 \\ 0 & 0 & 0 & 0 & 0 & 4 & 0 & 0 \\ 0 & 0 & 0 & 0 & 0 & 0 & 2 & 0 \\ 0 & 0 & 0 & 0 & 0 & 0 & 0 & 2 \end{bmatrix}.$$

An edge between two nodes can involve a measuring cost called *weight* that may represent quantities such as distance, time, force, and monetary value. A graph where the edges have weights is called a *weighted graph*. Otherwise, the graph is an *unweighted graph*. Figure 1.3a shows a weighted graph with weights shown on the edges. The graph may represent a transportation network consisting of cities (nodes) and roads (edges) linking the cities. In this case, the weights on the edges are the traveling time in hours between the cities. Figure 1.3b is a directed and weighted graph. In this graph, the unidirectional weight w_{uv} applies to the directed edge from nodes u to v, not from v to u.

For a weighted graph, the adjacency matrix $A = [a_{ij}]$ is a symmetric matrix with elements $a_{ij} = w_{ij}$ representing the weight between v_i and v_j. The degree matrix is a diagonal matrix

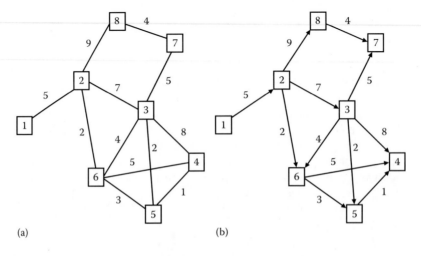

(a)   (b)

FIGURE 1.3
Weighted graphs, with (a) undirected and (b) directed.

$D = [d_{ij}]$ computed in the same manner as $d_{ii} = \sum_{j=1}^{n} w_{ij}$. Referring to Figure 1.3, the adjacency and degree matrices of the graph are

$$A = \begin{bmatrix} 0 & 5 & 0 & 0 & 0 & 0 & 0 & 0 \\ 5 & 0 & 7 & 0 & 0 & 2 & 0 & 9 \\ 0 & 7 & 0 & 8 & 2 & 4 & 5 & 0 \\ 0 & 0 & 8 & 0 & 1 & 5 & 0 & 0 \\ 0 & 0 & 2 & 1 & 0 & 3 & 0 & 0 \\ 0 & 2 & 4 & 5 & 3 & 0 & 0 & 0 \\ 0 & 0 & 5 & 0 & 0 & 0 & 0 & 4 \\ 0 & 9 & 0 & 0 & 0 & 0 & 4 & 0 \end{bmatrix} D = \begin{bmatrix} 5 & 0 & 0 & 0 & 0 & 0 & 0 & 0 \\ 0 & 23 & 0 & 0 & 0 & 0 & 0 & 0 \\ 0 & 0 & 26 & 0 & 0 & 0 & 0 & 0 \\ 0 & 0 & 0 & 14 & 0 & 0 & 0 & 0 \\ 0 & 0 & 0 & 0 & 6 & 0 & 0 & 0 \\ 0 & 0 & 0 & 0 & 0 & 14 & 0 & 0 \\ 0 & 0 & 0 & 0 & 0 & 0 & 9 & 0 \\ 0 & 0 & 0 & 0 & 0 & 0 & 0 & 13 \end{bmatrix}$$

In a directed graph, an edge has its tail in one node and head on another node to indicate the order of flow from the tail node to the head. For example, a directed edge from nodes u to v may be interpreted as the process must go through u first before v. In an undirected edge the flow can be in either way, from u to v and v to u.

A directed edge can also carry a weight, and the graph with weights on its directed edges is a directed and weighted graph. The *in-degree* of a node in a directed graph refers to the number of heads of edges incident on the node, while the *out-degree* is the number of tail edges originating from the nodes. From Figure 1.3b, v_3 has a degree of 5 with an in-degree and out-degree of 1 and 4, respectively.

The graph $G' = \{V', E'\}$ is a *subgraph* of G or $G' \subseteq G$ if $V' \subseteq V$ and $E' \subseteq E$. It can be seen from Figure 1.1 that $G' = \{V', E'\}$ for $V' = \{v_2, v_3, v_8\}$ and $E' = \{e_2, e_5\}$ is a subgraph of the graph G. The *complement* $\bar{G}$ of the graph G is a graph having the same nodes as G but the edges are opposite to that in G. The edges $\bar{G}$ are constructed from nonadjacent pairs in G, while pairs with edges in G are eliminated in $\bar{G}$.

1.1.1 Paths in a Graph

The *distance* $d(u, v)$ between two nodes u and v of a graph is defined as the number of edges or the number of hops in the shortest path between the two nodes. Obviously, a pair of adjacent nodes have a distance of 1. The *eccentricity* of a node v or $ecc(v)$ in graph G is defined as the maximum distance from the node to any other node in the graph. The *radius* of G or $r(G)$ is the minimum eccentricity of G while *diameter* or $d(G)$ is the maximum eccentricity of G. The *center* of G is the set of nodes where $ecc(G) = R(G)$. The center of a graph always lies in the middle position of the graph as it minimizes the maximum distance to any other nodes. It can be seen from Figure 1.1 that $D(v_1, v_5) = 3$, $ecc(v_4) = 3$, $r(G) = 1$, and $d(G) = 3$.

A *walk* is an alternating sequence of nodes and edges from the source to its destination. A *trail* in G is a walk from a node to another node. A *path* is a trail where no internal node is repeated. From Figure 1.1a, there are many paths from v_1 to v_4 and they include

$$v_1 \rightarrow v_2 \rightarrow v_3 \rightarrow v_4$$

$$v_1 \rightarrow v_2 \rightarrow v_8 \rightarrow v_7 \rightarrow v_3 \rightarrow v_4$$

A *cycle* is a path that starts and ends in the same node; an example is $v_2 \rightarrow v_3 \rightarrow v_6 \rightarrow v_2$ for the undirected graph in Figure 1.1. The directed graph in Figure 1.3b has no cycle as none of the nodes have directed edges that start and end in the same node. A graph that has at least one cycle is called a *cyclic graph*. If the graph does not have a cycle, the graph is an *acyclic graph*. It can be seen that the graph in Figure 1.3b is acyclic because no cycle exists in it. It is important to note that in an acyclic graph any path between a pair of nodes is unique, whereas in a cyclic graph more than one path may exist between a pair of nodes.

A *circuit* is a cycle that allows the repetitions of nodes but not edges. In Figure 1.1, there are many circuits. An example is

$$v_2 \rightarrow v_3 \rightarrow v_4 \rightarrow v_6 \rightarrow v_3 \rightarrow v_7 \rightarrow v_8 \rightarrow v_2$$

A Hamiltonian path of a graph consists of a path that visits every node in the graph exactly once. An example of the Hamiltonian path of the graph in Figure 1.1 is

$$v_1 \rightarrow v_2 \rightarrow v_8 \rightarrow v_7 \rightarrow v_3 \rightarrow v_4 \rightarrow v_5 \rightarrow v_6$$

The end points of the Hamiltonian path above are v_1 and v_6. If the two nodes are adjacent, then the Hamiltonian path becomes a *Hamiltonian cycle*. Obviously, the graph in Figure 1.1 does not have any Hamiltonian cycle.

Related to the Hamiltonian path is the *Eulerian path*, which is a path that visits every edge in the graph exactly once. It follows that the *Eulerian cycle* is a Eulerian path that starts and ends in the same node. It can be seen from Figure 1.1 that a Eulerian path is not possible in the graph.

A graph can be connected or disconnected. A *connected graph* has paths from a node to any other node in the graph. If a path between a pair of nodes does not exist, then the graph is a *disconnected graph*. The number of components of a disconnected graph is the number of disconnected subgraphs where each component is a connected subgraph of G.

In a connected graph, a *bridge* is an edge where if it is removed it will cause the graph to be disconnected. In a similar manner, an *articulation point* is a node in the graph where its removal will cause the graph to be disconnected. From Figure 1.1b, e_1 is the only bridge and v_2 is the only articulation point in G.

A *crossing* is a point in the graph other than nodes where two edges meet. A graph without any crossings is said to be a *planar graph*. It can be seen that the graph G in Figure 1.1 is not a planar graph as there is one crossing between e_8 and e_9. Finding the planar graph contributes to problems such as routing in printed circuit boards and transportation network design. For example, in designing a network of roads linking places in a city, certain parts of the network require reduction into one or more planar subgraphs. This is necessary in order to reduce the overall cost because any crossing requires an expensive structure, such as a bridge, to be constructed, which incurs additional costs.

1.1.2 Subgraph

The subgraph of G where all the nodes are adjacent to each other is called a *clique*. In Figure 1.1, several cliques exist, including

$\{v_7, v_8\}$

$\{v_2, v_3, v_6\}$

$\{v_3, v_4, v_5, v_6\}$

The maximum clique of the graph is the subgraph $G'(V', E')$ with the most number of nodes. The graph in Figure 1.1 has the maximum clique $G'(V', E')$, where $V' = \{v_3, v_4, v_5, v_6\}$ and $E' = \{e_6, e_7, e_8, e_9, e_{10}, e_{11}\}$.

A *complete graph* is a graph where all the nodes in the graph are adjacent to each other. A complete graph with n nodes is denoted as K_n and it has $\frac{n(n-1)}{2}$ nodes. A complete digraph is a digraph where every pair of nodes has a pair of edges, one going from the first node to the second and another from the second to the first. It can be seen that a complete graph has $(n-1)!$ Hamiltonian circuits in it.

An *independent set of a graph* is a set of nodes where none of the nodes are adjacent to themselves. The maximum independent set of G is the independent set with the maximum number of nodes in the graph. There are seven maximum independent sets of G in Figure 1.1; among them are

$\{v_1, v_6, v_8\}$

$\{v_1, v_5, v_7\}$

$\{v_1, v_3, v_8\}$

Related to a maximum independent set is the k-partite graph. A graph is said to be *k-partite* if the nodes can be partitioned into k sets in such a way that none of the nodes in the same sets are adjacent to themselves. A two-partite graph is also called a bipartite graph while a three-partite graph is tripartite. Figure 1.4a shows a bipartite graph with $\{(v_1, v_2), (v_3, v_4, v_5)\}$ while Figure 1.4b shows a tripartite graph with $\{(v_1, v_2), (v_3, v_6), (v_4, v_5)\}$.

A *regular graph* is a specialized graph where all the nodes have the same degree. A k-regular graph is a regular graph of degree k. Figure 1.5 shows two-regular (a), two-regular directed (b), and three-regular (c) graphs. A complete graph with n nodes is a good example of a regular graph with degree of $n-1$. A regular directed graph has the same number of in-degrees and out-degrees in every node.

1.1.3 Tree

A *tree* is a connected graph that has no cycle (acyclic). A set of trees form a *forest*. A tree is characterized as a graph where every path connecting any two pairs of nodes in it is unique. It is also easy to verify that every tree with n nodes has exactly $n-1$ nodes. The nodes with degrees greater than 1 in a tree are *internal nodes*. The nodes with degree 1 in

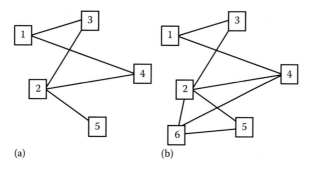

(a)   (b)

FIGURE 1.4
(a) Bipartite and (b) tripartite graphs.

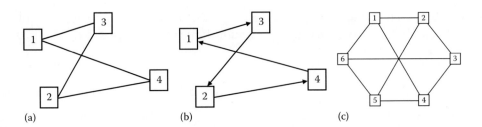

FIGURE 1.5
Examples of regular graphs. (a and c) Undirected graphs where each node has a degree of 2 and 3, respectively.
(b) Directed graph with each node having in-degree and out-degree of 1 each.

a tree are leaves and their removal will still retain the status of the graph as a tree. Every
internal node in a tree is an articulation point, and its removal will split the tree into a disconnected graph. Every edge of a tree is a bridge and its removal will also break the tree
into a disconnected graph.

Figure 1.6a is a tree while Figure 1.6b is a directed tree. Both graphs have $n = |V| = 8$ and
$m = |E| = 7$. A *directed tree* is a digraph that has its orientation originating from one or more
nodes called a *root*. A root can be a leaf or internal node. The maximum distance from the
root to a leaf in the tree is the number of levels of the tree. The level number starts with
1 at the root and increases toward the leaves. The total number of levels is the maximum
distance from the root to the leaves plus one. The rooted directed tree in Figure 1.6b has
v_2 as its root. The tree has three levels with $\{v_2\}$ at level 0, $\{v_8, v_1, v_6\}$ at level 1, and $\{v_7, v_3, v_4, v_5\}$ at level 2.

A *binary tree* is a tree where every node has a maximum degree of 2. A binary tree has a
root whose degree is 1 or 2. A *complete binary tree* is one where the degrees of the root and
leaf are 2 and 1, respectively, while the internal nodes have degrees of 3. It can be shown
that a complete binary tree with r levels has $\sum_{i=0}^{r-1} 2^i$ nodes where 2^{r-1} of them are leaves.

The tree in Figure 1.6a is a spanning tree of the graph in Figure 1.1. Spanning tree G'
is a tree subgraph of the graph G that maintains the same nodes. G' is derived from G by

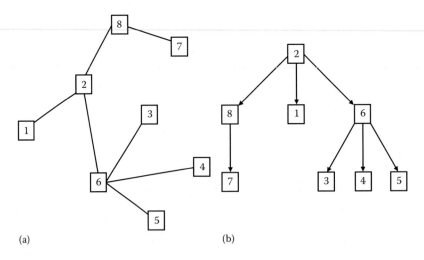

FIGURE 1.6
(a) Undirected tree and (b) rooted directed tree.

deleting some edges in G to remove cycles. There are many spanning trees in the graph. For the weighted graph in Figure 1.3, finding the *minimum spanning tree* becomes one tricky problem in graph theory. The problem is about finding the spanning tree of the given graph where the total weights are minimum.

Understanding the properties of a tree contributes to understanding an area of computer science known as data structure. This field of study is needed for relating objects in a graph by organizing them into structured and hierarchical entries. A good application of data structure is in a database management system that organizes raw data into columns (fields) and rows (records). A tree is also a model commonly used in routing, transportation, and scheduling problems. For example, broadcasting a message from a node requires finding the minimum spanning tree of the graph (network) so that the message reaches all the nodes at the quickest time possible.

1.2 Spectral Graph Theory

Spectral graph theory is an area of mathematics that relates the characteristic polynomial and eigenvalues and their corresponding eigenvectors of a matrix with the properties of a graph. In other words, spectral graph theory analyzes the spectrum of the matrix representing the graph. *Spectrums* are the eigenvectors of a graph ordered by the magnitude of their corresponding eigenvalues. Eigenvalues and eigenvectors provide global information about the structure of the graph. Therefore, these items play an important role in providing solutions to several problems in graph theory, among them graph coloring, graph partitioning, graph clustering, and minimum spanning tree.

A square matrix A of size $n \times n$ is always associated with eigenvalues and eigenvectors. *Eigenvalues* λ_i of the matrix for $i = 1, 2, \ldots, n$ are real or complex scalar values that form a system of linear equations from the following relationship:

$$A v = \lambda v \tag{1.5}$$

In the above equation, A is a matrix, $\lambda = [\lambda_i]$ is the set of eigenvalues, and v is the corresponding set of eigenvectors. It can be seen that the system of linear equations produces a characteristic polynomial through the identity matrix I according to the following steps:

$$A v = \lambda I v$$
$$A v - \lambda I v = 0$$
$$(A - \lambda I) v = 0$$

$$|A - \lambda I| = 0 \tag{1.6}$$

The characteristic polynomial is obtained from Equation 1.6 to produce a set of real and complex eigenvalues λ_i and the corresponding eigenvectors v_i.

One important feature of a simple graph is that its adjacency matrix is always symmetric. This means the matrix has real eigenvalues and a complete set of orthonormal eigenvectors. In an adjacency matrix, the set of graph eigenvalues is called the *spectrum* of the graph. If all the eigenvalues are integers, then the graph is said to be an *integral graph*.

If the eigenvalues λ_i are in the ordered sequence $\lambda_1 \geq \lambda_2 \geq \ldots \geq \lambda_n$ then $\lambda = \{\lambda_1, \lambda_2, \ldots, \lambda_n\}$ is the spectrum of matrix A with the corresponding eigenvectors $v_1, v_2, \ldots, v_n$, such that

$$A = \sum_{i+1}^{n} \lambda_i v_i v_i^T \tag{1.7}$$

where $v_1, v_2, \ldots, v_n$ are the orthogonal basis, $|v_1| = |v_2| = \ldots = |v_n| = 1$.

If λ_i and λ_j are two distinct eigenvalues of a symmetric matrix A, then their corresponding eigenvectors should be $< v_i, v_j > = 0$. The largest eigenvalue is $\lambda_{\max} = \max \dfrac{v^T A v}{v^T v}$ while the smallest eigenvalue is $\lambda_{\min} = \min \dfrac{v^T A v}{v^T v}$. A is positive-definite if all the eigenvalues are greater than 0. In a regular graph, the degree of the nodes is an eigenvalue of its matrix adjacency graph.

The eigenvalues give rise to a special matrix called a *Laplacian matrix*, L. This type of matrix is a square and symmetric matrix formed from the following relationship:

$$L = D - A \tag{1.8}$$

The Laplacian L has some unique properties. First, L is symmetric and a positive-definite, which suggests all eigenvalues of the matrix are real. In addition, 0 is one of the eigenvalues of the matrix while all other eigenvalues of L are positive. Another property is for every vector v,

$$v^T L v = \frac{1}{2} \sum_{i,j=1}^{n} w_{ij}(v_i - v_j)^2 \tag{1.9}$$

As well, the multiplicity of eigenvalue 0 is the trace of the matrix which gives the number of connected components of graph

$$trace(A) = \sum_{i=1}^{n} a_{ii} = \sum_{i=1}^{n} \lambda_i \tag{1.10}$$

2

Visualization with MFC

2.1 Windows Programming

Computers from the 1950s until the late 1970s did not have Windows. Therefore, solutions to mathematical problems were displayed as text and numbers only. Only a few specialized computers could display the output in the form of user-friendly interfaces, so only technical people related to the given application were able to understand the output.

The emergence of Windows in the 1980s changed the scenario drastically. Windows allows both problems and their solutions to be illustrated as text, numbers, graphs, charts, images, and other objects. Interfaces on Windows are easy to understand and use, which has contributed to making mathematics accessible to the general public.

Visualization is the graphical presentation of the solution to a problem in the form of numbers, text, graphics, images, and other objects. In other words, visualization provides a way for the average person on the street to understand the problem and its solution. Windows provides an effective approach for visualization by providing resources for creating user-friendly interfaces that connect a problem and its solution effectively. A good visualization can portray the given problem clearly as Windows objects on a computer screen.

Graphical objects in Windows connect visual objects and their relationships through curves, charts, and figures. A good visualization program should include the following features:

- Good illustration of the problem in the form of mathematical equations and their initial values or boundary values
- Good illustration of the mathematical model for its solution
- Sufficient data to support the computation
- Input section to allow interaction
- Clear display of the output in the form of numbers, graphs, and charts
- Other user-friendly actions for allowing changes by the user
- Flexible interface for linking to external programs

In providing a solution, the problem is first reduced into its mathematical model where the governing equations and their initial or boundary values are defined. This problem is then solved using the given mathematical method, and its solution is presented as numbers, graphs, and charts. To arrive at the solution, some long and tedious work on the simulation is needed. The main tool often involves software development through programming or the use of standard computer packages.

Figure 2.1 shows the steps in developing the solution to a typical problem using C++. The first step is the formulation of the mathematical model that governs the problem. This step involves formulating the mathematical equations and the initial or boundary values. In many cases, this step may involve getting the required data experimentally or through observation. The second step is to create a small version of the problem by reducing the size of the problem without losing its general features. For example, if the original problem involves 200 nodes in the graph, then it may be sufficient to work on a model of 6 to 10 nodes first.

The third step is to work on the small model manually using a standard mathematical method. It is important to derive the manual solution because all the fundamental steps involved will lead to the development of the related algorithms for solving the problem. It is easy to see the flow of the program when the size is small. The program flow leads to

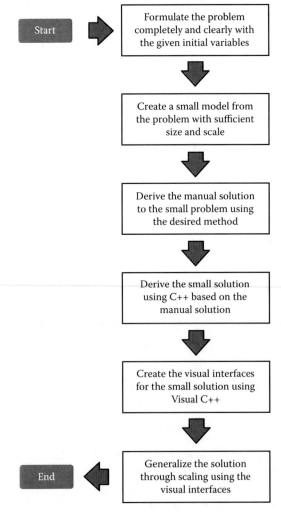

FIGURE 2.1
Work plan for designing a computer program.

the derivation of the algorithm. One important feature of the algorithm is that it should be language-independent and can easily be understood by anybody with expertise in different languages. As well, the algorithms should be designed so that the solution can be scaled up to any size for generalization. This means that if the algorithm works on a small-sized problem, then it should also apply to any other sizes larger or smaller than the worked model.

Successful completion of the manual solution leads to the next step, namely, C++ coding. The codes can easily be created if the algorithms in the previous step are clear and complete. The codes require hours of testing at runtimes with several variations of the data to verify the robustness of the solution.

The fifth step is to add visual interfaces to the solution of the small problem. Basically, every successful program must have each of the three main components of the solution: input, process, and output. Some useful objects are the edit boxes and mouse-click controls for input, a button for processing, and static boxes and list view windows for the output. The edit boxes serve to collect direct input from the keyboard, and this facility can be used to test for the output from runtime. Different entries will produce different output, which allows the developer to verify the correctness of the program.

The final step is the generalization of the solution to the original problem. The generalization can easily be achieved if the program has upward and downward scalability capabilities. This means if the program works on 6 to 10 nodes, then increasing the number of nodes to 200 nodes or other sizes should produce the desired results. Scalability should involve very few lines of code in the program only (between one to 10 lines of codes normally) without the need to rewrite the whole code.

References in [1–3] provide a good start to programming using Visual C++ using Microsoft Foundation Class libraries. The work in [2] discusses the application of Visual C++ in some selected numerical and graph theoretical problems. In [3], Visual C++ serves as a useful tool for solving problems in numerical methods and analysis.

2.2 Microsoft Foundation Classes

In order to utilize the full power of the computer, a good knowledge of the machine hardware is needed. Figure 2.2 shows a schematic drawing of the hardware in relation to the user interface. Central to the computer is the *kernel* that stores all the resources. However, access to the kernel is not easy. Traditionally, communication with the kernel can be achieved using low-level languages in the form of the machine or assembly language that provides access to the real locations of the resources inside the computer for input, processing, and output. The resources can be explored through a layer called the application programming interface (API). However, access to API through these languages is difficult and not many people use it today.

Several means to access the kernel have been developed without going through the low-level languages. One quick way is through the *graphics device interface* (GDI) that provides a path to the kernel through API. The Microsoft Foundation Classes™ (MFC) were developed as a high-level language alternative to handle GDI. Through the MFC, the concept of Windows programming that originated from UNIX can now be deployed on personal computers.

The main feature in the MFC is Windows programming. Standard C++ does not deal with the resources in Windows in order to make it portable across all computing platforms.

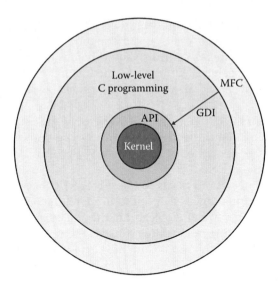

FIGURE 2.2
MFC path to Windows resources.

Windows programming has C++ as its engine to explore the resources in Windows. The resources are necessary in order to provide user-friendly interfaces for visualization. The MFC provides a platform for accessing the Windows resources through C++, which is known as Visual C++.

The MFC has a library consisting of hundreds of classes, and each in turn has dozens of functions for supporting high-level text and graphics interfaces with Windows. The functions simplify a lot of programming tasks particularly in getting input from the user, displaying the output, and controlling Windows resources. There are functions to collect input in the form of user-friendly edit boxes, displaying output as list view windows, and drawing graphs in Windows.

The classes in the MFC are arranged in a hierarchical order with *CObject* as its highest order as the base class. A good reference to the MFC is Microsoft Software Developer Network (MSDN™), which is provided by Microsoft for applications developers. The service is available online on the Microsoft website and is very helpful for finding references to classes, functions, and resources available in the MFC.

2.2.1 Windows Interface Design

Windows™ is a proprietary name belonging to Microsoft™. The generic name is *window*, which refers to a rectangular area on the computer screen. A window represents a work area for one application. At any instance of time, a computer can have several windows open, each catering to one unique application. Today's computers with multicore and multithreading features allow multitasking capabilities involving several open windows for different applications at the same time.

A window can be configured according to the screen resolution, for example, 800 × 600, which has 800 columns and 600 rows of tiny rectangular units called *pixels*. A high-resolution screen can be configured to approximately 1340 × 1280 for finer pixels to produce a bigger work area and sharper graphics. Figure 2.3 shows a typical window of 800 × 600. The window has 800 columns that are numbered from left to right as 0 to

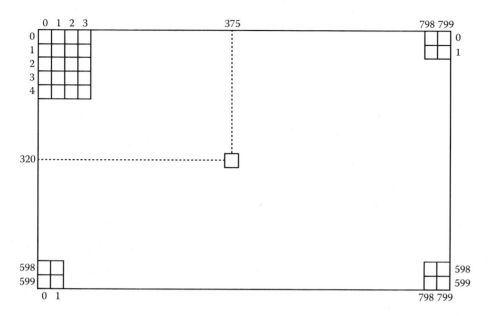

FIGURE 2.3
The main window made up of pixels in a 800 × 600 display.

799. There are 600 rows, numbered from top to bottom as 0 to 599. A pixel is formed from the intersection of the column and row. For example, the pixel shown in the middle of the figure has Windows coordinates of (375,320). Each pixel becomes active when it is assigned a color value to represent a small portion of text, number, graphic, and image.

2.2.2 Color Management

The color of an object in Windows is actually the color of a segregated set of pixels that make up the object. In the MFC, the color of each pixel is controlled using the function *RGB*() from its class called class of Device context objects (CDC) class. $RGB(r, g, b)$ consists of three integer arguments r, g, and b ordered from left to right as the red, green, and blue components, respectively. These three primary colors generate other colors through combinations of the arguments in the function. Each component in *RGB*() is an integer that represents the monotone scale from 0 to 255, with 0 as the darkest value and 255 as the lightest. Therefore, there are 256^3 or 16,777,216 combinations of colors possible for a pixel in Windows.

The monotone scales for red, green, and blue are easily obtained by blanking the other two color components, as follows:

Red, r	$RGB(r, 0, 0)$	for $0 \leq r \leq 255$
Green, g	$RGB(0, g, 0)$	for $0 \leq g \leq 255$
Blue, b	$RGB(0, 0, b)$	for $0 \leq b \leq 255$

Figure 2.4 shows a hypercube that represents color combinations with axes at red, green, and blue. Yellow is obtained by setting $r = g$ and $b = 0$, while $r = 0$, $g = 255$ and $b = 255$ produces cyan. A solid black color is obtained by setting $r = g = b = 0$, while $r = g = b = 255$ produces pure white. It is obvious from this hypercube that a gray-scale color is obtained

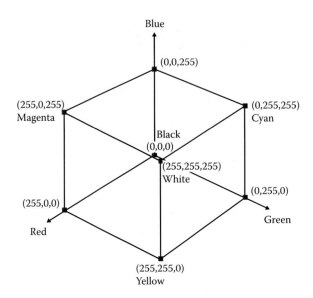

FIGURE 2.4
RGB color scheme in Windows.

by setting $r = g = b$ in $RGB(r, g, b)$, which corresponds to a position along the diagonal line from (0, 0, 0) to (255, 255, 255) in the hypercube.

In Windows, a graphical object is formed from a rectangular composition of pixels of varying intensity. Each pixel displayed in Windows is represented by a string consisting of 24 bits of binary digits that is represented in the MFC by $RGB()$. The first 8 bits in the string starting from the right form the red component. This is followed by 8 bits of green in the middle and the remaining bits make up blue. The alignment of a set of pixels in a rectangle makes up an image, a graphical object, and a text character. One unit of the graphical object is represented by 24 bits of data, so there are a total of 2^{24} or 16,777,126 color combinations possible for supporting various graphical output requirements.

Figure 2.5 shows a pixel in the shaded square having its value defined as $RGB(217, 75, 154)$ represented as a 24-bit string. The string is arranged with the red argument occupying the eight right-most bits, while the green argument is in the middle and blue in the eight left-most bits. The figure also shows the corresponding values in binary: 11011001 for 217, 01001011 for 75 and 10011010 for 154. With this arrangement, the value returned by $RGB(217, 75, 154)$ is one long string of the 24-bit binary number 100110100100101111011001.

2.2.3 Device Context

Device context is a Windows data structure that defines the drawing attributes of a device object for producing the output. The objects in Windows include a pen for drawing a line, a brush for spraying a region with a color, a bitmap for copying parts of the window, a region for clipping, and a path for drawing or painting.

Device context provides a device-independent approach to creating text and graphics in Windows. This capability is inherited by referencing to GDI every time when accessing the Windows resources. For example, to use a red pen for drawing a line the application must create the pen and select the object beforehand; otherwise, the default color, which is black,

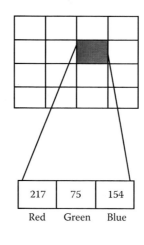

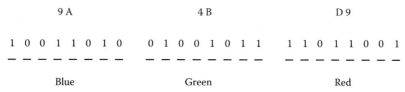

FIGURE 2.5
RGB(217, 75, 154) expressed as a 24-bit binary string.

is used. The same thing also applies to formatting and displaying text. The correct type-faces, colors, and their sizes must be specified by telling the GDI through the device context.

2.3 Writing the Simplest Windows Program

Unlike standard C++, a Windows program using C++ is not easy to write because the coding must take into consideration the Windows resources that must be declared and linked with the program. Every Windows program must involve tasks such as declaring, registering, and updating of the resources.

There are hundreds of functions available for use in the MFC. The best way to find out about the functions is to refer to the MSDN, which can be freely accessed from the Microsoft website. Most of the functions are also explained in various other websites, and they can be searched through many Internet search engines. Visual Studio also provides help for writing codes for Windows programming. The help comes in the form of generated codes for linking several files related to Windows resources. Unfortunately, some of these codes take time to get used to as they are very long and not easy to understand. Furthermore, the codes may confuse and distract a user from writing the actual codes for the application.

A window can be created from several MFC classes. Among the common means is the *CFrameWnd* class, which will be discussed throughout this book. *CFrameWnd* provides an empty window that requires the prototype file *afxwin.h*. This prototype file needs to be included in the application because it has all the function and macro declarations with regard to the use of the main window.

2.3.1 Initializing an Application with Windows

Every Windows program must be initialized, registered, and updated. For an application with the class name *ClassName* initialization, registration and updating can be performed by creating an application class from MFC's *CWinApp* class and including the following codes in the header file of the application:

```
class CMyWinApp:public CWinApp
{
      public:
              virtual BOOL InitInstance();
};
CMyWinApp MyApplication;
BOOL CMyWinApp::InitInstance(void)
{
      m_pMainWnd=new ClassName;
      m_pMainWnd->ShowWindow(m_nCmdShow);
      return TRUE;
}
```

2.3.2 Creating the Main Window

The main window can be created from the *CFrameWnd* class using the function *Create()*. As the main window is the foremost object before anything else, *Create()* should be applied in the application class constructor. The window consists of a work area whose size and other attributes can be configured through the arguments inside *Create()*. The main window is referred to as *this* by other objects and functions in the application. Several options for the window are available; among them are

```
Create(0, string)
```

Which creates a window using the default settings and caption expressed as *string*. This function can be extended by adding several other parameters in order to customize the window's appearance. This will be seen in all upcoming chapters.

2.3.3 Displaying a Text Message

Text can be displayed in the main window through the device context. The basic function is *TextOutW()*, which displays text according to the following format:

```
dc.TextOutW(col, row, string)
```

In the above function, *col* and *row* refer to the column and row number, respectively. The string is the text message, which may also include formatted integer and floating point values, as will be seen later.

2.3.4 *Code2A*: Skeleton Program

Our approach is to start writing C++ code with empty source files to produce a skeleton program. We call it a skeleton program because the code in the related files is kept to the minimum, and the program produces the bare-minimum output. This skeleton program

FIGURE 2.6
Skeleton output from *Code2A.*

may serve as the starting point for writing all other applications as the code can easily be expanded from here.

Our skeleton program is called *Code2A.* Figure 2.6 shows the output from this program, which is a small window with one line of message and caption. The program consists of two files: the C++ source file *Code2A.cpp* and its prototype file *Code2A.h.* A single class called *CCode2A* is used in the application.

Windows runtime is expressed as events. An *event* is an interrupt that is attended to by Windows immediately. Some common events on Windows include the left and right clicks of the mouse and a key stroke on the keyboard. As a rule, every event must be declared and mapped in Windows. The event must also be responded to by an *event handler*, which is a function.

There is only one event in *Code2A*: the display of a text message in the main window. This event is coded as *ON_WM_PAINT*(), which is a built-in function in the MFC. By default, the event handler for *ON_WM_PAINT*() is a function called *OnPaint*(). The event is mapped by including *ON_WM_PAINT*() inside the message map body, as follows:

```
BEGIN_MESSAGE_MAP
      ON_WM_PAINT()
END_MESSAGE_MAP
```

Figure 2.7 is the organization of *Code2A* showing the class, files, and functions. There are two main functions in the application: the constructor and the event handler. The constructor is *Code2A*(), which creates the main window using *Create*(). The constructor has the basic role of creating the class and allocates its memory. The constructor is also an ideal place to initialize variables, as will be seen in the applications of all upcoming chapters. The opposite of the constructor is the destructor ~ *Code2A*(), which destroys the class and returns its allocated memory to the computer.

OnPaint() is the default event handler for *ON_WM_PAINT*(). Any text message or graphics can only be displayed in the main window through a device context. In this application, the device context is *dc*, which is an object from the class *CPaintDC*. The text message is displayed through *TextOutW*().

The full listings of the codes are displayed below:

```
//Code2A.h
#include <afxwin.h>
class CCode2A: public CFrameWnd
{
public:
      CCode2A();
      ~CCode2A();
```

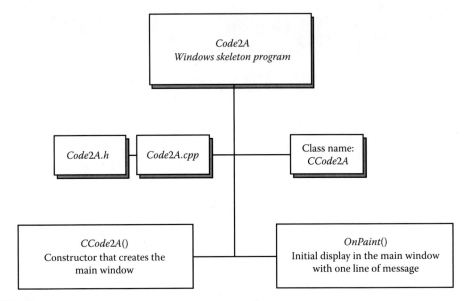

FIGURE 2.7
Organization of *Code2A*.

```
        afx_msg void OnPaint();
        DECLARE_MESSAGE_MAP()
};

class CMyWinApp:public CWinApp
{
        public:
                virtual BOOL InitInstance();
};
CMyWinApp MyApplication;

BOOL CMyWinApp::InitInstance(void)
{
        m_pMainWnd=new CCode2A;
        m_pMainWnd->ShowWindow(m_nCmdShow);
        return TRUE;
}

//Code2A.cpp
#include "Code2A.h"
BEGIN_MESSAGE_MAP(CCode2A,CFrameWnd)
        ON_WM_PAINT()
END_MESSAGE_MAP()

CCode2A::CCode2A()
{
        Create(0,L"Code2A: My skeleton Windows");
}

CCode2A::~CCode2A()
{
```

```
}
void CCode2A::OnPaint()
{
        CPaintDC dc(this);
        dc.TextOut(100,50,L"Welcome to Windows...");
}
```

2.4 Windows Resources for Text and Graphics

A single point in Windows with its own unique coordinates is represented as a pixel. It follows that all text and graphics objects in Windows are constructed from a group of pixels. For example, a line is produced from a set of successive pixels aligned according to the direction of the line. A circle is obtained from pixels that are equidistant from their center. A rectangle is constructed from two pairs of matching lines. A curve is obtained from the successive placement of pixels whose shape is governed by a mathematical function.

What tools are needed in graphics? Pens, brushes, paints, and so forth, just like what an artist requires. A programmer will need to have previously applied difficult low-level routines using C or the assembly language in order to display graphics in Windows. Obviously, a number of ready-made tools from GDI bypass these tedious steps and cut the development time for displaying graphics on Windows.

Table 2.1 summarizes some of the most common graphical *GDI* functions in MFC. They include primitive functions for drawing a point, a line, an arc, a circle, an ellipse, a rectangle, and a polygon. The primitives can be applied for drawing most other sophisticated graphics objects. For example, a mathematical curve can be drawn by plotting straight lines between points that are close to each other successively.

TABLE 2.1

Common Functions for Displaying Graphics in the *CDC* Class

Function	Description
Arc()	Draws an arc
BitBlt()	Copies a bitmap to the current device context
CRect()	Defines the rectangle position and size
Ellipse()	Draws an ellipse (including a circle)
FillRect()	Fills a rectangular region with the indicated color
FillSolidRect()	Creates a rectangle using the specified fill color
GetClientRect()	Returns the size of the window to a *CRect* object
GetPixel()	Gets the pixel value at the current position
LineTo()	Draws a line to the given coordinates
MoveTo()	Sets the current pen position to the coordinates
PolyLine()	Draws a series of lines passing through the given points
Rectangle()	Draws a rectangle according to the given coordinates
RGB()	Creates color from the red, green, and blue palettes
SelectObject()	Selects the indicated *GDI* drawing object
SetPixel()	Draws a pixel according to the chosen color

We list some of the most fundamental text and graphical primitives based on the MFC functions in Table 2.1. The following tasks are discussed, assuming a device context called *dc* has been created in the program.

2.4.1 Setting Up the Device Context

A device context is needed to access resources such as writing text, drawing graphics, and displaying images in the main window. The device context is created as an object from the *CPaintDC* if the function is *OnPaint*() or *CClientDC* if the function is other than *OnPaint*(). The following example creates *dc* as a device context object in the normal *CFrameWnd* window (designated as *this*):

$$CPaintDC\ dc(this);\quad \text{or}\quad CClientDC\ dc(this).$$

2.4.2 Formatted Text Output

By default, text is displayed as a string in Windows. Therefore, any other form of data requires conversion to a string before it can be displayed. The data include common variables: integer, floating-point, and character. In converting to their types, the data are identified according to their identifiers.

Formatted text is a string of message that includes data that have been formatted from their original type in *int*, *double*, and so forth. Text can be formatted for display as a *CString* object according to the identifiers using the function *Format*(). This function allows text to be formatted according to the correct data type using several identifiers in the same manner as the famous C function *printf*(). Table 2.2 lists some of the most common data types supported in *Format*().

The example in Figure 2.8 illustrates how the data *a* and *x*, whose types are *int* and *double*, respectively, are recognized through their identifiers, formatted as a string, and displayed at (50,85).

TABLE 2.2

Common Identifiers in *Format*()

Identifier	Variable Type
%c	*char*
%s	*char**
%d	*int*
%f	*float*
%lf	*double*

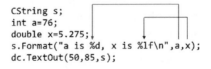

```
CString s;
int a=76;
double x=5.275;
s.Format("a is %d, x is %lf\n",a,x);
dc.TextOut(50,85,s);
```

    a is 76, x is 5.275

FIGURE 2.8
Formatted string output.

2.4.3 Setting the Pen Color

A pen is needed to draw an object with color attributes in Windows. By default, the pen color is black with a default thickness of 1 pixel. A pen can be created by declaring an object from the MFC's *CPen* class.

For example, the following code segment creates an object called *myPen* to represent a blue pen with a thickness of 3 pixels, and then selects this pen for drawing:

```
CPen myPen(PS_SOLID,3,RGB(0,0,200));
dc.SelectObject(myPen);
```

2.4.4 Defining a Point in Windows

A point in Windows is a pixel defined as an object from the class *CPoint*. The point has two components, the x and y coordinates, which represent the column and row numbers, respectively. The values for the point are integer numbers ranging from 0 to the maximum size of the window.

The following code segment creates a point *pt* with Windows coordinates of (70,50):

CPoint pt; pt = CPoint(70,50);	or	CPoint pt; pt.x = 70; pt.y = 50;

2.4.5 Plotting a Point in Windows

A point can be plotted in Windows using MFC's *SetPixel()*. The point is plotted as a pixel at the indicated Windows location using the current pen color. For example, the following statement plots a pixel at (340,200):

```
dc.SetPixel(340,200);
```

A single pixel may not appear to be visible on the screen, especially if the screen is high resolution. Therefore, in order in improve its visibility, a single point may be represented as a group of connected pixels.

2.4.6 Drawing a Line

Drawing a straight line requires moving the pen to the starting point through *MoveTo()*, and then dragging the pen to the ending point using *LineTo()*. For example, to draw a straight line from (250,280) to (100,340), as in Figure 2.9:

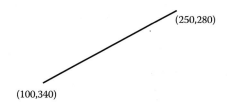

FIGURE 2.9
Drawing a line.

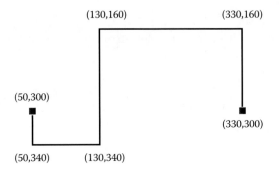

FIGURE 2.10
Drawing multiple connected lines.

```
dc.MoveTo(250,280);
dc.LineTo(100,340);
```

All other shapes involving lines are drawn using the above concept. For example, the following codes draw a track involving six points in the order from start to finish (50,300), (50,340), (130,340), (130,160), (330,160), and (330,300), as in Figure 2.10:

```
dc.MoveTo(50,300);
dc.LineTo(50,340);
dc.LineTo(130,340);
dc.LineTo(130,160);
dc.LineTo(330,160);
dc.LineTo(330,300);
```

2.4.7 Drawing a Polygon

A polygon can be drawn by placing the pen at the starting point through *MoveTo()*, and then dragging the pen along the other points in the polygon using *LineTo()*.

The following code segment draws a hexagon whose vertices are at (50,75), (100,25), (150,25), (200,75), (150,125), and (100,125), as in Figure 2.11:

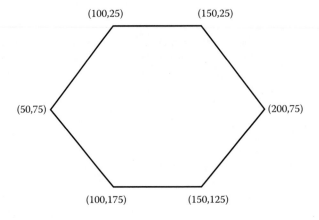

FIGURE 2.11
Hexagon constructed from lines.

```
dc.MoveTo(50,75);
dc.LineTo(100,25);
dc.LineTo(150,25);
dc.LineTo(200,75);
dc.LineTo(150,125);
dc.LineTo(100,175);
dc.LineTo(50,75);
```

2.4.8 Drawing an Empty Rectangle

Drawing an empty rectangle requires *CRect*() for its position and size, while *Rectangle*() displays the rectangle. The rectangle is represented by an object from the MFC's *CRect* class. *Rectangle*() only draws the boundaries of the rectangle; it does not color its inner region.

For example, to draw the rectangle in Figure 2.12 with its top left-hand corner coordinates and bottom right-hand corner coordinates at (50,70) and (300,350), respectively, and having a length of 100 pixels and width of 70 pixels:

```
CRect rct;
rct=CRect(50,70,150,140);
dc.Rectangle(rct);
```

or

```
CRect rct;
rct=CRect(CPoint(50,70),CSize(100,70));
dc.Rectangle(rct);
```

2.4.9 Drawing a Solid-Filled Color Rectangle

A solid-filled color rectangle is drawn in the same manner as drawing an empty rectangle. The function is *FillSolidRect*() and it requires two arguments: the rectangle specification and its color.

For example, to draw a solid blue rectangle with its top left-hand corner coordinates at (50,70), having a length of 100 pixels and width of 70 pixels:

```
CRect rct;
int color=RGB(0,0,200);
rct=CRect(CPoint(50,70),CSize(100,70));
dc.FillSolidRect(rct,color);
```

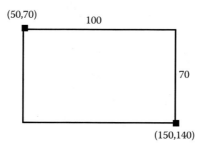

FIGURE 2.12
Drawing a rectangle.

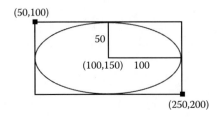

FIGURE 2.13
Drawing an ellipse.

2.4.10 Drawing an Ellipse

It follows that an ellipse is drawn in the same manner as in drawing a rectangle, using the function *Ellipse*(). For example, the following code segment draws the ellipse in Figure 2.13, whose top left-hand and bottom right-hand coordinates are (50,100) and (250,200), respectively:

```
CRect rct;
rct=CRect(50,100,250,200);
dc.Ellipse(rct);
```

or

```
CRect rct;
rct=CRect(CPoint(50,100),CSize(200,100));
dc.Ellipse(rct);
```

2.4.11 Drawing a Circle

A circle is an ellipse whose *x* and *y* radii are the same. Therefore, the same function *Ellipse*() can be used to draw a circle.

Figure 2.14 shows a circle with its center at (100,150) and with a radius of 50 pixels. In drawing this circle, simple deductions produce (50,100) and (150,200) as its top left-hand and bottom right-hand coordinates, respectively. Therefore, the code segment becomes

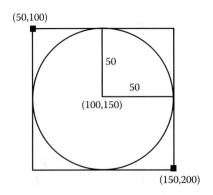

FIGURE 2.14
Drawing a circle.

```
CRect rct;
rct=CRect(50,100,150,200);
dc.Ellipse(rct);
```

or

```
CRect rct;
rct=CRect(CPoint(50,100),CSize(100,100));
dc.Ellipse(rct);
```

2.4.12 Clearing a Portion of Window

A rectangular region in Windows can be cleared simply by spraying the region with a color similar to its background. This can be achieved by defining an object for the brush from the MFC's *CBrush* class. The rectangular region is represented by a *CRect*() object. Spraying is done through *FillRect*().

The following code segment sprays white paint on a region specified by (150,250) and (250,350) as its top left-hand and bottom right-hand coordinates, respectively:

```
CBrush bkBrush(RGB(255,255,255));
CRect rct;
rct=CRect(150,250,250,350);
dc.FillRect(&rct,&bkBrush);
```

2.4.13 Clearing the Whole Window

The main window can be cleared in the same manner as for a portion. However, it is not necessary to measure the size of the window in this case. An MFC function called *GetClientRect*() detects the size and position of the main window automatically and returns the value to a *CRect*() object.

The following code segment clears the main window by spraying black paint onto the window:

```
CBrush bkBrush(RGB(0,0,0));
CRect rct;
GetClientRect(&rct);
dc.FillRect(&rct,&bkBrush);
```

2.4.14 *Code2B*: Graphics Drawing

We discuss an example of graphics drawing involving lines, rectangles, and circles. The program is called *Code2B*, and it includes the source files *Code2B.h* and *Code2B.cpp*. Figure 2.15 shows the output from *Code2B* that includes three columns of drawings. The first column is fairly straightforward as it has a rectangle, an ellipse, and a line. The second column has a set of rectangles and a set of concentric circles, while the last column displays a variation of colors in small rectangles on Windows (shown as shades of gray in the figure).

The rectangles in the second column are arranged as a table in a rectangular grid, and they are labeled according to their row and column numbers. Each rectangle is drawn iteratively with width and length using a nested loop, and its top-left point is marked as the *CPoint* object called *home*.

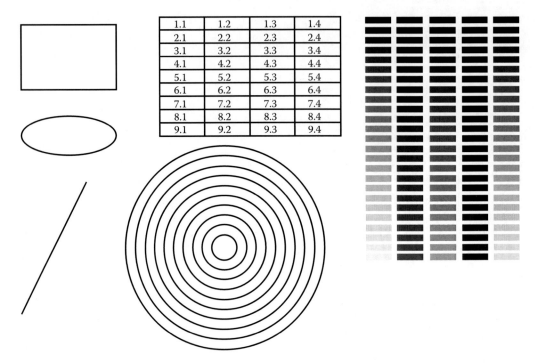

1.1	1.2	1.3	1.4
2.1	2.2	2.3	2.4
3.1	3.2	3.3	3.4
4.1	4.2	4.3	4.4
5.1	5.2	5.3	5.4
6.1	6.2	6.3	6.4
7.1	7.2	7.3	7.4
8.1	8.2	8.3	8.4
9.1	9.2	9.3	9.4

FIGURE 2.15
Output from *Code2B* showing primitive graphics.

In the second column, 10 circles with the same center are drawn starting with the largest to the smallest. A circle is drawn using *Ellipse*() by setting the value of the width equals the height of its rectangle.

The third column displays five columns of rectangles with varying colors. Each rectangle is drawn with a solid color through *FillSolidRect*(). This function differs from *Rectangle*() as the latter colors the sides of the rectangle, omitting the inner part of the rectangle. The color in each rectangle is generated by varying the arguments in *RGB*().

Figure 2.16 is the organization of *Code2B* that shows the related class, files, and functions. The codes are given as follows:

```
//Code2B.h
#include <afxwin.h>
class CCode2B: public CFrameWnd
{
public:
        CCode2B();
        ~CCode2B()      {}
        afx_msg void OnPaint();
        DECLARE_MESSAGE_MAP()
};
class CMyWinApp:public CWinApp
{
        public:
                virtual BOOL InitInstance();
};
CMyWinApp MyApplication;
```

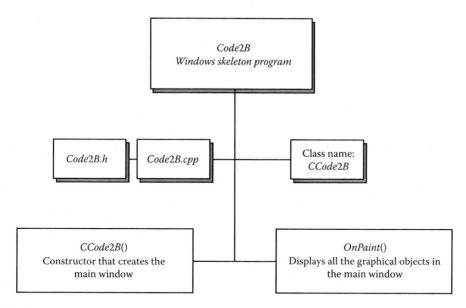

FIGURE 2.16
Organization of *Code2B*.

```
BOOL CMyWinApp::InitInstance(void)
{
      m_pMainWnd=new CCode2B;
      m_pMainWnd->ShowWindow(m_nCmdShow);
      return TRUE;
}

//Code2B.cpp
#include "Code2B.h"
BEGIN_MESSAGE_MAP(CCode2B,CFrameWnd)
      ON_WM_PAINT()
END_MESSAGE_MAP()

CCode2B::CCode2B()
{
      Create(NULL,L"Code2B: Drawing Graphics",
            WS_OVERLAPPEDWINDOW,CRect(0,0,1000,800));
}

void CCode2B::OnPaint()
{
      CPaintDC dc(this);
      CRect rct;
      CPen pen1(PS_SOLID,2,RGB(200,0,0));          //red
      CPen pen2(PS_SOLID,1,RGB(150,0,255));        //purple
      CPen pen3(PS_SOLID,3,RGB(200,200,0));        //gold
      //drawing a line
      dc.SelectObject(pen3);
      dc.MoveTo(150,300); dc.LineTo(50,500);
      //drawing a rectangle & ellipse
      dc.SelectObject(pen2);
```

```
rct=CRect(50,60,200,160);
dc.Rectangle(rct);
rct=CRect(50,200,200,260);
dc.Ellipse(rct);

//drawing empty rectangles
int i,j;
int M=10,N=5,width=70,height=20;
CString s;
CFont font;
font.CreatePointFont(100,L"Helvetica");
CPoint home;
home=CPoint(300,50);
dc.SelectObject(pen1);
dc.SelectObject(font); dc.SetTextColor(RGB(0,0,200));
for (i=1;i<=M;i++)                      //i=row
{
        //horizontal lines
        dc.MoveTo(home.x,home.y+height*(i-1));
        dc.LineTo(home.x+width*(N-1),home.y+height*(i-1));
        for (j=1;j<=N;j++)              //j=column
        {
                //vertical lines
                dc.MoveTo(home.x+width*(j-1),home.y);
                dc.LineTo(home.x+width*(j-1),home.y+height*(M-1));
                if (i<M && j<N)
                {
                        //captions
                        s.Format(L"%d,%d",i,j);
                        dc.TextOut(home.x+20+width*(j-1),home.
                                y+5+height*(i-1),s);
                }
        }
}

//drawing circles
int radius;
CPoint center=CPoint(home.x+100,home.y+350);
dc.SelectObject(pen2);
for (i=10;i>=1;i--)
{
        radius=20+15*(i-1);
        rct=CRect(CPoint(center.x-radius,center.y-radius),
                CSize(2*radius,2*radius));
        dc.Ellipse(rct);
}

//drawing solid rectangles
int color;
width=40; height=10; M=25, N=5;
for (i=1;i<=M;i++)
        for (j=1;j<=N;j++)
        {
                switch(j)
```

```
{
case 1:
        color=RGB(10*(i-1),10*(i-1),10*(i-1));
                break;//bw monotone
case 2:
        color=RGB(10*(i-1),0,0); break; //red
                monotone
case 3:
        color=RGB(0,10*(i-1),0); break; //green
                        monotone
case 4:
        color=RGB(0,0,10*(i-1)); break; //blue
                monotone
case 5:
        color=RGB(10*(i-1),10*(i-1),0); break; //
                yellow monotone
}
home=CPoint(650+(width+10)*(j-1),50+(height+5)*
        (i-1));
rct=CRect(CPoint(home),CSize(width,height));
dc.FillSolidRect(rct,color);
        }
}
```

2.5 Event and Event Handler

In object-oriented programming, an *event* is defined as an interrupt or a happening during runtime that requires immediate attention and response. The response to an event is the *event handler,* which is a function. An event can be regarded as an interrupt where a call to a specific task in the computer is immediately performed. Some obvious examples of events are the left click of the *mouse,* a key stroke, a choice of an item in a menu, and a push-button click.

Table 2.3 lists some of the most common events in Windows. Of particular interest is *ON_WM_PAINT*(), which triggers the initial display of the application in the main window using a default function called *OnPaint*(). The display is also updated through an

TABLE 2.3

Common Events and Their Event Handlers

Event	Description	Event Handler
ON_WM_PAINT()	Display in main window	*OnPaint*()
ON_WM_LBUTTONDOWN()	Mouse's left click	*OnLButtonDown*()
ON_WM_LBUTTONDBLCLK()	Mouse's double left click	*OnLButtonDblClk*()
ON_WM_LBUTTONUP()	Mouse's left button release	*OnLButtonUp*()
ON_WM_RBUTTONDOWN()	Mouse's right click	*OnRButtonDown*()
ON_WM_MOUSEMOVE()	Mouse move	*OnMouseMove*()
ON_WM_KEYDOWN()	Key press on the keyboard	*OnKeyDown*()
ON_BN_CLICKED()	Click on *CButton* object	to be named by the user
ON_COMMAND()	Menu item activation	to be named by the user

interrupt specified by a function called *Invalidate*() or *InvalidateRect*(). We will come across *ON_WM_PAINT*() in almost all applications in this book.

An event handler is a special function that has an advantage over other normal functions because it becomes activated immediately when its event is called. By default, the names of the functions are provided by MFC automatically and the names cannot be changed, for example, *OnPaint*() for *ON_WM_PAINT*(). Some events, for example *ON_BN_CLICKED*(), require the name of the event handler to be provided by the user.

In MFC, a message map is provided to detect the occurrence of an event during the run-time. The message map is written in the.cpp file according to the following format:

```
BEGIN_MESSAGE_MAP(ClassName,CFrameWnd)
       <List of events, their ids and event handlers>
END_MESSAGE_MAP()
```

The event is declared in the header file according to

```
afx_msg void EventHandler;
```

The example in Figure 2.17 illustrates a message map that shows three events: window display update, button click, and the mouse's left click, in a class called *CCode2*. Every event handler must be declared as a function of type void preceded by *afx_msg*. The event handlers for the example above are declared in the header file as follows:

```
afx_msg void_OnPaint();
afx_msg void_NewGraph();
afx_msg void_OnLButtonDown(UINT,CPoint);
```

From this example, *ON_WM_PAINT* updates the display in the main window by calling the default function *OnPaint*(), which is its event handler. The second event is *ON_BN*_CLICKED(), which is triggered when its *CButton* object is clicked with the mouse. The event handler in this case is *NewGraph*(). The third event is *ON_WM_LBUTTONDOWN*(), which is activated when a point in Windows is clicked. The default function called *OnLButtonDown*() is immediately called to act at the point of click.

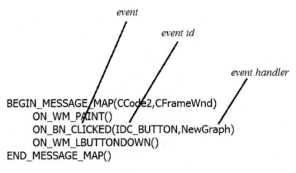

FIGURE 2.17
Mapping events and their handlers.

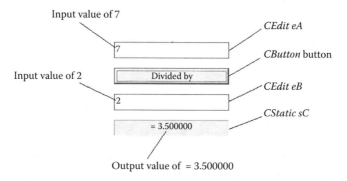

FIGURE 2.18
Edit box, static box, and button objects.

2.6 Windows Control Resources

The main objective of Windows is to provide user-friendly and easy-to-use interfaces that allow interaction between the user and the computer. These interfaces are represented as control objects that can be classified as input, output, and process objects. Several objects from each category can be prepared using the ready-made MFC functions through their respective classes. For mathematical modeling and simulation purposes, not all of the resources need to be discussed here. Only a few are relevant and are sufficient for producing good simulation models.

Three things happen within a computer: input of data and information, processing of the data, and output of the results. The input includes edit boxes, mouse clicks, the keys on the computer keyboard, and external files. For output, some of the common facilities include text and graphics output on the main window, static boxes, list view windows, and external files. The processes in Windows are commonly triggered by button clicks and menus.

Figure 2.18 displays a typical output from an application involving a simple division calculator. The calculator has two edit boxes for input, a static box for output, and a button for triggering the process. Input in the edit boxes produces the result in the static box, a process triggered by the push button. The idea from the calculator can be expanded to more sophisticated applications that share the common mechanism for input, process, and output.

2.6.1 Edit Box

The edit box represents the most common form of an input object in Windows applications. An edit box is a rectangular box that provides direct input from the keyboard. It is declared as an object from the *CEdit* class and created using the function *Create()*. The input is read as a *CString* object using a function called *GetWindowText()*. The input value can then be converted to *int* or *double* using *_ttoi()* and *_ttof()*. An edit box is destroyed using *DestroyWindow()*.

An edit box can be created as an object with its declaration according to the following format:

```
CEdit ObjectName;
```

The edit box is created by linking its object with Create() according to

```
ObjectName.Create(DisplayOptions, RectangularRegion, HostWindow, Id);
```

Referring to Figure 2.18, there are two edit boxes called *eA* and *eB*. The size of each box is 80 by 25. *eA* has its top left position at (250,50) while *eB* is at (250,250). The id for both objects is the integer variable *idc*, which changes their values through the increment of one unit.

```
CEdit eA,eB;
eA.Create(WS_CHILD | WS_VISIBLE | WS_BORDER,
      CRect(CPoint(250,50),CSize(80,25)),this,idc++);
eB.Create(WS_CHILD | WS_VISIBLE | WS_BORDER,
      CRect(CPoint(250,250),CSize(80,25)),this,idc++);
```

Input from the user is collected as a string through *GetWindowText*() according to the following format:

```
ObjectName.GetWindowText(string);
```

For example, *eA* reads an input string called *s*. The string is converted into a double value and stored as a variable called *A* using the function *_ttof*().

```
CString s;
eA.GetWindowText(s);
double A=_ttof(s);
```

2.6.2 Static Box

A static box is a very common way to display output. It consists of a rectangular space that stores the string for display using only a function called *SetWindowText*(). The object for a static box is declared from the *CStatic* class and created using *Create*(). For displaying the data, the value in *int* and *double* needs to be formatted to the string using *Format*(). As with all other Windows objects, a static box is destroyed using *DestroyWindow*().

A static box is created from a class called *CStatic*. As with an edit box, each static box is identified through a control id that can be defined in its creating function. The global declaration of a static box is made in the header file, as follows:

```
CStatic ObjectName;
```

This is followed by its creation in the constructor according to

```
ObjectName.Create(Title, DisplayOptions, RectangularRegion, HostWindow,
Id);
```

By default, a static box is displayed as a shaded rectangle. The shape and style of a static box can be set and modified through *DisplayOptions*. Referring to the example in Figure 2.18, the following statement creates the *CStatic* object called *sC*:

```
      sC.Create(L"",WS_CHILD | WS_VISIBLE | SS_SUNKEN
            | SS_CENTER,CRect(CPoint(250,350),CSize(80,25)),this, idc++);
```

Data in the form of a string can be displayed in a static box using the function *SetWindowText*() according to

```
ObjectName.SetWindowText(string);
```

For example, to display the content of the variable *C* into a static box, *sC*, the value must be formatted into a string first, as shown below:

```
double C=6.05;
CString s;
s.Format(L"%lf",C);
sC.SetWindowText(s);
```

2.6.3 Push Button

A push button is normally used to trigger an action. For example, once all the data from the edit boxes have been typed in by the user, the calculating process can start by clicking the push button. A push button is an object declared from the *CButton* class. It is created using *Create*() and destroyed using *DestroyWindow*().

A click of the push button is an event in Windows. The event is mapped as *ON_BN_CLICKED* and handled by an event handler that is the named function in the message map.

A push button is a resource in the form of a child window created from the *CButton* class. An object for a push button is normally declared in the header file, as follows:

```
CButton ObjectName;
```

The object is created in the constructor according to a format given by

```
ObjectName.Create(Title,DisplayOptions, RectangularRegion, HostWindow,
Id);
```

In the above format

Title: the title in the title bar

DisplayOptions: defines the shape of the button

RectangularRegion: defines the coordinates and size of the button

HostWindow: the host or parent window

Id: the control id

From Figure 2.18, the button is created at (250,150) as an object called *button* that has the id *IDC_BUTTON* and is captioned as *Divided By*:

```
CButton button;
button.Create(L"Multiply",WS_CHILD | WS_VISIBLE
     | BS_DEFPUSHBUTTON, CRect(CPoint(250,150),CSize(80,25)),this,
     IDC_BUTTON);
```

A click of a push button is an event. Therefore, a message handler for this event needs to be mapped according to

```
ON_BN_CLICKED(ObjectId,EventHandler)
```

In the above message handler, *ObjectId* is the control id and *EventHandler* is the name of the function that will respond to this event. For example, a function called OnButton() below responds to a click of a button with id IDC_BUTTON:

```
ON_BN_CLICKED(IDC_BUTTON,OnButton)
```

2.6.4 List View Window

A list view window is a child window within the main window for displaying a large table of data within its limited rectangular display area. The window is an object created from the MFC class called *CListCtrl*. The window has automatic vertical and horizontal scrolling bars that allow viewing beyond its physical space. A large matrix of size 50 × 50, for example, can be viewed in this window by scrolling vertically and horizontally through the bars. Figure 2.19 shows a typical list view window having 20 rows and 20 columns of data. The window has a horizontal bar for scrolling the columns (1–20) but no vertical bar because the size is enough for accommodating 20 rows of data.

In utilizing the resources in the list view window, a header file in MFC called *afxcmn.h* needs to be included. A list view box can have a flexible size and can be placed anywhere in the main window. More than one list view box can be created and displayed in the main window, which allows viewing of several different sets of data simultaneously.

Three steps are involved in creating a list view window. The first step is creating the window, the second step is creating the columns, and the third step is creating the rows for inserting the data. In designing a list view box, a list view object is first created from the *CListCtrl* class, as follows:

```
CListCtrl TableName;
```

i	x	y	f(x,y)	g(x,y)
0	1.000000	2.500000	5.269910	-3.002938
1	1.100000	2.550000	4.810978	-3.014445
2	1.200000	2.600000	4.333225	-3.051803
3	1.300000	2.650000	3.839211	-3.116449
4	1.400000	2.700000	3.331593	-3.209401
5	1.500000	2.750000	2.813101	-3.331246
6	1.600000	2.800000	2.286518	-3.482133
7	1.700000	2.850000	1.754657	-3.661761
8	1.800000	2.900000	1.220341	-3.869385
9	1.900000	2.950000	0.686379	-4.103819
10	2.000000	3.000000	0.155546	-4.363445
11	2.100000	3.050000	-0.369440	-4.646228
12	2.200000	3.100000	-0.885938	-4.949734
13	2.300000	3.150000	-1.391403	-5.271156
14	2.400000	3.200000	-1.883406	-5.607337
15	2.500000	3.250000	-2.359653	-5.954806
16	2.600000	3.300000	-2.817997	-6.309811
17	2.700000	3.350000	-3.256458	-6.668356
18	2.800000	3.400000	-3.673232	-7.026243
19	2.900000	3.450000	-4.066707	-7.379117

FIGURE 2.19
List view window showing a table of data.

The window is created using *Create*() according to the following format:

```
TableName.Create(DisplayOptions,RectangularRegion,HostWindow, Id);
```

Figure 2.19 shows a list view window for displaying the contents of the values of five variables. The list view object is called *table* with an identification number of 802. The list view window starts with (50,50) at the top left-hand corner to (430,400) at the bottom right:

```
CListCtrl table;
CRect rct=CRect(50,50,430,400);
table.Create(WS_VISIBLE | WS_CHILD | WS_DLGFRAME | LVS_REPORT
                | LVS_NOSORTHEADER,rct,this,802);
```

Once the table has been created the next logical step is to divide the table into columns according to the fields of the data. Each data *field* occupies a column in the window and it denotes the category of the data. The columns in the window are created using *InsertColumn*() with the column number starting from 0. For example, the following codes insert five columns of data labeled as 0, 1, 2, 3, 4, each with the width of 30 pixels:

```
CString S[]={"i","x","y","f(x,y)","g(x,y)"};
for (int j=0;j<=4;j++)
                table.InsertColumn(j,S[j],LVCFMT_CENTER,70);
```

The final step is to insert the rows of data according to their columns (fields). A new row is created in the first column (column 0) using *InsertItem*(). The row number always starts with 0. The second column and onward do not require a new row; therefore, the data are inserted using *SetItemText*(). The following example shows how 51 rows of data from five variables called i, $x[i]$, $y[i]$, $f(x[i], y[i])$ and $g(x[i], y[i])$ are inserted:

```
x[0]=1.0;  y[0]=2.5;
for (i=0;i<=50;i++)
{
        s.Format(L"%d",i);  table.InsertItem(i,s,0);
        s.Format(L"%lf",x[i]);  table.SetItemText(i,1,s);
        s.Format(L"%lf",y[i]);  table.SetItemText(i,2,s);
        s.Format(L"%lf",z=f(x[i],y[i]));  table.SetItemText(i,3,s);
        s.Format(L"%lf",z=g(x[i],y[i]));  table.SetItemText(i,4,s);
        x[i+1]=x[i]+h;  y[i+1]=y[i]+k;
}
```

2.7 Displaying a Graph

Now that we have explored the user-friendly interfaces for Windows, we are ready to create a computer program for displaying a graph visually on Windows using all the tools discussed in the last section.

A graph $G(V, E)$ with its nodes $V = \{v_i\}$ and edges $E = \{e_{ij}\}$ for $i, j = 1, 2, \ldots, n$ can be drawn in Windows quite easily using the primitive tools discussed earlier. The graph should occupy an area about one third or half of the Windows area. The nodes of the graph should be labeled with node numbers from 1 to n and they can be represented as rectangles or circles. The edges are represented as straight lines or sometime as arcs. The whole graph with n nodes should be displayed within a single window to illustrate the problem clearly.

2.7.1 Designing the Work Area

As already mentioned, the work area inside Windows consists of three main components: input, output, and process. The input area should include edit boxes while the output area normally has a drawing area, a text area for displaying the results, and small list view windows for displaying results in the form of tables. The process part is one that triggers actions that connect input with output, such as buttons and menus. Other essential objects include file read and save, check boxes, radio buttons, combo boxes, and slider controls, which may be added according to the program requirements.

Figure 2.20 shows a typical design of the drawing area of a graph, and several input, output, and process objects. This design is reasonably good for most small applications involving these three components. However, the real design will depend on the program requirements. For example, some applications may emphasize the graphical displays of the output. In this case, two or three drawing areas are needed in the window, replacing some other noncritical objects.

The drawing area for displaying the graph is normally rectangular and it does not have to be the whole window. A reasonable size should be between one third and three quarters

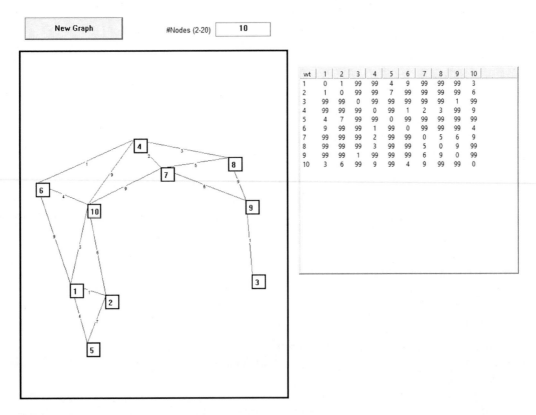

wt	1	2	3	4	5	6	7	8	9	10
1	0	1	99	99	4	9	99	99	99	3
2	1	0	99	99	7	99	99	99	99	6
3	99	99	0	99	99	99	99	99	1	99
4	99	99	99	0	99	1	2	3	99	9
5	4	7	99	99	0	99	99	99	99	99
6	9	99	99	1	99	0	99	99	99	4
7	99	99	99	2	99	99	0	5	6	9
8	99	99	99	3	99	99	5	0	9	99
9	99	99	1	99	99	99	6	9	0	99
10	3	6	99	9	99	4	9	99	99	0

FIGURE 2.20
Randomly scattered nodes of a weighted graph.

of the size of the window. Two *CPoint* objects called *home* and *end* are assigned to mark the top-left and bottom-right corners of the rectangular drawing area and confine the drawing to be within this area. The supporting objects such as buttons, edit boxes, static boxes, menus, and list view windows should be placed at good locations so that they can be viewed easily.

2.7.2 Data Structure of a Graph

In drawing a graph, structures are needed to organize the variables in nodes and edges. Here, we call *NODE* a structure for the nodes and edges. The typical data structures for *NODE* in a weighted graph with n nodes are

```
typedef struct
{
        CPoint home;        //top left-hand corner of the rectangle
        CRect rct;          //rectangle to represent the point
        double wt[n+1];     //weight of the edge
} NODE;
NODE v[n+1];                //the node v_i for i=1, 2, …, n
```

In *NODE*, *home* is the physical position of a node v_i in the graph. For example, $v[3].home = CPoint(50,170)$ denotes the assignment of the home coordinates of v_3 at $(50,170)$. In Windows, the node appears as a box represented by the rectangular object *rct* with *home* as its top left-hand point. A node may connect with another node through an edge. If the edge exists, then its value is represented by *wt*. Thus, $v[3].wt[7]$ is the weight between v_3 and v_7.

The members of the structures are the attributes belonging to the nodes and edges. Many more attributes can be added depending on the program requirements. In a directed graph, for example, the edges are vectors that denote directions. Each edge may then be characterized with a direction called *dir* which is a double variable in *NODE*.

2.7.3 Random Placement of Nodes

The nodes of a graph can be arranged at various locations in a structured or unstructured manner. In a structured arrangement, the nodes are placed in an organized manner to make them easy and convenient to access or communicate. The nodes may be placed in rows and columns or according to a fixed arrangement in a network. The nodes have fixed coordinates where the values are determined according to the system requirement. Interconnection networks, printed circuit boards, and computer networks are some good examples of structured networks.

On the other hand, some networks have nodes located in an unstructured manner. The nodes are scattered over the region with their locations randomly determined. Some examples of networks of this type are wireless sensor networks and transportation networks involving supply and demand centers. Unstructured data are very commonly encountered in statistics, economics, and engineering.

Random number generation has become an important useful feature in a program to support simulations. A computer can simulate randomly scattered nodes in a graph. Inside the computer is a clock that ticks at every microsecond. The ticks of the clock can be tapped to generate random numbers as integer numbers.

In Visual C++, the random number facility becomes available by including the prototype file *fstream.h*. This is followed by *srand(time(0))* for initializing the clock in order to start the random process. It is sufficient to initialize the clock only once, normally done at the beginning of the program. The ideal place to do this is in the constructor.

Random numbers are produced by *rand*(). The function returns a positive integer value between 0 and the constant *RAND_MAX*, which has a value of 32,767. A useful *operator%* (modulus) enhances *rand*() for producing a random number within a certain interval. This mathematical operator returns a number when a number is divided by another number. For example,

- *rand*()%10 produces a number between 0 and 10
- 50 + *rand*()%100 produces a number between 50 and 149
- (*double*)1/(1 + *rand*()) produces a double value between 0 and 1
- sin((double)*rand*()) produces a double value from 0 to 1

The function *rand*() will be used heavily in this book mainly for generating the (*x, y*) random coordinates of the nodes of a graph.

2.7.4 *Code2C*: Displaying a Graph

Code2C is a project for creating and displaying a graph with interfaces using the Windows resources discussed in Section 2.6. The project incorporates the use of an edit box, a button, a list view window, and a drawing area for supporting the display of the graph. The project produces a weighted graph whose size can be determined by the user through entry in an edit box. The graph is generated with randomly scattered nodes.

The edges between the nodes are drawn based on the adjacency status of the nodes: an edge exists if the distance between a pair of nodes is smaller than a predetermined threshold value called *LinkRange*. If the edge exists, then its weight is given randomly.

Figure 2.20 shows a sample output from the project showing a graph *G*(*V, E*) with *n* = |*V*| = 10 and the weight matrix of the graph shown as an output in the list view window. The number of nodes in the graph is an input value read from an edit box. The drawing area for the graph is bounded by the *CPoint* objects *home* and *end*, on the top-left and bottom-right corners, respectively.

There are two push buttons called *Draw Graph* and *New Graph*. The first button draws the initial graph based on the input value from the edit box when it is clicked. The second button destroys the first button when it is clicked and redraws the graph with new locations for the nodes and edges. Any Windows object can be destroyed through a function called *DestroyWindow*(). The weights of the edges are numbers from 1 to 9 that are generated randomly. An edge with the value of 99 is symbolic; this value denotes that the edge between the pair of nodes does not exist.

Project *Code2C* has *Code2C.h* and *Code2C.cpp* as the source files. The application class is *CCode2C*. Three events are associated with this application, namely, the display of the output in the main window, a push-button click for computing the graph, and another push-button click for redrawing the graph.

Table 2.4 describes important objects and variables in *Code2C* while the organization of the project is shown in Figure 2.21. The constructor *Code2C*() constructs the window and creates *eNodes* for collecting the number of nodes from the user. It also creates *button*1 for getting this value and proceeds with creating the graph through *Initialize*(). The latter function generates random locations for the nodes and the edges between the nodes. The graph is displayed through *OnPaint*(), which generates the weight matrix through

TABLE 2.4

Important Objects and Variables in *Code2C*

Variable	Type	Description
eNodes	*CEdit*	Input for number of nodes
*button*1, *button*2	*CButton*	Buttons for computing the graph and redrawing a new graph
home, end	*CPoint*	Top left and bottom right corners of the drawing area
table	*CListCtrl*	List view table to show the weight matrix
v	*NODE*	A node of the graph

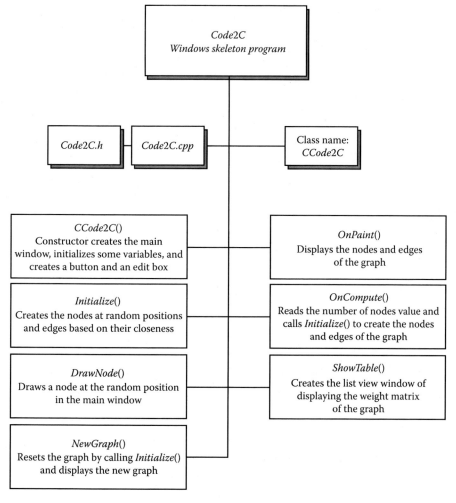

FIGURE 2.21
Organization of *Code2C*.

ShowTable(). The second button is *button2*, which refreshes the window by resetting the graph with new random locations.

The interface first asks the user to provide an input value for the number of nodes in the graph. With this value, a click on the *Draw Graph* button immediately produces the graph through *Initialize*() and *OnPaint*(). The function *Initialize*() creates the graph by generating the coordinates of the nodes and assigning weight values to the edges using random numbers. The graph is displayed through *OnPaint*(), which includes a call to *DrawNode*() to draw the nodes of the graph. The graph is refreshed whenever the *New Graph* button is clicked, which calls *Initialize*() and *Invalidate*(). The output function *ShowTable*() creates a list view window for displaying the weight matrix of the graph.

The complete code listings for *Code2C.h* and *Code2C.cpp* are given below:

```
//Code2C.h
#include <afxwin.h>
#include <afxcmn.h>
#include <fstream>
#define N 20                    //max #nodes
#define LinkRange 200           //transmission range
#define IDC_BUTTON1 501
#define IDC_BUTTON2 502

class CCode2C: public CFrameWnd
{
private:
        int n;
        CEdit eNodes;
        CButton button1,button2;
        CPoint home,end;
        CFont fHelvetica,fArial;
        CListCtrl table;
        typedef struct
        {
                CPoint home;
                CRect rct;
                int wt[N+1];
        } NODES;
        NODES v[N+1];
public:
        CCode2C();
        ~CCode2C()      {}
        void Initialize(),NewGraph(),DrawNode(int),ShowTable();
        afx_msg void OnPaint();
        afx_msg void OnCompute();
        DECLARE_MESSAGE_MAP()
};

class CMyWinApp:public CWinApp
{
        public:
                virtual BOOL InitInstance();
};
CMyWinApp MyApplication;
```

```
BOOL CMyWinApp::InitInstance(void)
{
      m_pMainWnd=new CCode2C;
      m_pMainWnd->ShowWindow(m_nCmdShow);
      return TRUE;
}

//Code2C: Graph
#include "Code2C.h"
BEGIN_MESSAGE_MAP(CCode2C,CFrameWnd)
      ON_WM_PAINT()
      ON_BN_CLICKED (IDC_BUTTON1,OnCompute)
      ON_BN_CLICKED (IDC_BUTTON2,NewGraph)
END_MESSAGE_MAP()

CCode2C::CCode2C()
{
      CString s;
      home=CPoint(30,100); end=CPoint(500,700);
      fArial.CreatePointFont(60,L"Arial");
      fHelvetica.CreatePointFont(100,L"Helvetica");
      Create(NULL,L"Graph",WS_OVERLAPPEDWINDOW,CRect(0,0,1000,800));
      button1.Create(L"Draw Graph",WS_CHILD | WS_VISIBLE |
                        BS_DEFPUSHBUTTON,
            CRect(CPoint(30,30),CSize(180,40)),this,IDC_BUTTON1);
      eNodes.Create(WS_CHILD | WS_VISIBLE | WS_BORDER | SS_CENTER,
            CRect(CPoint(home.x+350,home.y-60),CSize(100,25)),this,401);
}

void CCode2C::OnCompute()
{
      CString s;
      eNodes.GetWindowText(s); n=_ttoi(s);
      n=((n>N || n<=1)?N:n);
      Initialize();
}

void CCode2C::Initialize()
{
      int i,j;
      double distance;
      button1.DestroyWindow(); button2.DestroyWindow();
      button2.Create(L"New Graph",WS_CHILD | WS_VISIBLE |
                        BS_DEFPUSHBUTTON,
            CRect(CPoint(30,30),CSize(180,40)),this,IDC_BUTTON2);
      srand(time(0));
      for (i=1;i<=n;i++)
      {
            v[i].home.x=home.x+20+rand()%(end.x-home.x-50);
            v[i].home.y=home.y+20+rand()%(end.y-home.y-50);
            v[i].rct=CRect(v[i].home.x,v[i].home.y,v[i].home.
                              x+25,v[i].home.y+25);
      }
      for (i=1;i<=n;i++)
```

```
        {
                v[i].wt[i]=0;
                for (j=i+1;j<=n;j++)
                {
                        distance=sqrt(pow((double)(v[i].home.x-v[j].
                                        home.x),2)
                                        +pow((double)(v[i].home.y-v[j].
home.y),2));
                        if (distance<=LinkRange)
                                v[i].wt[j]=1+rand()%9;
                        else
                                v[i].wt[j]=99;
                        v[j].wt[i]=v[i].wt[j];
                }
        }
        Invalidate();
        ShowTable();
}

void CCode2C::OnPaint()
{
        CPaintDC dc(this);
        CString s;
        CRect rct;
        CPen rPen(PS_SOLID,3,RGB(0,0,0));
        CPen mPen(PS_SOLID,2,RGB(0,0,0));
        CPen qPen(PS_SOLID,1,RGB(100,100,100));
        CPoint mPoint,pt;
        int i,j;
        dc.SelectObject(fHelvetica);
        dc.TextOutW(home.x+260,home.y-55,L"#Nodes (2-20)");
        dc.SelectObject(rPen);
        dc.Rectangle(home.x-10,home.y-10,end.x+10,end.y+10);
        dc.SelectObject(qPen); dc.SelectObject(fArial);
        for (i=1;i<=n;i++)
                for (j=1;j<=n;j++)
                        if (v[i].wt[j]!=99)
                        {
                                dc.MoveTo(CPoint(v[i].home));
                                dc.LineTo(CPoint(v[j].home));
                                mPoint=CPoint((v[i].home.x+v[j].home.x)/2,
                                        (v[i].home.y+v[j].home.y)/2);
                                s.Format(L"%d",v[i].wt[j]);
                                dc.TextOut(mPoint.x,mPoint.y,s);
                        }
        for (i=1;i<=n;i++)
                DrawNode(i);
}

void CCode2C::DrawNode(int u)
{
        CClientDC dc(this);
        CString s;
        CPoint pt;
```

```
        CPen pBlack(PS_SOLID,2,RGB(0,0,0));
        dc.SelectObject(pBlack);
        pt=v[u].home;
        dc.Rectangle(v[u].rct);
        s.Format(L"%d",u); dc.TextOut(pt.x+5,pt.y+5,s);
}

void CCode2C::ShowTable()
{
        CString s;
        int i,j;
        table.DestroyWindow();
        CRect rct=CRect(CPoint(end.x+30,home.y+20),CSize(400,370));
        //creates the list view window
        table.Create(WS_VISIBLE | WS_CHILD | WS_DLGFRAME | LVS_REPORT
                | LVS_NOSORTHEADER,rct,this,802);
        for (j=0;j<=n;j++)
        {
                s.Format((j==0)?L"wt":L"%d",j);
                table.InsertColumn(j,s,LVCFMT_CENTER,30);
        }
        for (i=1;i<=n;i++)
        {
                s.Format(L"%d",i);
                table.InsertItem(i-1,s,0);
                for (j=1;j<=n;j++)
                {
                        s.Format(L"%d",v[i].wt[j]); table.
                                SetItemText(i-1,j,s);
                }
        }
}

void CCode2C::NewGraph()
{
        Initialize(); Invalidate();
}
```

2.8 Hexagonal Network

One interesting type of network based on a graph is the *hexagonal network,* sometimes called the honeycomb network. The hexagonal network has many applications, among them are in the wireless cellular network and multistage interconnection network. The wireless cellular network marks its area of coverage consisting of a set of regular hexagons. A single hexagon represents an area covered by a base station at its center. The circular transmission emitted from the base station overlaps with transmission circles from five other base stations (assuming all transmissions are the same size) to produce the regular hexagon.

A multistage interconnection network is a network that has processing nodes on the left and right boundaries and switches in its internal nodes. A common design has a hexagonal arrangement of the switches to enable communication between the nodes from the

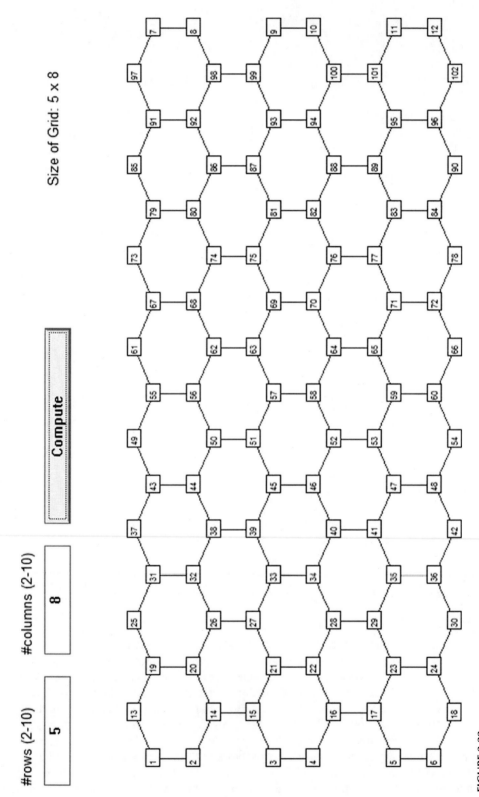

FIGURE 2.22
Hexagonal network of size 5 × 8.

opposing boundaries. The left nodes may serve as the input nodes, and they communicate with the nodes on the right boundary using the paths through the switches. Figure 2.22 shows a typical hexagonal network of size 5 × 8 for a multistage interconnection network. The nodes are labeled based on columns in the order of top to bottom. The numbering of nodes starts from the left boundary and continues with the right boundary followed by the interior nodes from left to right.

2.8.1 *Code2D*: Designing a Hexagonal Network

Code2D illustrates the design of a hexagonal network as shown in Figure 2.23. There are five rows and eight columns of hexagonal paths with the vertices of the hexagons forming the nodes of the network. A communication path can be drawn between pairs of left–right nodes. For example, v_5 can communicate with v_8 through the following path:

$$v_5 \rightarrow v_{17} \rightarrow v_{23} \rightarrow v_{29} \rightarrow v_{28} \rightarrow v_{34} \rightarrow v_{40} \rightarrow v_{46} \rightarrow v_{52} \rightarrow v_{58} \rightarrow v_{57}$$

$$\rightarrow v_{63} \rightarrow v_{69} \rightarrow v_{75} \rightarrow v_{81} \rightarrow v_{87} \rightarrow v_{93} \rightarrow v_{99} \rightarrow v_{98} \rightarrow v_8$$

Of course, this path may not be optimal. Finding the shortest path between the pairs of nodes is one of the most challenging problems. In many cases, matching the pairs of nodes with some constraints often reduces to graph theoretical problems.

In this section, we discuss the hexagonal network design problem. Fundamentally, any drawing on the computer requires good trigonometric knowledge and understanding. Lines and other graphical objects are drawn based on the alignment of pixels using the available Windows coordinates. In mathematics, the (x, y) Cartesian coordinates are treated differently: x increases from left to right, while y increases from bottom to top. Windows supports integer values only with (0,0) start at the top-left corner. The coordinates increase from left to right and top to bottom.

Figure 2.23 shows the Cartesian (a) and Windows (b) coordinates systems. We denote the coordinates of a point in the Cartesian system as (x, y) and a pixel in the Windows system as (X, Y). The Windows coordinate system starts with (0,0) in the top left-hand corner as its origin. The end points in Windows depend on the desktop resolution of the computer. A typical 800 × 600 screen resolution has 800 columns and 600 rows of pixels with the coordinates of (799,0), (0,599) and (799,599) in the top right-hand, bottom left-hand, and bottom right-hand corners, respectively. A higher resolution screen, such as 1280 × 1024, can be

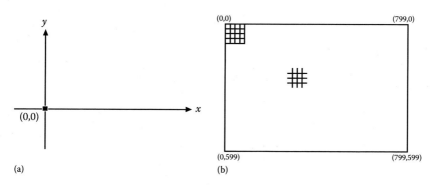

FIGURE 2.23
Cartesian system (a) and windows resolution of size 800 × 600 (b).

obtained by setting the properties in the desktop resolution. This setting produces finer pixels made from 1280 columns and 1024 rows for a crisper display. However, a very fine screen resolution also has some drawbacks. As the number of pixels increases, more memory is needed in displaying graphical objects. Some graphics-intensive applications may fail if the allocated amount of memory is not sufficient as a result of high-resolution displays.

Mapping a point (x, y) from the Cartesian system to its corresponding pixel (X, Y) in Windows is a straightforward procedure involving the linear relationships, $X = m_1 x + c_1$ and $Y = m_2 y + c_2$. In these equations, m_1 and m_2 are the gradients in the mapping $x \to X$ and $y \to Y$, respectively. The constants c_1 and c_2 are the y-axis intercepts of the lines in the Cartesian system. In solving for m_1, m_2, c_1, and c_2, a total of four equations are required that can be obtained from points — any two points in the Cartesian system and their range in Windows.

The mapping from (x, y) to (X, Y) is a type of mapping called *linear transformation*. In this transformation, all coordinates in the domain are related to their images in the range through a linear function. Figure 2.24 shows a linear transformation from a line in the Cartesian system to Windows. The line on the left is made up of the points (x_1, y_1) and (x_2, y_2) that map to the Windows coordinates, (X_1, Y_1) and (X_2, Y_2), respectively. The mapping involving $x \to X$ is linearly represented as

$$X_1 = m_1 x_1 + c_1, \tag{2.1}$$

$$X_2 = m_1 x_2 + c_1. \tag{2.2}$$

At the same time, the mapping $y \to Y$ is also linearly represented as

$$Y_1 = m_2 y_1 + c_2, \tag{2.3}$$

$$Y_2 = m_2 y_2 + c_2. \tag{2.4}$$

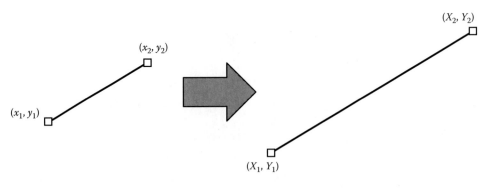

FIGURE 2.24
Conversion from Cartesian to Windows.

Solving the first two equations, we obtain

$$m_1 = \frac{X_2 - X_1}{x_2 - x_1}, \tag{2.5}$$

$$c_1 = X_1 - m_1 x_1. \tag{2.6}$$

The mapping equation in $y \rightarrow Y$ is solved in the same manner to produce the following results:

$$m_2 = \frac{Y_2 - Y_1}{y_2 - y_1}, \tag{2.7}$$

$$c_2 = Y_1 - m_2 y_1. \tag{2.8}$$

Equations 2.1 through 2.4 provide very useful conversion criteria for coordinates from Cartesian to Windows. The contribution is obvious especially in confining certain points in the real coordinate system to be within a rectangular region in Windows.

Cartesian points using floating-point numbers can be declared in C++ using a structure called *PT*, as follows:

```
typedef struct
{
        double x,y;
} PT;
```

From the above structure, conversion can be performed from Cartesian to Windows, and vice versa. First, the drawing area needs to have the direct mappings of its points inside from Cartesian to Windows. Let *home* and *end* be the top-left and bottom-right Windows *CPoint* objects of the drawing area, respectively. Also, let *CHome* and *CEnd* be their corresponding Cartesian points. It follows from Equations 2.5 through 2.8 that

$$m_1 = \frac{end.x - home.x}{CEnd.x - CHome.x}, \tag{2.9}$$

$$c_1 = home.x - m_1 CHome.x, \tag{2.10}$$

$$m_2 = \frac{end.y - home.y}{CEnd.y - CHome.y}, \tag{2.11}$$

$$c_2 = home.y - m_2 CHome.y. \tag{2.12}$$

A special function called *ConvertCW()* is introduced in *Code2D* to handle the conversion of coordinates from Cartesian to Windows using Equations 2.9 through 2.12.

A single hexagon is drawn using the Cartesian coordinates and then the coordinates are transformed for display in Windows using Equations 2.9 through 2.12. It follows that a tile of hexagons can be drawn by first drawing a single hexagon and then repeating the drawing using a single or nested loop.

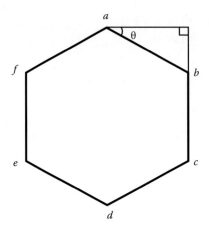

FIGURE 2.25
A regular hexagon and its vertices.

Figure 2.25 shows a regular hexagon with its vertices marked as *a*, *b*, *c*, *d*, *e*, and *f*. All these vertices can be declared with type *PT* as they will use the Cartesian coordinates. They can also be grouped to be the hexagonal members of the structure *HEX*, as follows:

```
typedef struct
{
        PT a,b,c,d,e,f;
} HEX;
```

It is also known that each interior angle of a regular hexagon is 120°. Therefore, θ in the figure is 30°, or π/6 radians. Assuming each side of the hexagon has length *L* and *a* has the coordinates (0, 0), then the following values are obtained relative to *a*:

$$a.x = 0, a.y = 0 \tag{2.13}$$

$$b.x = a.x + L\cos\frac{\pi}{6}, b.y = a.y - L\sin\frac{\pi}{6} \tag{2.14}$$

$$c.x = b.x, c.y = a.y - L - L\sin\frac{\pi}{6} \tag{2.15}$$

$$d.x = a.x, d.y = a.y - L - 2L\sin\frac{\pi}{6} \tag{2.16}$$

$$e.x = a.x - L\cos\frac{\pi}{6}, e.y = c.y \tag{2.17}$$

$$f.x = e.x, f.y = b.y. \tag{2.18}$$

TABLE 2.5

Important Objects and Variables in *Code2D*

Variable	Type	Description
bCompute	*CButton*	Buttons for computing the graph
CHome, CEnd	PT	Cartesian coordinates of the top-left and bottom-right points, respectively, of the drawing area
eR, eC	*CEdit*	Input boxes for the number of rows and columns, respectively
h[i][j]	*HEX* array	Hexagon where *i* is the row number and *j* is the column number
home, end	*CPoint*	Top left and bottom right corners of the drawing area
m, n	*int*	Number of rows and columns, respectively, of the hexagonal network
v[i]	*NODE*	A vertex of the hexagons labeled as *i*

In *Code2D*, the vertices are assigned with the values according to Equations 2.13 through 2.18 inside a function called *Initialize()*. The equations produce coordinates for a single hexagon. A nested loop is then applied to produce *m* rows and *n* columns of hexagons by repeatedly drawing the single hexagon. The coordinates are converted into Windows coordinates using *ConvertCW()* before they are displayed in the window.

Code2D has a single class called *CCode2D*. The source files are *Code2D.h* and *Code2D.cpp*. Table 2.5 lists important variables and objects in the program. The input has edit boxes for *eR* and *eC* for getting the number of rows and columns. A button called *bCompute* activates the input by displaying the hexagonal network in the drawing area. The drawing area is bounded by *home* and *end* at the top-left and bottom-right points, respectively. The corresponding Cartesian points are *CHome* and *CEnd*. The hexagons are labeled as a two-dimensional array called *h* while their vertices are represented as *v*.

Figure 2.26 is the organization of *Code2D* showing all the related functions. The constructor *CCode2D()* creates the main window, a button, and two edit boxes *eR* and *eC* for the input. The function also initializes some variables. *OnPaint()* displays the initial output that consists of the drawing area.

The input is activated by the button *bCompute* that calls *OnCompute()* to read the input values in the form of number of rows and columns for the network. This function then calls *Initialize()* to compute the coordinates of the vertices of the hexagons and *DrawHex()* to display the hexagons in the drawing area. A hexagon is identified as *h[i][j]*, where *i* is the row number and *j* is the column number. The vertices of the hexagons are labeled as array *v* starting from the top left according to columns and rows.

Initialize() computes the coordinates of the vertices based on Equations 2.13 through 2.18. The coordinates are converted to Windows through *ConvertCW()*. Each hexagon has its vertices arranged in a structured manner according to Figure 2.25. The labeling of the vertices starts in the first column in the order from top to bottom then continues at the last column before the interior nodes.

The full codes for *Code2D* are given as follows:

```
//Code2D.h
#include <afxwin.h>
#include <math.h>
#define IDC_COMPUTE 500
#define M 10        //max columns
#define N 10        //max rows
```

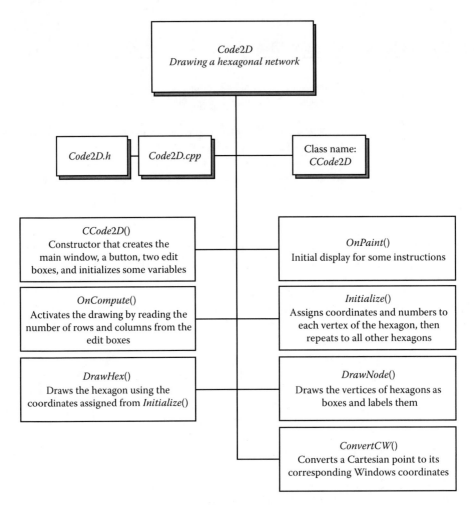

FIGURE 2.26
Organization of *Code2D*.

```
typedef struct               //Cartesian points
{
      double x,y;
      int p;

} PT;
typedef struct
{
      PT a,b,c,d,e,f;

} HEX;
typedef struct
{
      PT home;
      CRect rct;
      bool C[2*M*(2*N+1)];
} NODE;
```

```
class CCode2D: public CFrameWnd
{
private:
        NODE v[2*M*(2*N+1)];
        int idc,m,n;
        CEdit eR,eC;
        CButton bCompute;
        CPoint home,end;               //Windows home, end
        CFont fArial,fHelvetica;
        HEX h[M+1][N+1];
        PT CHome,CEnd;                 //CHome,CEnd=Cartesian home,end
public:
        CCode2D();
        ~CCode2D()               {}
        void Initialize(),DrawHex(),DrawNode(int);
        CPoint ConvertCW(PT);
        afx_msg void OnCompute();
        afx_msg void OnPaint();
        DECLARE_MESSAGE_MAP();
};

class CMyWinApp: public CWinApp
{
public:
        virtual BOOL InitInstance();
};
CMyWinApp MyApplication;

BOOL CMyWinApp::InitInstance()
{
        m_pMainWnd=new CCode2D;
        m_pMainWnd->ShowWindow(m_nCmdShow);
        return TRUE;
}

//Code2D.cpp
#include "Code2D.h"
#define PI 3.142
#define L (double)1.0

BEGIN_MESSAGE_MAP(CCode2D,CFrameWnd)
        ON_WM_PAINT()
        ON_BN_CLICKED(IDC_COMPUTE,OnCompute)
END_MESSAGE_MAP()
CCode2D::CCode2D()
{
        idc=400;
        Create(NULL,L"Hexagon",WS_OVERLAPPEDWINDOW,CRect(0,0,1000,800));
        home=CPoint(20,50); end=CPoint(600,500);
        CHome.x=-2; CEnd.x=10; CHome.y=2; CEnd.y=-10;
        eR.Create(WS_CHILD | WS_VISIBLE | WS_BORDER | SS_CENTER,
                CRect(CPoint(home.x+30,home.y-10),CSize(100,25)),this,idc++);
        eC.Create(WS_CHILD | WS_VISIBLE | WS_BORDER | SS_CENTER,
                CRect(CPoint(home.x+150,home.y-10),CSize(100,25)),this,idc++);
```

```
        bCompute.Create(L"Compute",WS_CHILD | WS_VISIBLE | BS_DEFPUSHBUTTON,
             CRect(CPoint(home.x+270,home.y-10),CSize(180,25))),
                   this,IDC_COMPUTE);
        fArial.CreatePointFont(60,L"arial");
        fHelvetica.CreatePointFont(100,L"Helvetica");
}

void CCode2D::OnPaint()
{
        CPaintDC dc(this);
        dc.SelectObject(fHelvetica);
        dc.TextOutW(home.x+30,home.y-35,L"#rows (2-10)");
        dc.TextOutW(home.x+150,home.y-35,L"#columns (2-10)");
}

void CCode2D::OnCompute()
{
        CString s;
        eR.GetWindowText(s); m=_ttoi(s);
        if (m>10)
            m=10;
        eC.GetWindowText(s); n=_ttoi(s);
        if (n>10)
            n=10;
        if (m%2==0)
            m--;
        Initialize();
        DrawHex();
}

void CCode2D::Initialize()
{
        int i,j,k,r;
        h[1][1].a.x=0; h[1][1].a.y=0;
        for (i=1;i<=m;i++)                    //i=row, j=column
            for (j=1;j<=n;j++)
                {
                    if (i%2==1)
                    {
                        h[i][j].a.x=h[1][1].a.x+2*L*cos(PI/6)*(j-1);
                        h[i][j].a.y=h[1][1].a.y-(2*L+2*L*sin(PI/6))*(i-1)/2;
                    }
                    else
                        if (j<n)
                            {
                                h[i][j].a.x=h[1][1].a.x+L*cos(PI/6)+2*L*
                                cos(PI/6)*(j-1);
                                h[i][j].a.y=(h[1]
[1].a.y-L-L*sin(PI/6))
                                -(2*L+2*L*sin(PI/6))*((i-1)/2);
                            }
                        h[i][j].b.x=h[i][j].a.x+L*cos(PI/6);
                        h[i][j].b.y=h[i][j].a.y-L*sin(PI/6);
```

```
                        h[i][j].c.x=h[i][j].b.x;
                        h[i][j].c.y=h[i][j].a.y-L*sin(PI/6)-L;
                        h[i][j].d.x=h[i][j].a.x;
                        h[i][j].d.y=h[i][j].a.y-2*L*sin(PI/6)-L;
                        h[i][j].e.x=h[i][j].a.x-L*cos(PI/6); h[i][j].e.y=h[i]
[j].c.y;
                        h[i][j].f.x=h[i][j].e.x; h[i][j].f.y=h[i][j].b.y;
            }

    k=1;
    for (j=1;j<=n;j++)
    {
            for (i=1;i<=m;i++)
                    if (i%2==1)
                    {
                            if (j==1)      //first column (f,e,a,d) and last
                            columns (b,c)
                            {
                                    r=k+(j-1)*(m+1);
                                    h[i][j].f.p=r; v[r].home=h[i][j].f;
                                    r++;
                                    h[i][j].e.p=r; v[r].home=h[i][j].e;
                                    r=k+(j+1)*(m+1);
                                    h[i][j].a.p=r; v[r].home=h[i][j].a;
                                    r++;
                                    h[i][j].d.p=r; v[r].home=h[i][j].d;
                                    h[i+1][j].f.p=h[i][j].d.p;
                                    r=k+j*(m+1);
                                    h[i][n].b.p=r; v[r].home=h[i][n].b;
                                    r++;
                                    h[i][n].c.p=r; v[r].home=h[i][n].c;
                                    k+=2;
                            }
                            if (j>1)       //intermediate columns (b,c,a,d)
                            {
                                    if (j==2 && i==1)
                                            k=1;
                                    r=k+(j+1)*(m+1); h[i][j].f.p=r;
                                    h[i][j-1].b.p=h[i][j].f.p;
                                    v[r].home=h[i][j].f;
                                    r++;
                                    h[i][j].e.p=r; v[r].home=h[i][j].e;
                                    h[i][j-1].c.p=h[i][j].e.p;
                                    r=k+(j+2)*(m+1); h[i][j].a.p=r;
                                    v[r].home=h[i][j].a;
                                    r++;
                                    h[i][j].d.p=r; v[r].home=h[i][j].d;
                                    k+=2;
                            }
                    }
    }
    for (i=1;i<=m;i+=2)
            for (j=1;j<=n-1;j++)
```

```
                {
                        h[i][j].b.p=h[i][j+1].f.p;
                        h[i][j].c.p=h[i][j+1].e.p;
                }
        for (i=2;i<=m-1;i+=2)
                for (j=1;j<=n-1;j++)
                {
                        h[i][j].f.p=h[i-1][j].d.p;
                        h[i][j].e.p=h[i+1][j].a.p;
                        h[i][j].a.p=h[i-1][j].c.p;
                        h[i][j].d.p=h[i+1][j].b.p;
                        h[i][j].b.p=h[i-1][j+1].d.p;
                        h[i][j].c.p=h[i+1][j+1].a.p;
                }
        int u[7],ii,jj;
        for (i=1;i<=m;i++)
                for (j=1;j<=n;j++)
                {
                        u[1]=h[i][j].a.p; u[2]=h[i][j].b.p; u[3]=h[i]
[j].c.p;
                        u[4]=h[i][j].d.p; u[5]=h[i][j].e.p;      u[6]=h[i]
[j].f.p;
                        if (i%2==1 || (i%2==0 && j<n))
                                for (ii=1;ii<=6;ii++)
                                        for (jj=1;jj<=6;jj++)
                                        {
                                                v[u[ii]].C[u[jj]]=0;
                                                if (ii!=jj)
                                                        v[u[ii]].C[u[jj]]=1;
                                        }
                }
}

void CCode2D::DrawHex()
{
        CClientDC dc(this);
        CPoint Q;
        CString s;
        int i,j,k,r;
        CPen pBlue(PS_SOLID,1,RGB(0,0,255));
        CPen pBlack(PS_SOLID,1,RGB(0,0,0));
        CPen pRed(PS_SOLID,1,RGB(255,0,0));
        dc.SelectObject(fHelvetica);
        s.Format(L"Size of Grid:%d x%d",m,n);
        dc.TextOutW(end.x,home.y-10,s);
        dc.SelectObject(fArial);dc.SelectObject(pBlue);
        for (i=1;i<=m;i++)
                for (j=1;j<=n;j++)
                {
                        Q=ConvertCW(h[i][j].a); dc.MoveTo(Q);
                        Q=ConvertCW(h[i][j].b); dc.LineTo(Q);
                        Q=ConvertCW(h[i][j].c); dc.LineTo(Q);
                        Q=ConvertCW(h[i][j].d); dc.LineTo(Q);
```

```
                          Q=ConvertCW(h[i][j].e); dc.LineTo(Q);
                          Q=ConvertCW(h[i][j].f);    dc.LineTo(Q);
                          Q=ConvertCW(h[i][j].a); dc.LineTo(Q);
                }
        k=1;
        for (j=1;j<=n;j++)
        {
                for (i=1;i<=m;i++)
                        if (i%2==1)
                        {
                                if (j==1)     //first column (f,e,a,d),last
                                                columns (b,c)
                                {
                                        r=k+(j-1)*(m+1); DrawNode(r);
                                        r++; DrawNode(r);
                                        r=k+(j+1)*(m+1); DrawNode(r);
                                        r++; DrawNode(r);
                                        r=k+j*(m+1); DrawNode(r);
                                        r++; DrawNode(r);
                                        k+=2;
                                }
                                if (j>1)       //intermediate columns (b,c,a,d)
                                {
                                        if (j==2 && i==1)
                                                k=1;
                                        r=k+(j+1)*(m+1); DrawNode(r);
                                        r++; DrawNode(r);
                                        r=k+(j+2)*(m+1); DrawNode(r);
                                        r++; DrawNode(r);
                                        k+=2;
                                }
                        }
        }
}

void CCode2D::DrawNode(int u)
{
        CClientDC dc(this);
        CString s;
        CPoint pt;
        CRect rct;
        CPen pBlack(PS_SOLID,1,RGB(0,0,0));
        dc.SelectObject(fArial); dc.SelectObject(pBlack);
        pt=ConvertCW(v[u].home);
        rct=CRect(pt.x-7,pt.y-7,pt.x+10,pt.y+7);
        dc.Rectangle(rct);
        s.Format(L"%d",u); dc.TextOut(pt.x-5,pt.y-5,s);
}

CPoint CCode2D::ConvertCW(PT p)           //converts C to W
{
        CPoint px;
        double m1,c1,m2,c2;
```

```
m1=(double)(end.x-home.x)/(CEnd.x-CHome.x);
c1=(double)(home.x-CHome.x*m1);
m2=(double)(end.y-home.y)/(CEnd.y-CHome.y);
c2=(double)(home.y-CHome.y*m2);
px.x=p.x*m1+c1; px.y=p.y*m2+c2;
return px;
}
```

3

Graph Coloring

3.1 Background

Graph coloring is a well-known NP-complete problem in discrete mathematics. The problem is classified as combinatorial optimization and has three main subproblems: node coloring, edge coloring, and face coloring. In *node coloring*, or vertex coloring, the problem is to minimize the number of colors needed to color the nodes of a connected graph in such a way that no two adjacent nodes share the same color. In a similar manner, *edge coloring* is the problem of coloring the edges in such a way that no two edges sharing the same node are assigned with the same color. In general, edge coloring becomes node coloring by mapping the graph to a new graph where the edges of the first graph are the nodes of the second graph. In *surface coloring*, the objective is to minimize the number of colors assigned to the faces (regions) of a planar graph in such a way that no two adjacent faces share the same color. Traditionally, this problem applies in map coloring where different colors are used to color bordering countries.

Graph coloring applies to many problems in everyday life. Some notable applications are scheduling, assignment, and partitioning. In a school timetable problem, for example, different colors are assigned to subjects offered to a group of students. The timetable is to be scheduled in such a way that the subjects are offered at different class times and in different rooms to the same students. Referring to Figure 3.1, three students marked as 1, 2, and 3 are taking classes u, v, and w in rooms Q and R. The node adjacency in the graph speaks for the relationship where student 1 is enrolled in u and v, Student 2 has u and w, while student 3 takes v and w. The classes u and v are to be conducted at rooms Q, while w is assigned to R.

With this arrangement, the next step is to schedule the students in such a way that each will take one class at a time in one room. There are three parameters in the problem: students, the classes they are taking, and the rooms for the classes. For scheduling, the time parameter t is added to produce the schedule shown in Figure 3.2.

This schedule looks simple as only three students are involved. Imagine a school having 30 rooms with 1000 students and 50 different subjects offered to the students. A timetable schedule is to be drawn using five timeslots, that is, from $t = 0$ (8 am) to $t = 4$ (12 pm). Scheduling does not only apply to school timetables. In factories, there are machine scheduling problems that involve the assignment of different machines for parts manufacturing according to an assembly requirement. The parts need to be scheduled to their corresponding machines and assembled in the minimum time so that they can be delivered on time. This is a typical optimization problem that involves a graph coloring problem.

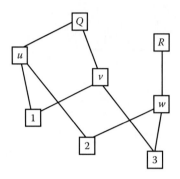

FIGURE 3.1
Example of graph for the time table.

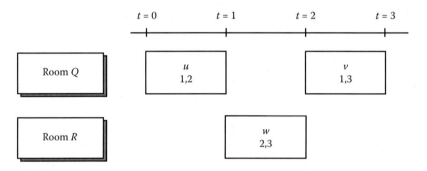

FIGURE 3.2
Timetable for the graph in Figure 3.1.

Graph coloring is very much related to *graph partitioning*, which is another branch of graph theory. The problem can be stated as follows:

> Given a weighted graph $G(V, E_w)$ set of nodes $V = \{v_i\}$ for $i = 1, 2,...n$ and edges e_{ij} for i, $j = 1, 2,...n$, how can the nodes be partitioned into k bins for $k = 2, 3,..., n$ in such a way that no nodes inside each bin are adjacent to one another? The main objective in this problem is to minimize the total interbin weights.

The relationship between graph partitioning and graph coloring can be seen from the nodes partitioned in each bin, which will be assigned with unique colors. Hence, the number of bins required in the graph partitioning problem is the same as the total number of colors. The only additional thing to do is to satisfy the objective.

Good literature on the graph partitioning problem and its solution using neural networks can be found in Van de Bout and Miller [1]. Given a weighted graph with n nodes, the problem of partitioning the nodes v_i into two bins, b_1 and b_2, can be stated as

$$\min \sum_{i=1}^{n} \sum_{j=1, j\neq i}^{n} c_{ij}(1-s_j)s_i - \lambda \sum_{i=1}^{n} \sum_{j=1, j\neq i}^{n} (1-s_j)s_i, \tag{3.1}$$

such that $s_i = \begin{cases} 0 & \text{if } v_i \in b_1 \\ 1 & \text{if } v_i \in b_2. \end{cases}$

In Equation 3.1, c_{ij} is the weight of the link between v_i and v_j, s_i is the state of the node v_i, and λ is the repulsion constant for balancing the equation. A simplified form of Equation 3.1 is obtained by disregarding the second term of the equation, and this form is commonly called the min-cut problem of graph partitioning.

In this chapter, the discussion will be confined to the node coloring problem only when its solution can also be applied as the solutions to the other two problems. This can be seen as edge and face coloring are both reduced to node coloring by reproducing a new graph in such a way the edges and the faces become the nodes of the new graph.

3.2 Node Coloring Problem

In a connected graph G with n nodes, n colors can be assigned to the nodes without any problems. But this is not optimum. The number of colors needed can be reduced by reusing the already assigned colors without breaking the general rule of no two adjacent nodes sharing the same color. Node coloring has been discussed in much of the literature; some excellent articles include Harary [2] and Gross and Yellen [3]. Among the methods proposed for solving the node coloring problem include [4] for the exact solution, [5] using tabu search, and [6] using a hybrid genetic algorithm.

The minimum number of colors required to color a graph in such a way that no two adjacent nodes share the same color is called the *chromatic number* of the graph, often denoted as $\chi(G)$. If the number of colors is fixed at k (which may not be the minimum number), the problem is called a k-coloring problem. A graph is said to be *k-colorable* if all the k colors can be assigned to the nodes, or $\chi(G) \leq k$. If $\chi(G) = k$, then the graph is k-chromatic. Figure 3.3

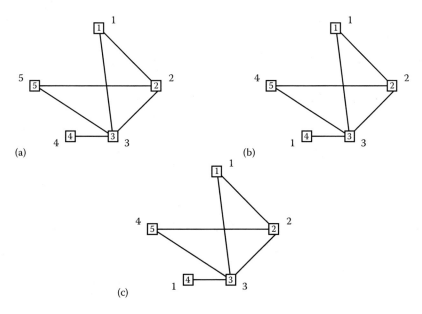

FIGURE 3.3
A graph that is (a) 5-colorable, (b) 4-colorable, and (c) 3-colorable.

shows three identical graphs that are (a) 5-colorable, (b) 4-colorable, and (c) 3-colorable. The color number for the nodes in each graph is shown in the boxes. This graph has $\chi(G) = 3$ as shown in Figure 3.3c.

The number of distinct ways the nodes of a graph can be colored using not more than a given number of colors is represented as a *chromatic polynomial*. The polynomial is also the number of *k*-colorings of the graph.

In general, finding $k = \chi(G)$ is the same as the *k*-partitioning problem. A bipartite graph has $\chi(G) = 2$, a tripartite graph has $\chi(G) = 3$, while a complete graph with $|V| = n$ has $\chi(G) = n - 1$.

The above node coloring is the simplest form of the problem that assigns different colors to pairs of nodes that are adjacent. Several other problems evolve from the node coloring problem. In another form of the problem called *T-coloring*, a total of *c* colors are assigned to nodes of the graph *G* such that if *u* and *v* are nodes of *G* and $u, v \in E(G)$, then $|c_u - c_v| \notin T$. Some interesting articles on this topic include Liu [7] and Graf [8].

3.2.1 Node Coloring and Eigenvalues

Graph coloring is very much related to eigenvalues and eigenvectors. This is obvious as a graph, either unweighted or weighted, can be represented as an adjacency matrix. Spectral graph theory is a branch of mathematics that relates a graph with eigenvalues and eigenvectors. As discussed in Chapter 1, eigenvalues λ and their corresponding eigenvectors *v* are formed from a given square matrix *A* through the relationship $Av = \lambda v$. This gives

$$(A - \lambda I)\, v = 0. \tag{3.2}$$

The above relationship leads to

$$|A - \lambda I| = 0, \tag{3.3}$$

which produces the characteristic polynomial $P(\lambda)$.

In general, a square matrix with size $n \times n$ has *n* real and complex eigenvalues λ_i for $i = 1, 2,..., n$. For a symmetric matrix, eigenvalues are real and distinct, and they can be arranged in terms of magnitude into the following form:

$$|\lambda_1| > |\lambda_2| >...> |\lambda_n|. \tag{3.4}$$

In the above ordered sequence, λ_1 is called the *most dominant eigenvalue* while λ_n is the *least dominant* eigenvalue. The corresponding eigenvectors are v_1 and v_n, respectively. The set $\{\lambda_n, \lambda_{n-1},..., \lambda_1\}$ is called the *spectrum* of the graph. The *spectral radius* $\rho(A)$ of the matrix *A* is the magnitude of the most dominant eigenvalue, or $\rho(A) = |\lambda_1|$.

Finding the most dominant and least dominant eigenvalues contribute in solving the node coloring problem where matrix *A* in the problem is the adjacency matrix of the graph. Assuming the graph is symmetric, which is represented by the adjacency matrix *A*, and has nonnegative weights, the following properties hold:

a. The eigenvalues λ_i for $i = 1, 2,..., n$ are the roots of $|A - \lambda I| = 0$ and have real values.

b. The adjacency matrix can be written as $A = \sum_{i=1}^{n} \lambda_i u_i u_i^T$, or $A = U^T D U$, where D is the diagonal matrix consisting of eigenvalues λ_i and $U = [u_i]$ is an orthogonal matrix with columns from eigenvectors of the corresponding eigenvalues λ_i. The eigenvectors u_i form the orthonormal basis vectors, or $u_i^T u_i = 1$.

c. The sum of the eigenvalues is zero, or $\sum_{i=1}^{n} \lambda_i = 0$.

d. λ_1 is the spectral radius of the graph, or $\lambda_1 \geq |\lambda_k|$ for all k.

e. The most dominant eigenvalue is simple (occurs once once), or $\lambda_1 > \lambda_2$. A normalized eigenvector has one of its elements with the largest value equaling unity, that is, 1 in this case.

It can be shown that the most dominant eigenvalue λ_1 is related to the chromatic number through the interval [2,3,9] defined by

$$\frac{n}{n - \lambda_1} \leq \chi(G) \leq 1 + \lambda_1, \tag{3.5}$$

which provides the lower and upper bounds for the chromatic number. The elements of the eigenvector of the most dominant vector then correspond to the number of colors of the graph.

3.2.2 Power Method for Finding the Most Dominant Eigenvalue

Both the most dominant and least dominant eigenvalues can be approximated using the numerical methods called the *power method* and *shifted power method*, respectively. The shifted power method is basically the same power method but applied on a slightly different matrix. The power method extension is the inverse power method that produces other eigenvalues based on an initial approximation. An excellent discussion of the numerical solution to this problem can be found in [10] while its implementation using C++ can be found in [11].

The power method is an iterative method that produces the most dominant eigenvalue through convergence after several iterations. The method starts with an initial eigenvector $v^{(0)}$, which is a normalized vector. A series of iterations k are performed to update the value of the eigenvalue λ and its corresponding eigenvector v using the following formula [10,11]:

$$v^{(k+1)} = \frac{1}{\lambda^{(k+1)}} v^{(k)}. \tag{3.6}$$

The power method is outlined in Algorithm 3.1. In the first iteration with $k = 0$, the product $Av^{(0)}$ produces a vector and the element of this vector with the most dominant value becomes $\lambda^{(0)}$. The eigenvector $v^{(1)}$ is then produced from Equation 3.1. The second iteration with $k = 1$ repeats the process to produce $\lambda^{(2)}$ and $v^{(2)}$. The error $|\lambda^{(2)} - \lambda^{(1)}|$ is computed and compared with the given tolerance ε, which is a small number close to 0. Subsequent iterations are performed with the same process applied until the error is smaller than ε. Convergence is achieved with this error and the final values of λ and the corresponding v become λ_1 and v_1, respectively.

Algorithm 3.1: Power Method for Finding the Most Dominant Eigenvalue [10,11]

Input: Matrix A.
Initial values:
    Maximum iterations k_{max}, $v^{(0)}$ and error tolerance ε.
Process:
For $k = 0$ to k_{max}
    Compute $Av^{(k)}$;
    Determine $\lambda^{(k+1)}$ = most dominant element in $v^{(k)}$;

    Update $v^{(k+1)} = \dfrac{1}{\lambda^{(k+1)}} v^{(k)}$;
    If $k > 0$
        Compute error = $|\lambda^{(k+1)} - \lambda^{(k)}|$;
        If error $< \varepsilon$
            $\lambda_1 \leftarrow \lambda^{(k+1)}$;
            $v_1 \leftarrow v^{(k+1)}$;
            Break from the loop;
        Endif
    Endif
Endfor
Output: λ_1 and v_1.

3.2.3 *Code3A*: Power Method for Estimating the Chromatic Number

The power method contributes to estimating the chromatic number of a graph. As given by Equation 3.4, the relationship requires the value of the most dominant eigenvalue. *Code3A* illustrates the power method using an example from the graph G from Figure 3.3. The graph produces the adjacency matrix A, as follows:

$$A = \begin{bmatrix} 0 & 1 & 1 & 0 & 0 \\ 1 & 0 & 1 & 0 & 1 \\ 1 & 1 & 0 & 1 & 1 \\ 0 & 0 & 1 & 0 & 0 \\ 0 & 1 & 1 & 0 & 0 \end{bmatrix}.$$

Figure 3.4 shows the output from *Code3A* for the above graph. The results are produced from the initial values given by $v^{(0)} = (0,1,0,0,0)^T$ and error tolerance of $\varepsilon = 0.05$. The iteration starts with $k = 0$ to produce $Av^{(0)} = (1,0,1,0,0)^T$, $\lambda^{(1)} = 1.0000$, and $v^{(1)} = \dfrac{1}{\lambda^{(1)}} Av^{(0)} = (1,0,1,0,0)^T$.

Next, with $k = 1$ we get $Av^{(1)} = (1,3,1,1,2)^T$, $\lambda^{(2)} = 3.0000$, $v^{(2)} = \dfrac{1}{\lambda^{(2)}} Av^{(1)} = (0.3333, 1.0000, 0.6667,$

$0.3333, 0.3333)^T$, and error = $|\lambda^{(2)} - \lambda^{(1)}| = 1 > \varepsilon$. Further iterations produce the results as shown in the list view of the figure.

The most dominant eigenvalue is $\lambda_1 = \lambda^{(15)} = 2.6821$ obtained after $k = 14$ iterations. The corresponding eigenvector is

$$v^{(15)} = (0.7078, 0.8993, 1.0000, 0.3728, 0.7078)^T.$$

From Equation 3.4, $2.1571 < \chi(G) < 3.6821$.

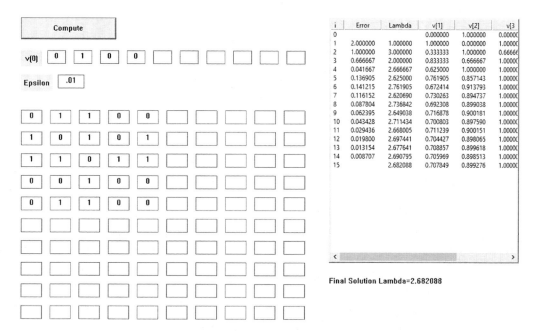

FIGURE 3.4
Output from *Code3A* for the most dominant eigenvector.

Code3A consists of the files *Code3A.h* and *Code3A.cpp*. A single class called *CCode3A* is used in this application. The main variables in this application are an eigenvalue and its corresponding eigenvector, which are represented as *lambda* and *v*, respectively. The variables are organized into a structure called *EIGEN*.

Table 3.1 lists some of the most important variables and objects in *Code3A*. For graph *G* with *n* nodes, the adjacency matrix is *A*. The power method computes *lambda* and *v* through a series of iterations that converge when the error is lower than *epsilon*.

The organization of *Code3A* is shown in Figure 3.5. The figure shows all the functions involved in the project. The processing starts with the constructor *CCode3A*(), which creates the main window and the *Compute* button. The function also creates the input edit boxes for matrix *A*, vector *v*, and the error tolerance *Epsilon* for the iterations in the power method.

TABLE 3.1

Important Objects and Variables in *Code3A*

Variable	Type	Description
$A[i][j]$	*double*	Matrix A with element A_{ij}
button	*CButton*	Buttons for computing the most dominant eigenvalue and its eigenvectors
table	*CListCtrl*	Displays the iterations for the most dominant eigenvalue and its eigenvectors
n	*int*	Matrix size for A; that is, $n \times n$
$eA[i][j]$, $eV[i]$, *eEpsilon*	*CEdit*	Input for A, v, and *Epsilon*
Epsilon	*double*	Error tolerance for the convergence

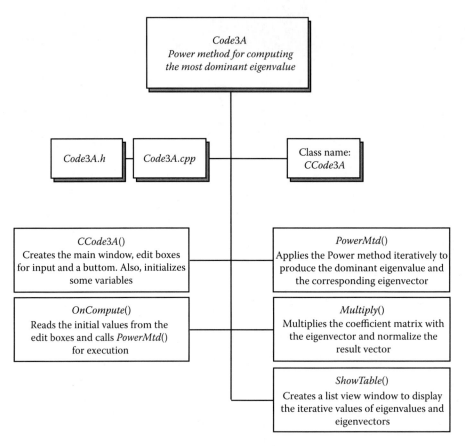

FIGURE 3.5
Organization of *Code3A*.

The power method is activated when the *Compute* button is pressed. The event is handled by *OnCompute*(), which first reads the input values from the edit boxes. The function calls *PowerMtd*(), which applies the power method in updating *lambda* and *v* values through a series of iterations. *PowerMtd*() calls *Multiply*() to perform multiplication on the matrix and the vector. Convergence is achieved when the computed error is lower than *epsilon*. With the convergence, the results from the iterations are displayed in the list view table through *ShowTable*().

The full codes from *Code3A* are listed as follows:

```
//Code3A.h
#include <afxwin.h>
#include <math.h>
#include <afxcmn.h>
#define N 10
#define M 200
#define IDC_BUTTON 501

class CCode3A: public CFrameWnd
{
private:
      typedef struct
```

```
        {
                double lambda,v[N+1];
        } EIGEN;
        EIGEN Eigen[M+1];
        int idc,n;
        double A[N+1][N+1],Epsilon;
        CEdit eA[N+1][N+1],eEpsilon,eV[N+1];
        CStatic sLambda[N+1],sV[N+1];
        CStatic sLabel1,sLabel2;
        CButton button;
        CListCtrl table;
public:
        CCode3A();
        ~CCode3A()      {}
        afx_msg void OnCompute();
        void PowerMtd();
        void Multiply(int),ShowTable(double [M+1],int);
        DECLARE_MESSAGE_MAP();
};

class CMyWinApp:public CWinApp
{
        public:
                virtual BOOL InitInstance();
};
CMyWinApp MyApplication;

BOOL CMyWinApp::InitInstance(void)
{
        m_pMainWnd=new CCode3A;
        m_pMainWnd->ShowWindow(m_nCmdShow);
        return TRUE;
}


//Code3A.cpp
#include "Code3A.h"

BEGIN_MESSAGE_MAP(CCode3A,CFrameWnd)
     ON_BN_CLICKED (IDC_BUTTON,OnCompute)
END_MESSAGE_MAP()

CCode3A::CCode3A()
{
     int i,j;
     idc=200; n=0;
     Create(NULL,L"Power Method",WS_OVERLAPPEDWINDOW,CRect(0,0,1000,900));
     button.Create(L"Compute",WS_CHILD | WS_VISIBLE | BS_DEFPUSHBUTTON,
         CRect(CPoint(20,50),CSize(180,40)),this,IDC_BUTTON);
     for (i=1;i<=N;i++)
        {
                for (j=1;j<=N;j++)
                        eA[j][i].Create(WS_CHILD | WS_VISIBLE | WS_BORDER |
SS_CENTER,
```

```
              CRect(CPoint(20+55*(j-1),220+40*(i-1)),CSize(40,25)),this,idc++);
                 eV[i].Create(WS_CHILD | WS_VISIBLE | WS_BORDER | SS_CENTER,
                      CRect(CPoint(70+50*(i-1),110),CSize(35,25)),this,
idc++);
        }
        eEpsilon.Create(WS_CHILD | WS_VISIBLE | WS_BORDER | SS_CENTER,
              CRect(CPoint(90,155),CSize(50,25)),this,idc++);
        sLabel1.Create(L"Epsilon", WS_VISIBLE | SS_CENTERIMAGE |
SS_CENTER,
              CRect(CPoint(20,155),CSize(60,30)),this,IDC_STATIC);
        sLabel2.Create(L"v(0)", WS_VISIBLE | SS_CENTERIMAGE | SS_CENTER,
              CRect(CPoint(20,110),CSize(40,30)),this,IDC_STATIC);
}

afx_msg void CCode3A::OnCompute()
{
        int i,j;
        CString s;
        bool flag=0;
        for (i=1;i<=N;i++)
        {
                for (j=1;j<=N;j++)
                {
                        eA[j][i].GetWindowTextW(s); A[i][j]=_tstof(s);
                        if (i==j && s=="")
                        {
                                flag=1; break;
                        }
                }
                eV[i].GetWindowTextW(s); Eigen[0].v[i]=_tstof(s);
                if (flag)
                {
                        n=i-1; break;
                }
        }
        eEpsilon.GetWindowTextW(s); Epsilon=_tstof(s);
        PowerMtd();
}

void CCode3A::PowerMtd()
{
        int k,m=0;
        double error[M+1];
        for (k=0;k<M;k++)
        {
                Multiply(k);
                if (k>0)
                {
                        error[k]=fabs(Eigen[k+1].lambda-Eigen[k].lambda);
                        if (error[k]<Epsilon)
                        {
                                ShowTable(error,m);
                                break;
                        }
                }
```

```
                }
                m++;
        }
}

void CCode3A::Multiply(int k)
{
        int i,j;
        double max,w[N+1];
        for (i=1; i<=n; i++)
        {
                w[i]=0;
                for (j=1;j<=n;j++)
                w[i] +=A[i][j]*Eigen[k].v[j];
        }
        max=w[1];
        for (i=1;i<=n;i++)
                if (fabs(max)<fabs(w[i]))
                        max=w[i];
        Eigen[k+1].lambda=max;
        for (i=1;i<=n;i++)
                Eigen[k+1].v[i]=w[i]/Eigen[k+1].lambda;
}

void CCode3A::ShowTable(double error[M+1],int m)
{
        CClientDC dc(this);
        CString s;
        CString S[]={"","v[1]","v[2]","v[3]","v[4]","v[5]","v[6]","v[7]",
"v[8]"};
        int i,k;
        CRect rcTable=CRect(600,50,960,500);
        table.DestroyWindow();

        //creates the list view window
        table.Create(WS_VISIBLE | WS_CHILD | WS_DLGFRAME | LVS_REPORT
                | LVS_NOSORTHEADER,rcTable,this,idc++);
        table.InsertColumn(0,L"i",LVCFMT_CENTER,25);
        table.InsertColumn(1,L"Error",LVCFMT_CENTER,70);
        table.InsertColumn(2,L"Lambda",LVCFMT_CENTER,70);
        for (i=1;i<=n;i++)
        {
                table.InsertColumn(i+2,S[i],LVCFMT_CENTER,70);
        }

        for (k=0;k<=m+1;k++)
        {
                s.Format(L"%d",k); table.InsertItem(k,s,0);
                if (k>0)
                {
                        s.Format(L"%lf",Eigen[k].lambda); table.
SetItemText(k,2,s);
                        if (k<m+1)
                        {
```

```
                              s.Format(L"%lf",error[k]); table.
SetItemText(k,1,s);
                         }
                         if (k==m+1)
                         {
                              s.Format(L"Final Solution Lambda=%.6lf",
Eigen[k].lambda);
                              dc.TextOutW(600,525,s);
                         }
                    }
                    for (i=1;i<=n;i++)
                    {
                              s.Format(L"%lf",Eigen[k].v[i]); table.
SetItemText(k,i+2,s);
                         }
               }
}
```

3.3 Greedy Algorithm

Coloring the nodes of a graph is an NP-hard problem. Fundamentally, a greedy algorithm can be designed by taking one node v_i of the graph $G(V, E)$ at a time. This node's color is compared with the colors of the rest of the nodes in the graph. Let $c_i = k$ be the color assigned to v_i. The idea starts with the first node v_1. As initial values, assign the first color $k = 1$ to this node and $k = 0$ to all other nodes; that is, $c_0 = 1$ and $c_i = 0$. Next, select the second node v_2. Check for the adjacency between (v_1, v_2). If v_1 and v_2 are not adjacent, then v_2 will share the same color $k = 1$ as v_1. Otherwise, if v_2 is adjacent to v_1, then v_2 cannot be assigned with $c_2 = 1$. This node will definitely get the next color, or $c_2 = 2$.

Continuing with the next node v_3, again let $k = 1$. A check is made on the adjacencies of the pairs (v_3, v_1) and (v_3, v_2). If (v_3, v_1) is adjacent, then k increases by one to make $c_3 = 2$; otherwise, $c_3 = 1$. Similarly, if (v_3, v_2) is adjacent, then k increases by one again; otherwise, this value will not change. With that, the assignment becomes $c_3 = 1$, $c_3 = 2$, or $c_3 = 3$, depending on the adjacency status on (v_3, v_1) and (v_3, v_2).

This iterative process is repeated with $i = 4,5,...,n$. The color k starts with one at each i iteration, and increases by one for every adjacency in the pairs (v_i, v_j) for $j = 1,2,...,n$. The last value of k at the end of the iterations becomes the final color for v_i.

The whole idea of the greedy algorithm can be summarized as follows:

//Greedy algorithm for node coloring
Input:
   n = number of nodes,
   Coordinates of v_i for $i = 1,2,...,n$,
   Adjacency range R.
Initialize: let the node colors $c_1 = 1$, $c_k = 0$, for $k = 2,3,...,n$.
   Determine the adjacency of (v_i, v_j) for $i, j = 1,2,...,n$.
Process:
   For $i = 1$ to n

```
Let kMax = 1;
For j = 1 to n
    If (vᵢ, vⱼ) and are adjacent
        If kMax = cⱼ
            Update kMax++;
        Endif
    Endif
Endfor
Update cᵢ = kMax;
Endfor
```

Output: color assignment for v_i for $i = 1,2,...,n$.

3.3.1 *Code3B*: Node Coloring Using a Greedy Algorithm

A greedy algorithm is illustrated through *Code3B*. Figure 3.6 shows a sample output from this program. The output consists of two edit boxes for input, a drawing area for displaying the graph, a list view window for displaying the adjacency matrix, and the assignment of colors for the nodes. The coloring results are shown in the drawing area where the nodes

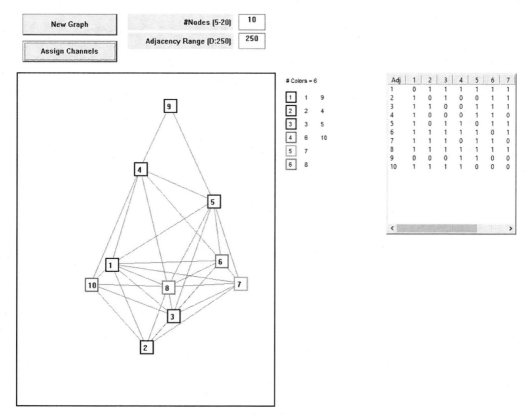

FIGURE 3.6
Sample output from *Code3B*.

TABLE 3.2

Important Objects and Variables in *Code3B*

Variable	Type	Description
adjRange	*int*	The range for two nodes to be adjacent
bCompute, bNGraph	*CButton*	Buttons for computing the graph and redrawing a new graph
eN, eAdj	*CEdit*	Input for the number of nodes and adjacency range
home, end	*CPoint*	Top-left and bottom-right corners of the drawing area
idc	*int*	Ids of objects in the application
kMax	*int*	Maximum number of colors assigned to the nodes
n	*int*	Number of nodes in the graph
table	*CListCtrl*	List view window for displaying the adjacency matrix
vColor[i]	*int*	Color assigned to v_i

are assigned with colors from the greedy algorithm. The output also displays the adjacency matrix of the graph in the list view window and the list of nodes assigned to the colors.

Input for *Code3B* consists of two edit boxes. The first edit box asks for the number of nodes for the graph, while the second asks for the adjacency range. The value in the adjacency range becomes the threshold value for the adjacency relationship between the pairs of nodes. This threshold value is compared with the Euclidean distance between a pair of nodes: if the distance is lower than the threshold value, then the two nodes are adjacent; otherwise, they are not. There are also two buttons, one for refreshing the drawing area for a new graph and another for assigning the colors to the nodes in the graph.

Code3B has the application class *CCode3B* in two files, *Code3B.h* and *Code3B.cpp*. A structure called *NODE* describes the members belonging to v_i including the coordinates (*home*), the adjacency relationship with other nodes (*adj*), and the assigned color (*color*).

The main objects and variables are listed in Table 3.2. Input is obtained from the edit boxes *eN* and *eAdj*, which represent the number of nodes and the adjacency range for the nodes, respectively. There are two buttons, *bCompute* for the *Compute* button and *bNGraph* for the *New Graph* button. The main output is the drawing area that is bounded by *home* and *end* as its top-left and bottom-right coordinates. The color assigned to v_i is denoted as *vColor[i]*.

Figure 3.7 is the organization of *Code3B*. There are seven application functions. The constructor *CCode3B()* creates the main window, two buttons, and two edit boxes for the input. The function also initializes some variables in the application. The initial node coordinates and edges linking them are defined in *Initialize()*. These values define the graph that is displayed and updated through *OnPaint()*. The graph is refreshed with new coordinates for the nodes and edges through *OnNewGraph()*.

The greedy algorithm is implemented in *OnCompute()*. The function starts by initializing the colors of all nodes to 0. A series of iterations is then performed in assigning the colors to the nodes. The iteration starts by assigning the first node v_1 with the first color. The color k for v_i is then compared to other nodes v_j in the iterations, and if v_j has the same color as v_i and both nodes are adjacent then k is increased by one. The color value k is assigned to v_i following this rule, and its maximum value is updated as *kMax*. With the assignment, the nodes are shown according to their assigned colors and this updates the graph in the drawing area.

The full codes for *Code3B* are listed as follows:

```
//Code3B.h
#include <afxwin.h>
#include <math.h>
```

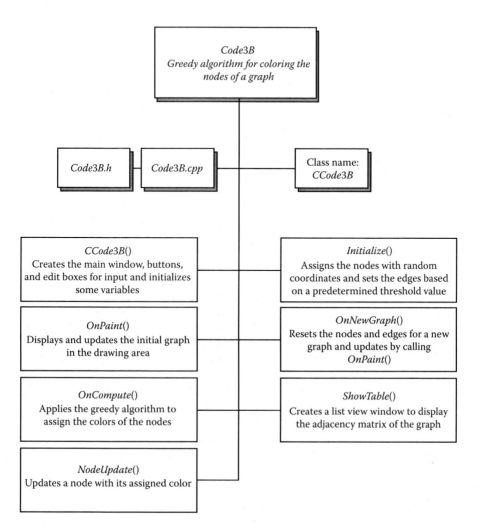

FIGURE 3.7
Organization of *Code3B*.

```
#include <afxcmn.h>
#define N 20
#define IDC_NGRAPH 500
#define IDC_COMPUTE 501

class CCode3B:public CFrameWnd
{
private:
        int idc,vColor[N+1];
        int n,kMax,adjRange;
        CPen pColor[N+1];
        CButton bNGraph,bCompute;
        CPoint home,end;
        CFont fArial8,fArial10;
        CListCtrl table;
        CEdit eN,eAdj;
        CStatic sN,sAdj;
```

```
        typedef struct
        {
                CPoint home;
                CRect rct;
                bool adj[N+1];
                int color;
        } NODE;
        NODE v[N+1];

public:
        CCode3B();
        ~CCode3B()              {}
        void Initialize(),NodeUpdate(int),ShowTable();
        afx_msg void OnPaint();
        afx_msg void OnLButtonDown(UINT,CPoint);
        afx_msg void OnNewGraph(),OnCompute();
        DECLARE_MESSAGE_MAP()
};

class CMyWinApp:public CWinApp
{
public:
        virtual BOOL InitInstance();
};
CMyWinApp MyApplication;

BOOL CMyWinApp::InitInstance()
{
        m_pMainWnd=new CCode3B;
        m_pMainWnd->ShowWindow(m_nCmdShow);
        return TRUE;
}

//Code3B.cpp
#include "Code3B.h"

BEGIN_MESSAGE_MAP(CCode3B, CFrameWnd)
        ON_WM_PAINT()
        ON_BN_CLICKED(IDC_NGRAPH,OnNewGraph)
        ON_BN_CLICKED(IDC_COMPUTE,OnCompute)
END_MESSAGE_MAP()

CCode3B::CCode3B()
{
        CPoint bHome=CPoint(30,20);
        idc=800;
        home=CPoint(30,140); end=CPoint(500,740);
        Create(NULL, L"Vertex coloring: Greedy algorithm",
                WS_OVERLAPPEDWINDOW,CRect(0,0,1000,800));
        bNGraph.Create(L"New Graph",WS_CHILD | WS_VISIBLE |
BS_DEFPUSHBUTTON,
                CRect(CPoint(bHome),CSize(180,40)),this,IDC_NGRAPH);
        bCompute.Create(L"Assign Colors",WS_CHILD | WS_VISIBLE |
BS_DEFPUSHBUTTON,
```

```
                      CRect(CPoint(bHome.x,bHome.y+50),CSize(180,40)),this,
IDC_COMPUTE);
        sN.Create(L"#Nodes (5-20) ", WS_VISIBLE | SS_CENTERIMAGE |
SS_RIGHT,
                  CRect(CPoint(bHome.x+200,bHome.y),CSize(200,30)),this,
IDC_STATIC);
        eN.Create(WS_CHILD | WS_VISIBLE | WS_BORDER | SS_CENTER,
                  CRect(CPoint(bHome.x+410,bHome.y),CSize(50,30)),this,idc++);
        sAdj.Create(L"Adjacency Range (D:250) ", WS_VISIBLE |
SS_CENTERIMAGE
                  | SS_RIGHT, CRect(CPoint(bHome.x+200,bHome.y+35),
                  CSize(200,30)),this,IDC_STATIC);
        eAdj.Create(WS_CHILD | WS_VISIBLE | WS_BORDER | SS_CENTER,
                  CRect(CPoint(bHome.x+410,bHome.y+35),CSize(50,30)),this,
idc++);
        fArial10.CreatePointFont(100,L"Arial");
        fArial8.CreatePointFont(80,L"Arial");
        int color[]={RGB(150,150,150),
                RGB(0,0,0),RGB(0,0,200),
                RGB(200,0,0),RGB(0,200,0),
                RGB(200,200,0),RGB(0,200,200),
                RGB(200,0,200),RGB(200,50,150),
                RGB(100,100,255),RGB(250,50,255),
                RGB(100,200,50),RGB(250,50,0),
                RGB(0,50,250),RGB(100,100,255),
                RGB(250,0,50),RGB(50,250,0),
                RGB(50,0,250),RGB(0,250,50),
                RGB(255,100,100),RGB(100,255,100)};
        for (int i=0;i<=N;i++)
                if (i==0)
                        pColor[i].CreatePen(PS_SOLID,1,color[i]);
                else
                        pColor[i].CreatePen(PS_SOLID,2,color[i]);
        Initialize();
}

void CCode3B::Initialize()
{
        int i,j;
        double distance;
        CString s;
        s=""; eN.GetWindowText(s); n=_ttoi(s); n=((n<=N)?n:N);
        s=""; eAdj.GetWindowText(s); adjRange=_ttoi(s);
        adjRange=((adjRange== 0)?250:adjRange);
        srand(time(0));
        for (i=1;i<=n;i++)
        {
                v[i].home.x=home.x+20+rand()%(end.x-home.x-50);
                v[i].home.y=home.y+20+rand()%(end.y-home.y-50);
                v[i].rct=CRect(CPoint(v[i].home.x-10,v[i].home.y-10),
CSize(25,25));
        }
        for (i=1;i<=n;i++)
        {
```

```
                v[i].adj[i]=0;
                for (j=i+1;j<=N;j++)
                {
                        distance=sqrt(pow(double(v[i].home.x-v[j].
home.x),2)
                            +pow(double(v[i].home.y-v[j].home.y),2));
                        v[i].adj[j]=((distance<=adjRange)?1:0);
                        v[j].adj[i]=v[i].adj[j];
                }
        }
}

void CCode3B::OnPaint()
{
        CPaintDC dc(this);
        CString s;
        int i,j;
        dc.SelectObject(pColor[1]);
        dc.Rectangle(home.x-10,home.y-10,end.x+10,end.y+10);
        dc.SelectObject(pColor[0]);
        for (i=1;i<=n;i++)
                for (j=1;j<=n;j++)
                        if (v[i].adj[j])
                        {
                                dc.MoveTo(CPoint(v[i].home));
                                dc.LineTo(CPoint(v[j].home));
                        }
        dc.SelectObject(fArial10); dc.SelectObject(pColor[1]);
        for (i=1;i<=n;i++)
        {
                dc.Rectangle(v[i].rct);
                s.Format(L"%d",i);
                dc.TextOutW(v[i].home.x-5,v[i].home.y-5,s);
        }
}

void CCode3B::OnNewGraph()
{
        Initialize(); Invalidate();
}

void CCode3B::OnCompute()
{
        int i,j,k;
        for (k=1;k<=n;k++)
        {
                vColor[k]=0; v[k].color=0;
        }
        for (i=1;i<=n;i++)
        {
                k=1;
                if (i==1)
                        kMax=k;
                else
```

```
                       for (j=1;j<=n;j++)
                            if (v[i].adj[j])
                                 if (k==v[j].color)
                                      k++;
              v[i].color=k; vColor[k]++;
              if (kMax<k)
                   kMax=k;
              NodeUpdate(i);
     }
     ShowTable();
}

void CCode3B::NodeUpdate(int i)
{
     CClientDC dc(this);
     CRect rct;
     CString s;
     int k=v[i].color;
     dc.SelectObject(pColor[k]);
     dc.Rectangle(v[i].rct);
     s.Format(L"%d",i);                         //display nodes in graph

     //display color info
     dc.TextOutW(v[i].home.x-5,v[i].home.y-5,s);
     rct=CRect(CPoint(end.x+30,home.y+25*k),CSize(20,20));
     dc.Rectangle(rct);
     dc.SelectObject(fArial8);
     s.Format(L"%d",k);           //display channel number in rct
     dc.TextOutW(end.x+35,home.y+3+25*k,s);
     s.Format(L"%d",i);           //display nodes for given channel
     dc.TextOutW(end.x+35+30*vColor[k],home.y+3+25*k,s);
     s.Format(L"# Colors=%d",kMax);
     dc.TextOutW(end.x+30,home.y-3,s);
}

void CCode3B::ShowTable()
{
     CString s;
     int i,j;
     CRect rct;
     table.DestroyWindow();

     //creates the list view window
     rct=CRect(CPoint(end.x+220,home.y-10),CSize(250,300));
     table.Create(WS_VISIBLE | WS_CHILD | WS_DLGFRAME | LVS_REPORT
          | LVS_NOSORTHEADER,rct,this,idc++);
     rct=CRect(CPoint(end.x+220,home.y+300),CSize(250,300));

     for (j=0;j<=n;j++)
     {
          s.Format((j==0)?L"adj":L"%d",j);
          table.InsertColumn(j,s,LVCFMT_CENTER,((j==0)?35:30));
     }
     for (i=1;i<=n;i++)
```

```
        {
                s.Format(L"%d",i); table.InsertItem(i-1,s,0);
                for (j=1;j<=n;j++)
                {
                        s.Format(L"%d",v[i].adj[j]); table.
SetItemText(i-1,j,s);
                }
        }
}
```

3.4 Channel Assignment on Wireless Mesh Networks

Another interesting application of graph coloring is the channel assignment problem for *wireless mesh networks* (WMNs). A WMN is a multihop wireless network consisting of nodes called mesh routers, mesh clients, and gateways connected in a mesh topology. In their standard form, WMNs include Wi-Fi, ad hoc networks, cellular telephone systems, and wireless sensor networks. The nodes in WMNs can be in a static position or be mobile, and they are capable of communicating with other nodes through the built-in radio transmitters and receivers.

Most communication in the network is performed wirelessly although cables are still used among some nodes. Each node transceiver (transmitter and receiver) has a limited transmission range that allows it to send and receive messages with other nodes. Mesh routers are normally in a stationary position and they provide coverage for the mesh clients that can be in static or mobile positions. The routers function to forward traffic messages, while the gateways connect the network to the Internet. The mesh clients are common electronic devices such as laptops and cellular phones that provide input and output to the system.

WMNs are deployed in many applications due to their current and potential applications such as in community networks, broadband home networking, video on demand (VoD), last-mile Internet access, and video surveillance. In the most common form, the Internet is a good example of a WMN that provides an efficient network to connect people all over the world. Part of the network includes the Iridium satellite system, which has 66 satellites that communicate with each other as a single WMN. Smaller WMNs can be seen on university campuses and in hospitals and hotels.

The international standard for wireless communications is provided by an organization called the Institute of Electrical and Electronics Engineers (IEEE), an organization based in the United States. Communication in the multihop WMN is provided through nonoverlapping channels operating under a protocol provided by the IEEE coded as 802.11. This protocol is the IEEE standard for the Medium Access Control (MAC) and physical layer for wireless local area network (WLAN). There are many versions of 802.11 standards; among the popular ones are 802.11, 802.11a, 802.11b, and 802.11g. The standard used in many Wi-Fi and other mesh networks including 802.11a and 802.11g, which support 2.4-GHz transmission. In IEEE 802.11b/g, there are 14 overlapping and nonoverlapping channels labeled as Channel 1 (2.412 GHz at its center), Channel 2 (2.417 GHz), and so forth until Channel 14 (2.484 GHz). Each channel has a width of 22 MHz and the throughput decreases with a smaller channel spacing. Three channels 1, 6, and 11 are orthogonal to each other as they do not overlap.

A radio channel is required before communication between two nodes in a WMN can be established. A single channel allows one-way delivery of a message or data, and it may not be sufficient to satisfy the high volume of communication demand between the nodes in the network. Therefore, WMN relies on multiple channels on each node in the network in order to fulfill this requirement. To achieve this goal, every node in a WMN is equipped with multiple network interface cards (NICs) that allow tuning to one of the available channels. A pair of nodes can communicate with each other if their NICs are tuned to the same channel in the transmission range of each other and the channels are not in the interference state. The use of the same channel by several nodes at the same time when they are within the transmission range of each other may result in electromagnetic interferences. Therefore, different channels are assigned to links that originate from the same node in order to avoid these interferences.

WMNs provide fast service and convenience by providing an unlimited area of coverage through flexible and scalable connections of nodes. The network size is easily increased according to demand by adding new clients, routers, and gateways. As the number of nodes increases, a proper management system is needed to organize the nodes for problems involving routing, data transmission, network management, and fault tolerance.

Channel assignment in a WMN can be modeled as a graph coloring application associated with the assignment of limited resources to the mesh routers and mesh clients. The resources consist of radio channels that are assigned according to frequencies. The number of channels, or the bandwidth, in a particular node is limited according to the average demand in the node. Therefore, an efficient system for managing the assignment of channels is important in order to optimize their assignments.

3.4.1 Problem Statement

A channel assignment problem for WMNs can be stated as follows: Given a network in the form of the connected graph $G(V, E)$ with n nodes and m edges, where $V = \{v_i\}$ for $i = 1,2,\ldots,n$ and $E = \{e_{ij}\}$ for $i, j = 1,2,\ldots,m$, how can the channels be assigned in order to support data communication between the pairs of nodes in the network so that they use the minimum number of channels?

Channels are limited resources in the network and they are assigned from the allocated bandwidths. The main objective in channel assignment is to assign the channels with minimum bandwidth, that is, to assign as few channels as possible but satisfy the demand in the network.

Channels are assigned to the nodes in such a way as to avoid electromagnetic interferences. There are three types of interferences and together they form the constraints in the problem: adjacency, cochannel, and cosite channel constraints. The *adjacency constraint* exists when two adjacent nodes share the same channel. The two nodes should use two noninterfering channels in order to transmit or receive messages at the same time. The *cochannel constraint* arises when two nodes sharing the same channel are located too close to each other. The nodes should be separated by some distance in order to share the same channel. The *cosite* constraint arises from the use of two or more channels in the same location. Although the allocated channels are different, they must be separated by some length in order to function properly.

//Greedy algorithm for channel assignment
Given:
    Graph $G(V, E)$with n nodes and m edges representing a wireless mesh network;
    Channels c_i for $k = 1,2,\ldots,kMax$;

Initialization:
    Assign every node with the first color c_1;
Process:
    For $i = 1$ to n
        Let $k = 1$
        For $j = 1$ to n
            If the adjacency or co-channel constraint exists between (v_i, v_j)
                Update $k \leftarrow k + 1$;
            Endif
        Endfor
        Assign k to v_i;
        Update the graph;
    Endfor
Output: Assigned channels for v_i for $i = 1,2,\ldots,n$.

3.4.2 *Code3C*: Constrained Single-Channel Assignment

Code3C is a program for a single-channel assignment of a mesh network with adjacency and cochannel constraints. The basic idea for this program is the greedy algorithm presented in *Code3B,* which is improved to support the two constraints. The solution is obtained through a new greedy algorithm that compares the colors of the nodes by checking on the two constraints before assigning the nodes with channels. The anticipated output from this program is a graph that shows the randomly scattered nodes with assigned colors to denote their assigned channels.

The program supports up to 20 nodes where the adjacencies between the pairs of nodes are determined from their Euclidean distances from one another: if the distance is less than a given threshold value, then they are adjacent; otherwise, they are not. The cochannel relationship between a pair of nodes is also determined from their Euclidean distance: the two nodes can share the same channel if they are at least some distance away from each other.

Figure 3.8 is a sample output from *Code3C.* There are three edit boxes for collecting input as the number of nodes, adjacency range, and cochannel range. There are also two buttons: the *New Graph* button creates a new graph by refreshing the coordinates of the nodes while *Assign Channel* assigns the channels to the nodes in the graph by applying the greedy algorithm.

The output is presented as a graph in the drawing area with nodes colored according to the assigned channels. A color table to the right of the drawing area displays the channels and their assigned nodes. Further to the right are two tables, one showing the adjacency matrix of the nodes, while the other is the cochannel matrix.

Figure 3.8 simulates a mesh network having 20 nodes that are scattered randomly in the drawing area. With the given positions and parameter values, the results produced are 13 channels with channel 4 shared the most with three nodes. The output will definitely be different with a different graph and parameters.

Code3C has *Code3C.h* and *Code3C.cpp* as its source files. A structure called *NODE* organizes the variables concerning the attributes of the nodes of the graph. Its object is array v that represents a node in the graph. *NODE* has five members for describing the attributes of each node: *home* is its position, *rct* is the appearance as a rectangular box, *adj* is the adjacency relationship with the neighboring node, *coc* is the cochannel relationship with the stated node, and *channel* is the channel assigned to the node using the greedy algorithm.

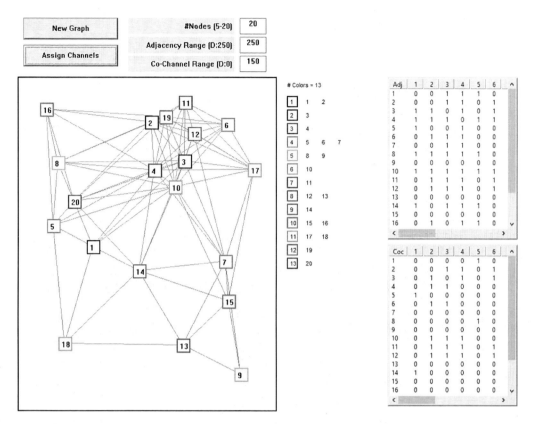

FIGURE 3.8
Output from *Code3C*.

Table 3.3 lists the variables and objects in *Code3C*. The input area has *eN*, *eAdj*, and *eCoc* edit boxes for collecting information on the number of nodes, the adjacency range, and the cochannel range, respectively. The input values are read once the push buttons *bNGraph* and *bCompute* are activated, with *bNGraph* refreshing the display with a new graph and *bCompute* computing the channels for the nodes. The main output is the graph, which is displayed inside the drawing area with *home* and *end* as its top-left and bottom-right

TABLE 3.3

Important Objects and Variables in *Code3C*

Variable	Type	Description
home, *end*	*CPoint*	Top-left and bottom-right points of the drawing area
bNGraph, *bCompute*	*CButton*	Buttons for redrawing a new graph and computing the channels
eN, *eAdj*, *eCoc*	*CEdit*	Input boxes for the number of nodes, adjacency range, and cochannel range
adjTable, *cocTable*	*CListCtrl*	List view tables for the adjacency and cochannel matrices
vColor	*int*	Array for the node numbers for the assigned color
kMax	*int*	Maximum number of assigned channels
n	*int*	Number of nodes in the graph
vColor[*i*]	*int*	Node number assigned to the given color

boundary points. The output is also supported by *adjTable* and *cocTable,* which display the adjacency and cochannel matrices, respectively, in their list view windows.

Figure 3.9 is the organization of *Code3C* showing all the functions. The constructor *Code3C()* creates the main window and eight child windows consisting of edit boxes, static boxes, and buttons. The function also initializes variables and objects for display and calls *Initialize()* to initialize the graph variables.

The initial display is the graph produced by *OnPaint()* based on the initial values of the nodes and edges defined in *Initialize().* The initial display also has the adjacency and cochannel matrices in the list view windows through *ShowTable().* Another function *OnNewGraph()* refreshes the graph by calling *Initialize()* to set new values for the nodes and edges.

The main engine is *OnCompute(),* which implements the greedy algorithm for assigning the channels. This function is activated when the *bCompute* push button is clicked.

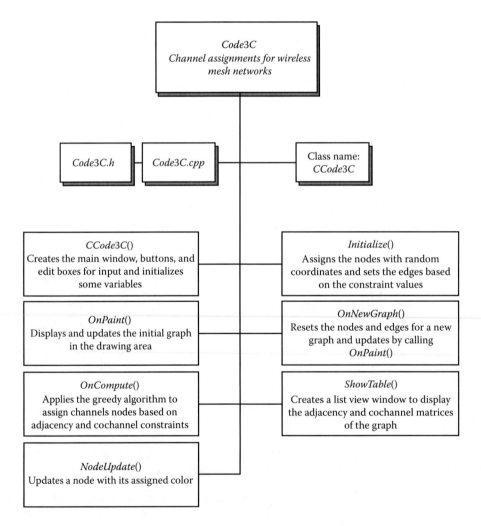

FIGURE 3.9
Organization of *Code3C.*

OnCompute() initially assigns the value of 1 to all nodes. Next, each node v_i for $i = 1,2,...,N$ is compared to every other node v_j for $j = 1,2,...,i-1$ in the graph. If v_i and v_j are adjacent and the cochannel distance between them is above the given threshold value, then a number k is matched to the color of v_j starting at 1 upward. The first value of k that differs from the color of j becomes the assigned channel for v_i. With this assignment, all other information regarding the number of nodes assigned to each channel is updated. The total number of colors is given by *kMax*, which is obtained by counting the number of different channels assigned to the nodes.

The full listing of the codes is given as follows:

```
//Code3C.h
#include <afxwin.h>
#include <time.h>
#include <math.h>
#include <afxcmn.h>
#define N 20
#define IDC_NGRAPH 501
#define IDC_COMPUTE 502

class CCode3C: public CFrameWnd
{
private:
        CPoint home,end;
        CButton bNGraph,bCompute;
        CStatic sN,sCoc,sAdj;
        CEdit eN,eCoc,eAdj;
        CFont fArial10,fArial8;
        CPen pColor[N+1];
        CListCtrl adjTable,cocTable;
        int idc,n,adjRange,cocRange;      //n=#nodes
        int vChannel[N+1],kMax;           //kMax=max #channels
        typedef struct
        {
                int channel;              //allocated channels
                bool adj[N+1];            //adjacency: 0=no, 1=yes
                bool coc[N+1];            //co-channel: 0=no, 1=yes
                CPoint home;
                CRect rct;
        } NODE;
        NODE v[N+1];
public:
        CCode3C();
        ~CCode3C()      {}
        afx_msg void OnPaint();
        afx_msg void OnCompute();
        afx_msg void OnNewGraph();
        void Initialize(),ShowTable();
        void NodeUpdate(int);
        DECLARE_MESSAGE_MAP();
};
```

```
class CMyWinApp:public CWinApp
{
      public:
            virtual BOOL InitInstance();
};
CMyWinApp MyApplication;

BOOL CMyWinApp::InitInstance(void)
{
      m_pMainWnd=new CCode3C;
      m_pMainWnd->ShowWindow(m_nCmdShow);
      return TRUE;
}

//Code3C.cpp
#include "Code3C.h"

BEGIN_MESSAGE_MAP(CCode3C,CFrameWnd)
      ON_WM_PAINT()
      ON_BN_CLICKED (IDC_NGRAPH,OnNewGraph)
      ON_BN_CLICKED (IDC_COMPUTE,OnCompute)
END_MESSAGE_MAP()

CCode3C::CCode3C()
{
      CPoint bhome=CPoint(30,20);
      idc=800;
      home=CPoint(30,140); end=CPoint(500,740);
      Create(NULL,L"Wireless mesh network: one channel per node",
            WS_OVERLAPPEDWINDOW,CRect(0,0,1000,800));
      bNGraph.Create(L"New Graph",WS_CHILD | WS_VISIBLE |
BS_DEFPUSHBUTTON,
            CRect(CPoint(bhome),CSize(180,40)),this,IDC_NGRAPH);
      bCompute.Create(L"Assign Channels",WS_CHILD | WS_VISIBLE |
BS_DEFPUSHBUTTON,
            CRect(CPoint(bhome.x,bhome.
y+50),CSize(180,40)),this,IDC_COMPUTE);
      sN.Create(L"#Nodes (5-20) ", WS_VISIBLE | SS_CENTERIMAGE |
SS_RIGHT,
            CRect(CPoint(bhome.x+200,bhome.y),CSize(200,30)),this,
IDC_STATIC);
      eN.Create(WS_CHILD | WS_VISIBLE | WS_BORDER | SS_CENTER,
            CRect(CPoint(bhome.x+410,bhome.y),CSize(50,30)),this,idc++);
      sAdj.Create(L"Adjacency Range (D:250) ", WS_VISIBLE |
SS_CENTERIMAGE
            | SS_RIGHT,CRect(CPoint(bhome.x+200,bhome.y+35),
            CSize(200,30)),this,IDC_STATIC);
      eAdj.Create(WS_CHILD | WS_VISIBLE | WS_BORDER | SS_CENTER,
            CRect(CPoint(bhome.x+410,bhome.y+35),CSize(50,30)),this,
idc++);
      sCoc.Create(L"Co-channel Range (D:0) ", WS_VISIBLE |
SS_CENTERIMAGE
            | SS_RIGHT,CRect(CPoint(bhome.x+200,bhome.y+70),
            CSize(200,30)),this,IDC_STATIC);
```

```
        eCoc.Create(WS_CHILD | WS_VISIBLE | WS_BORDER | SS_CENTER,
                CRect(CPoint(bhome.x+410,bhome.y+70),CSize(50,30)),this,
idc++);
        fArial10.CreatePointFont(100,L"Arial");
        fArial8.CreatePointFont(80,L"Arial");
        int Color[]={RGB(150,150,150),
                RGB(0,0,0),RGB(0,0,200),
                RGB(200,0,0),RGB(0,200,0),
                RGB(200,200,0),RGB(0,200,200),
                RGB(200,0,200),RGB(200,50,150),
                RGB(100,100,255),RGB(250,50,255),
                RGB(100,200,50),RGB(250,50,0),
                RGB(0,50,250),RGB(100,100,255),
                RGB(250,0,50),RGB(50,250,0),
                RGB(50,0,250),RGB(0,250,50),
                RGB(255,100,100),RGB(100,255,100)};
        for (int i=0;i<=N;i++)
                if (i==0)
                        pColor[i].CreatePen(PS_SOLID,1,Color[i]);
                else
                        pColor[i].CreatePen(PS_SOLID,2,Color[i]);
        Initialize();
}


afx_msg void CCode3C::OnNewGraph()
{
        Initialize(); Invalidate();
}


void CCode3C::Initialize()
{
        int i,j;
        double distance;
        CString s;
        s=""; eN.GetWindowText(s); n=_ttoi(s); n=((n<=N)?n:N);
        s=""; eAdj.GetWindowText(s); adjRange=_ttoi(s);
        adjRange=((adjRange==0)?250:adjRange);
        s=""; eCoc.GetWindowText(s); cocRange=_ttoi(s);
        cocRange=((cocRange==0)?0:cocRange);
        srand(time(0));
        for (i=1;i<=n;i++)
        {
                v[i].home.x=home.x+20+rand()%(end.x-home.x-50);
                v[i].home.y=home.y+20+rand()%(end.y-home.y-50);
                v[i].rct=CRect(CPoint(v[i].home.x-10,v[i].home.y-10),
CSize(25,25));
                v[i].channel=0;        //initialize the cells with channel 0
        }
        for (i=1;i<=n;i++)
        {
                v[i].adj[i]=0; v[i].coc[i]=0;
                for (j=i+1;j<=n;j++)
                {
```

```
                        distance=sqrt(pow((double)(v[i].home.x-v[j].
home.x),2)
                              +pow((double)(v[i].home.y-v[j].home.y),2));

                    v[i].adj[j]=((distance<=adjRange)?1:0);
                    v[j].adj[i]=v[i].adj[j];
                    v[i].coc[j]=((distance>=cocRange)?0:1);
                    v[j].coc[i]=v[i].coc[j];
            }
        }
}

afx_msg void CCode3C::OnPaint()
{
        CPaintDC dc(this);
        CString s;
        int i,j;
        dc.SelectObject(pColor[1]);
        dc.Rectangle(home.x-10,home.y-10,end.x+10,end.y+10);
        dc.SelectObject(pColor[0]);
        for (i=1;i<=n;i++)
                for (j=1;j<=n;j++)
                        if (v[i].adj[j])
                        {
                                dc.MoveTo(CPoint(v[i].home));
                                dc.LineTo(CPoint(v[j].home));
                        }
        dc.SelectObject(fArial10); dc.SelectObject(pColor[1]);
        for (i=1;i<=n;i++)
        {
                dc.Rectangle(v[i].rct);
                s.Format(L"%d",i);
                dc.TextOutW(v[i].home.x-5,v[i].home.y-5,s);
        }
}

void CCode3C::OnCompute()
{
        int i,j,k,r,cMax;
        bool w[N+1];

        //assign the channels
        kMax=0;
        for (k=1;k<=n;k++)
            vChannel[k]=0;
        for (i=1;i<=n;i++)
        {
                if (i==1)
                {
                    v[i].channel=1;
                    k=1; vChannel[k]=0;
                }
                else
                    while (k<=n)
```

```
                        {
                                w[k]=0;
                                for (j=1;j<=i-1;j++)
                                        if (v[i].adj[j] && abs(k-v[j].
channel)<1)
                                        {
                                                v[i].channel=k+1;
                                                w[k]=1;
                                        }
                                if (!w[k])
                                {
                                        r=0;
                                        for (j=1;j<=i-1;j++)
                                        {
                                                v[i].channel=k;
                                                if ((v[i].channel==v[j].channel)
&& v[i].coc[j])
                                                {
                                                        v[i].channel=k+1;
                                                        r++;
                                                }
                                        }
                                        if (r==0)
                                                break;
                                }
                                k++;
                        }
                vChannel[k]++;
                if (kMax<k)
                        kMax=k;
                NodeUpdate(i);
        }
        ShowTable();
}


void CCode3C::NodeUpdate(int i)
{
        CClientDC dc(this);
        CRect rct;
        CString s;
        dc.SelectObject(pColor[v[i].channel]);
        dc.Rectangle(v[i].rct);
        s.Format(L"%d",i);                              //display nodes in graph

        //display color info
        dc.TextOutW(v[i].home.x-5,v[i].home.y-5,s);
        rct=CRect(CPoint(end.x+30,home.y+25*v[i].channel),CSize(20,20));
        dc.Rectangle(rct);
        dc.SelectObject(fArial8);
        s.Format(L"%d",v[i].channel);   //display channel number in rct
        dc.TextOutW(end.x+35,home.y+3+25*v[i].channel,s);
        s.Format(L"%d",i);                              //display nodes for given channel
```

```
        dc.TextOutW(end.x+35+30*vChannel[v[i].channel],home.y+3+25*v[i].
channel,s);
        s.Format(L"# Colors=%d",kMax); dc.TextOutW(end.x+30,home.y-3,s);
}

void CCode3C::ShowTable()
{
        CString s;
        int i,j;
        cocTable.DestroyWindow(); adjTable.DestroyWindow();
        CRect rct;

        //creates the list view window
        rct=CRect(CPoint(end.x+220,home.y-10),CSize(250,300));
        adjTable.Create(WS_VISIBLE | WS_CHILD | WS_DLGFRAME | LVS_REPORT
                | LVS_NOSORTHEADER,rct,this,idc++);
        rct=CRect(CPoint(end.x+220,home.y+300),CSize(250,300));
        cocTable.Create(WS_VISIBLE | WS_CHILD | WS_DLGFRAME | LVS_REPORT
                | LVS_NOSORTHEADER,rct,this,idc++);
        for (j=0;j<=n;j++)
        {
                s.Format((j==0)?L"Adj":L"%d",j);
                adjTable.InsertColumn(j,s,LVCFMT_CENTER,((j==0)?35:30));
                s.Format((j==0)?L"Coc":L"%d",j);
                cocTable.InsertColumn(j,s,LVCFMT_CENTER,((j==0)?35:30));
        }
        for (i=1;i<=n;i++)
        {
                s.Format(L"%d",i); adjTable.InsertItem(i-1,s,0);
                s.Format(L"%d",i); cocTable.InsertItem(i-1,s,0);
                for (j=1;j<=n;j++)
                {
                        s.Format(L"%d",v[i].adj[j]); adjTable.SetItemText
(i-1,j,s);
                        s.Format(L"%d",v[i].coc[j]); cocTable.SetItemText
(i-1,j,s);
                }
        }
}
```

4

Computing the Shortest Path

4.1 Problem Description

In a weighted graph, a pair of nodes can be linked in one or more unique paths. The objective here is to find the path with the minimum sum of the weights along its path. The *shortest path SP* between two nodes, the source, and its destination in a graph $G(V, E)$ is defined as follows:

> Given a weighted and connected graph $G(V, E)$ with n nodes, find the minimal cost linking a pair of nodes in the graph.

Cost in this problem is normally measured as the total sum of the weights along the path between the two nodes. The *source* in this path is the starting node while the *destination* is the ending node. There may exist m different paths from the source to the destination, and the shortest path always refers to the one with the minimal cost.

The shortest path is a fundamental problem in network design. Several references discuss the topic for many applications in network design. The problem of finding more than one shortest path or k shortest paths of a graph is discussed in [1]. This problem applies in the design of alternative routes in the shortest path problem that is widely implemented in the Global Positioning System (GPS). In [2], a constraint is imposed in the shortest path problem for planning the routing model for heavy goods vehicles traveling in a city since they are only allowed on the city streets during certain hours of the working day. Also, the graph theoretical models for interconnection network models are discussed in detail in [3,4].

In discrete optimization the shortest path problem can be represented by the following equation:

$$SP(\text{source, destination}) = \min \sum_{ij} x_{ij} w_{ij}$$

$$\text{subject to } x_{ij} \geq 0 \text{ and } \sum_{j} x_{ij} - \sum_{j} x_{ji} = \begin{cases} 1 & i = \text{source} \\ -1 & i = \text{destination} \\ 0 & \text{otherwise.} \end{cases}$$

In the above equation, x_{ij} is a variable, and w_{ij} is the cost or weight from the source v_i to its destination v_j.

The longest path LP is the opposite of the shortest problem defined in a similar manner: Given a weighted and connected graph G with n nodes, find the maximal cost linking a pair of nodes in the graph.

$$LP(\text{source}, \text{destination}) = \max \sum_{ij} x_{ij} w_{ij}.$$

Finding the shortest path between the two nodes is a very common problem as it has many real-world applications. For example, the shortest path is needed whenever a driver navigates his or her car to a destination and would like to take the quickest route, or a truck driver needs to distribute the gasoline he or she is carrying to a number of gas stations in a city in a way that the whole job can be completed in the shortest time before moving on to another city.

The shortest path is also important in the wiring between pairs of pins in a compact printed circuit board, which contributes toward making the circuits organized and keeping congestion in the board low. A compact but complete design is also important to minimize the heat generated from of the friction between the components in the circuitry.

The longest path problem also has some real-world applications. For example, in a transportation problem the maximal cost for a path applies to the maximum profit to be collected from various supply centers in a geographical region by visiting the most number of centers. In this case, the maximal cost may refer to the longest path that connects most nodes in the network. Very often, the longest path may end up as finding the Hamiltonian path of the graph as the maximum distance results from visiting all the nodes of the graph.

The bold lines in Figure 4.1 show the shortest path between v_1 and v_4 in a weighted graph. The radius of the graph is 2 while the diameter is 3. The shortest or longest paths of the graph are not necessarily on the paths through the radius and diameter of the graph.

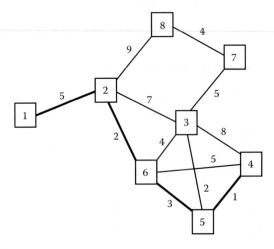

FIGURE 4.1
Shortest path between v_1 and v_4 in a weighted graph.

There are many paths between v_1 and v_4 with their costs indicated as $C(v_1, v_4)$ as the sum of the weights along the paths. Among them are

$$C(v_1, v_4) = C(5:1 \rightarrow 2,7:2 \rightarrow 3,8:3 \rightarrow 4) = 20$$

$$C(v_1, v_4) = C(5:1 \rightarrow 2,7:2 \rightarrow 3,4:3 \rightarrow 6,5:6 \rightarrow 4) = 21$$

$$C(v_1, v_4) = C(5:1 \rightarrow 2,9:2 \rightarrow 8,4:8 \rightarrow 7,5:7 \rightarrow 3,8:3 \rightarrow 4) = 31$$

$$C(v_1, v_4) = C(5:1 \rightarrow 2,2:2 \rightarrow 6,5:6 \rightarrow 4) = 12$$

$$C(v_1, v_4) = C(5:1 \rightarrow 2,2:2 \rightarrow 6,3:6 \rightarrow 5,1:5 \rightarrow 4) = 11.$$

Since the size of the graph is small, it can easily be seen that the shortest path $SP(v_1, v_4)$ is 11, which is along the path $v_1 \rightarrow v_2 \rightarrow v_6 \rightarrow v_5 \rightarrow v_4$, or

$$SP(v_1, v_4) = C(5:1 \rightarrow 2, 2:2 \rightarrow 6, 3:6 \rightarrow 5, 1:5 \rightarrow 4) = 11.$$

At the same time, the longest path $LP(v_1, v_4)$ of the graph is 31, which is the path $v_1 \rightarrow v_2 \rightarrow v_8 \rightarrow v_7 \rightarrow v_3 \rightarrow v_4$, or

$$LP(v_1, v_4) = C(5:1 \rightarrow 2, 9:2 \rightarrow 8, 4:8 \rightarrow 7, 5:7 \rightarrow 3, 8:3 \rightarrow 4) = 31.$$

In most cases, the shortest path problem has many more practical and relevant applications in the real world compared to the longest path problem. Therefore, in this chapter we will concentrate our discussion on the shortest path problem.

We now present four different programs related to the shortest path problems: *Code4A, Code4B, Code4C,* and *Code4D. Code4A* is the greedy algorithm solution using Dijkstra's algorithm for the single-source shortest path problem, while *Code4B* is the dynamic programming approach to the all-pairs shortest path problem using the Floyd-Warshall's algorithm. *Code4C* is an extension of *Code4A* with a specific application to designing a simple GPS. Finally, *Code4D* is another derivation of *Code4A* for routing a multistage interconnection network (MIN).

4.2 Single-Source Shortest Path Problem

The shortest path problem has its solutions in the form of greedy algorithms, the most common being the Dijkstra, Bellman-Ford, and Floyd-Warshall algorithms. The first two algorithms compute the shortest paths from a single source, while the last computes the shortest paths between all pairs of nodes in a graph. Dijkstra's algorithm is a graph search algorithm that computes the shortest paths on a weighted graph whose weights are nonnegatives. The algorithm will not function properly if one or more edges have negative weight values. For edges with negative weights, a good solution is Bellman-Ford's algorithm.

4.2.1 Dijkstra's Algorithm

Dijkstra's algorithm is the most widely used search algorithm for the shortest path problem. The algorithm was proposed by the Dutch computer scientist Edsger Dijkstra in 1959. Dijkstra's algorithm computes the shortest path by recursively selecting the unvisited vertex with the lowest distance to each unvisited neighbor. For a graph with n nodes having nonnegative weights on the edges, the method computes the path with the least cost between a pair of nodes with a complexity of $O(n^2)$. Dijkstra's algorithm computes the shortest path from the source node to its destination by recursively computing the shortest paths from the source node to all other nodes in the graph first. The last node is the destination node whose minimum distance from the source node is obtained after the shortest paths to other nodes have been computed.

The algorithm starts with the source node as the initial node s_0. It then searches for all nodes adjacent to s_0 and the costs from s_0 to the nodes are noted. The vertex s_1 with the minimum cost is marked. Next, the costs from s_0 to the adjacent nodes of s_1 through s_1 are calculated. The costs are compared to the costs from the other adjacent nodes of s_0 and the lowest values for the nodes are marked as the costs from s_0 to the nodes. The process is repeated to other nodes to produce the shortest paths s_j for $j = 2, 3,..., n - 1$ until the destination node is reached. The minimum cost for the destination node (the last node) is marked as s_n, and this value stands as the final solution.

Dijkstra's algorithm involves a loop for updating the shortest distances from the source node to other nodes in the recursive process. It is summarized as follows:

> **Given:**
>   Graph $G(V, E)$ for $V = [v_i]$, $E = [Wt(i, j)]$ with $|V| = n$ for $i, j = 1, 2,..., n$.
> **Problem:**
>   To find the shortest path from the source Sv to destination Dv, for $Sv, Dv \in V$.
> **Initialization:**
>   Let $count = 0$;
>   Unmark v_i by setting the flag $fv_i = 0$ for $i = 1, 2, ..., n$;
>   Let the cost from Sv to v_i, or $D[i] = \infty$ and $D[Sv] = 0$;
>   Let the $Pv_i = -1$ be the predecessor node to v_i;
> **Process:**
>   While ($count < n$)
>   Find v_{cun}, the closest unmarked node from Sv;
>   Mark v_{cun} by setting the flag $fv_i = 1$;
>   For $i = 1$ to n
>   If v_i is unmarked, or $fv_i = 0$
>   If $D[i] > D[v_{cun}] + Wt(i, v_{cun})$
>   Update $D[i] = D[v_{cun}] + Wt(i, v_{cun})$;
>   Assign $Pv_i = v_{cun}$;
>   Endif
>   Endfor
> $count$++;
>   Endwhile
> **Output:** the shortest path $SP(Sv, Dv) = D[Dv]$.

Figure 4.2 shows the implementation of Dijkstra's algorithm on the weighted graph $G(V, E)$ with $n = |V| = 8$. The process starts by marking the source node v_1 and the minimum cost

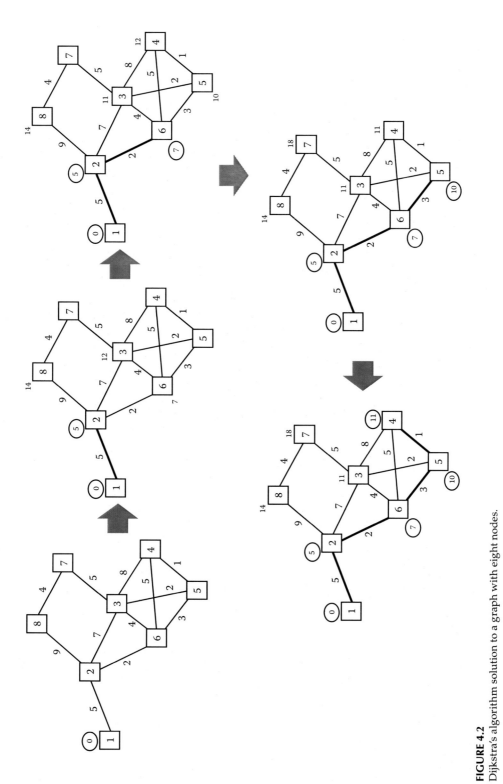

FIGURE 4.2
Dijkstra's algorithm solution to a graph with eight nodes.

with itself is set to 0. Since v_2 is its only neighbor, this node is marked with the minimum cost of 5, which is the weight in (v_1, v_2). Next, the minimum costs from v_2 to its neighbors $\{v_8, v_3, v_6\}$ are computed to produce $C(v_2, v_8) = 14$, $C(v_2, v_3) = 12$ and $C(v_2, v_6) = 7$ with $C(v_2, v_6) = 7$ being the lowest. This marks v_6 and the costs to its neighbors $\{v_3, v_4, v_5\}$ are computed to produce $C(v_6, v_3) = 11$, $C(v_6, v_4) = 12$ and $C(v_6, v_5) = 10$. In this case, $C(v_6, v_3) = 11$ overrides $C(v_2, v_3) = 12$ as the former is lower. Finally, the costs $C(v_5, v_3) = 12$ and $C(v_5, v_4) = 11$ bring the path to v_4 with the final solution of $SP(v_1, v_4) = 11$ through $v_1 \rightarrow v_2 \rightarrow v_6 \rightarrow v_5 \rightarrow v_4$.

4.2.2 *Code4A*: Implementing Dijkstra's Algorithm

Code4A is the implementation of Dijkstra's algorithm for finding the shortest path between a pair of nodes in a graph. The output of *Code4A* is shown in Figure 4.3, which displays 20

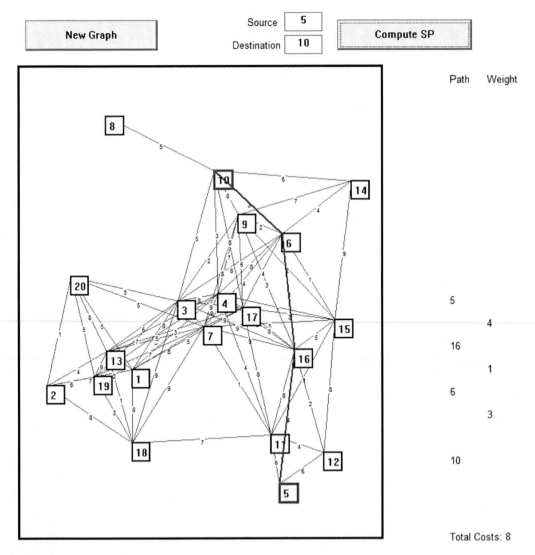

FIGURE 4.3
Output from *Code4A*.

randomly scattered nodes inside a box. Two edit boxes for collecting the source and destination nodes serve as the input. The shortest path between the two nodes, $SP(v_5, v_{10})$ in this case, is computed and displayed in the graph once the *Compute SP* button is clicked. The interface also allows a new graph to be generated through the *New Graph* button.

Code4A has a single class called *CCode4A* with two source files, *Code4A.h* and *Code4A.cpp*. In *Code4A.h*, *Sv* and *Dv* represent the source and destination nodes, respectively, and their values are obtained from the two edit boxes. The nodes and edges of the graph are represented by the structure *NODE* with the contents. There are also two button objects called *bNGraph* for *New Graph* and *bCompute* for *Compute SP*. Other important objects and variables are briefly described in Table 4.1.

Figure 4.4 is the design organization showing all the functions in this program. The application starts with the constructor *Code4A()*, which draws the main window and the *New Graph* button as well as initializes some variables. The constructor calls *Initialize()*, which creates a new graph with *n* nodes by assigning random coordinates and weights to the edges. Each node is drawn using *DrawNode()* and the whole graph is drawn inside the drawing area through *OnPaint()*.

Input for this program are the source and destination nodes collected from the edit boxes *eSv* and *eDv*, respectively. The edit boxes are created inside *Initialize()* and their values are read inside *OnCompute()*.

The main engine in this application is *Dijkstra()*, which applies Dijkstra's algorithm for computing the shortest path. The shortest path from the source to the destination nodes is displayed through *DrawPath()*. In computing the shortest path *Dijkstra()* calls *GetCUN()* repeatedly to get the closest unmarked nodes from the current node.

A new graph with the randomly located positions in Windows and randomly determined weights on the edges is produced by *Initialize()*. This graph is later drawn in Windows through *OnPaint()*. Another function *DrawNode()* serves is to draw the node specified in its argument.

The full listing of the codes for *Code4A* are given below.

```
//Code4A.h
#include <afxwin.h>
#include <math.h>
#define N 20
#define LinkRange 200
#define IDC_NGRAPH 500
#define IDC_COMPUTE 501
```

TABLE 4.1

Important Objects/Variables in *Code4A* and Their Descriptions

Variable	Type	Description
n	constant	Number of nodes in the graph
LinkRange	constant	Range for adjacency between two nodes in the graph
Sv, Dv	*int*	Source, destination nodes
$v[i].wt[j]$	*int*	Weight between (v_i, v_j)
$v[i].sp[j]$	*int*	Shortest path between (v_i, v_j)
Pv	*int* array	Predecessor node to the currently marked node
fv	*bool* array	Marking status of nodes, 0 = unmarked, 1 = marked
home, end	*CPoint*	Top-left and bottom-right corners of the drawing area
bNGraph	*CButton*	Button for generating a new graph
bCompute	*CButton*	Button for computing the shortest path

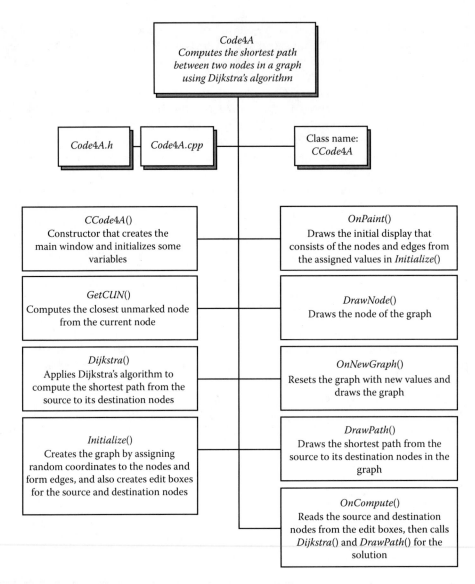

FIGURE 4.4
Organization of *Code4A* showing all the functions.

```
class CCode4A:public CFrameWnd
{
private:
        int idc;
        int Sv,Dv;
        int Pv[N+1];
        bool fv[N+1];
        CButton bNGraph,bCompute;
        CPoint home,end;
        CFont fHelvetica,fArial;
        CEdit eSv,eDv;

        typedef struct
```

```
        {
                CPoint home;
                CRect rct;
                int wt[N+1];
                int sp[N+1];
        } NODE;
        NODE v[N+1];

public:
        CCode4A();
        ~CCode4A()                {}
        int GetCUN();
        void Dijkstra();
        void DrawPath(),Initialize(),DrawNode(int);
        afx_msg void OnPaint();
        afx_msg void OnNewGraph(),OnCompute();
        DECLARE_MESSAGE_MAP()
};

class CMyWinApp:public CWinApp
{
public:
        virtual BOOL InitInstance();
};
CMyWinApp MyApplication;

BOOL CMyWinApp::InitInstance()
{
        m_pMainWnd=new CCode4A;
        m_pMainWnd->ShowWindow(m_nCmdShow);
        return TRUE;
}

//Code4A.cpp
#include "Code4A.h"

BEGIN_MESSAGE_MAP(CCode4A, CFrameWnd)
ON_WM_PAINT()
        ON_BN_CLICKED(IDC_NGRAPH,OnNewGraph)
        ON_BN_CLICKED(IDC_COMPUTE,OnCompute)
END_MESSAGE_MAP()

CCode4A::CCode4A()
{
        idc=800;
        home=CPoint(30,100); end=CPoint(500,700);
        Create(NULL, L"Dijkstra's Shortest Path Solution",
                WS_OVERLAPPEDWINDOW,CRect(0,0,1000,800));
        bNGraph.Create(L"New Graph",WS_CHILD | WS_VISIBLE |
BS_DEFPUSHBUTTON,CRect(CPoint(30,30),CSize(180,40)),this,IDC_NGRAPH);
        Initialize();
        fArial.CreatePointFont(60,L"Arial");
        fHelvetica.CreatePointFont(100,L"Helvetica");
}
```

```
void CCode4A::Initialize()
{
        int i,j,k;
        double distance;
        eSv.DestroyWindow(); eDv.DestroyWindow(); bCompute.DestroyWindow();
        bCompute.Create(L"Compute SP",WS_CHILD | WS_VISIBLE
                | BS_DEFPUSHBUTTON,CRect(CPoint(end.x-50,30),CSize(180,40)),
                this,IDC_COMPUTE);
        eSv.Create(WS_CHILD | WS_VISIBLE | WS_BORDER | SS_CENTER,
CRect(CPoint(home.x+350,20),CSize(50,25)),this,idc++);
        eDv.Create(WS_CHILD | WS_VISIBLE | WS_BORDER | SS_CENTER,
CRect(CPoint(home.x+350,50),CSize(50,25)),this,idc++);

        srand(time(0));
        for (i=1;i<=N;i++)
        {
                v[i].home.x=home.x+20+rand()%(end.x-home.x-50);
                v[i].home.y=home.y+20+rand()%(end.y-home.y-50);
                v[i].rct=CRect(v[i].home.x,v[i].home.y,v[i].home.x+25,v[i].
home.y+25);
        }
        for (i=1;i<=N;i++)
        {
                v[i].wt[i]=0;
                for (j=i+1;j<=N;j++)
                {
                        distance=sqrt(pow((double)(v[i].home.x-v[j].
home.x),2)+ pow((double)(v[i].home.y-v[j].home.y),2));
                        if (distance<=LinkRange)
                                v[i].wt[j]=1+rand()%9;
                        else
                                v[i].wt[j]=99;
                        v[j].wt[i]=v[i].wt[j];
                }
        }
        Sv=0; Dv=0;
}


void CCode4A::DrawNode(int u)
{
        CClientDC dc(this);
        CString s;
        CPoint pt;
        CPen pBlack(PS_SOLID,2,RGB(0,0,0));
        CPen pRed(PS_SOLID,3,RGB(200,0,0));
        if (u==Sv || u==Dv)
                dc.SelectObject(pRed);
        else
                dc.SelectObject(pBlack);
        pt=v[u].home;
        dc.Rectangle(v[u].rct);
        s.Format(L"%d",u); dc.TextOut(pt.x+5,pt.y+5,s);
}
```

```
void CCode4A::OnPaint()
{
        CPaintDC dc(this);
        CString s;
        CRect rct;
        CPen rPen(PS_SOLID,3,RGB(0,0,0));
        CPen mPen(PS_SOLID,2,RGB(0,0,0));
        CPen qPen(PS_SOLID,1,RGB(100,100,100));
        CPoint mPoint,pt;
        int i,j;
        dc.SelectObject(rPen);
        dc.Rectangle(home.x-10,home.y-10,end.x+10,end.y+10);
        dc.SelectObject(qPen); dc.SelectObject(fArial);
        for (i=1;i<=N;i++)
                for (j=1;j<=N;j++)
                        if (v[i].wt[j]!=99)
                        {
                                dc.MoveTo(CPoint(v[i].home));
                                dc.LineTo(CPoint(v[j].home));
                                mPoint=CPoint((v[i].home.x+v[j].home.x)/2,
                                        (v[i].home.y+v[j].home.y)/2);
                                s.Format(L"%d",v[i].wt[j]);
                                dc.TextOut(mPoint.x,mPoint.y,s);
                        }
        dc.SelectObject(fHelvetica); dc.SelectObject(mPen);
        for (i=1;i<=N;i++)
                DrawNode(i);
        dc.TextOut(home.x+300,25,L"Source");
        dc.TextOut(home.x+280,55,L"Destination");
}


int CCode4A::GetCUN()
{
        int minDistance=99;
        int cun;
        for (int i=1;i<=N;i++)
                if ((!fv[i]) && (minDistance>=v[Sv].sp[i]))
                {
                        minDistance=v[Sv].sp[i];
                        cun=i;
                }
        return cun;
}

void CCode4A::Dijkstra()
{
        int cun;        //closest unmarked node
        int count=0;

        //initialize
        for (int i=1;i<=N;i++)
        {
                fv[i]=false;
```

```
                Pv[i]=-1;
                v[Sv].sp[i]=99;
        }
        v[Sv].sp[Sv]=0;

        //compute
        while (count<N)
        {
                cun=GetCUN();
                fv[cun]=true;
                for (int i=1;i<=N;i++)
                        if ((!fv[i]) && (v[cun].wt[i]>0))
                                if (v[Sv].sp[i]>v[Sv].sp[cun]+v[cun].wt[i])
                                {
                                        v[Sv].sp[i]=v[Sv].sp[cun]+v[cun].wt[i];
                                        Pv[i]=cun;
                                }
                count++;
        }
}

void CCode4A::DrawPath()
{
        CClientDC dc(this);
        CString s;
        CPen rPen(PS_SOLID,2,RGB(200,0,0));
        int i,u,w;

        //display the source
        dc.SelectObject(fHelvetica);
        dc.SetTextColor(RGB(0,0,0));
        dc.TextOut(end.x+100,home.y,L"Path");
        dc.TextOut(end.x+150,home.y,L"Weight");
        s.Format(L"Total Costs:%d",v[Sv].sp[Dv]);
        dc.TextOut(end.x+100,end.y,s);

        w=Dv;
        dc.SelectObject(rPen);
        dc.MoveTo(v[w].home);
        DrawNode(w);
        s.Format(L"%d",Dv); dc.TextOutW(end.x+100,end.y-100,s);
        for (i=1;i<=N;i++)
        {
                u=Pv[w];
                if (u!=-1)
                {
                        dc.LineTo(v[u].home);
                        s.Format(L"%d",v[w].wt[u]);
                        dc.TextOutW(end.x+150,end.y-100-60*i,s);
                        w=u;
                        DrawNode(u);
                        s.Format(L"%d",w);
                        dc.TextOutW(end.x+100,end.y-130-60*i,s);
                }
        }
}
```

```
void CCode4A::OnCompute()
{
        CString s;
        eSv.GetWindowText(s); Sv=_ttoi(s);
        eDv.GetWindowText(s); Dv=_ttoi(s);
        Dijkstra(); DrawPath();
}

void CCode4A::OnNewGraph()
{
        Initialize(); Invalidate();
}
```

4.3 Floyd-Warshall's Method for the All-Pairs Shortest Paths

Floyd-Warshall's algorithm computes the shortest paths between all pairs of nodes in the graph. The approach is dynamic programming and it has the complexity of $O(n^3)$ for computing the shortest paths of the graph $G(V, E)$ having n nodes. Floyd-Warshall's algorithm is related to Dijkstra's by considering n multiplied by Dijkstra's $O(n^2)$ solution to produce Floyd-Warshall's $O(n^2)$.

Consider α_{ij} as the shortest path between (v_i, v_j) for $i, j = 1, 2,..., n$. The intermediate nodes between them are $(v_1, v_2,..., v_k)$. If v_k is an intermediate node, then $\alpha_{ij}^{(k)} = \alpha_{ik}^{(k-1)} + \alpha_{kj}^{(k-1)}$; otherwise, $\alpha_{ij}^{(k)} = \alpha_{ij}^{(k-1)}$. That gives $\alpha_{ij}^{(k)} = \min\left(\alpha_{ij}^{(k-1)}, \alpha_{ik}^{(k-1)} + \alpha_{kj}^{(k-1)}\right)$ in the case of $k > 0$ and $\alpha_{ij}^{(0)} = w_{ij}$ for $k = 0$. The idea summarizes the algorithm as follows:

Given:
    The graph $G(V, E)$ and weight matrix $W = [w_{ij}]$ for $i, j = 1, 2,..., n$.
Initialization:
$$D = \left[\alpha_{ij}^{(0)}\right] = [w_{ij}].$$

Process:
    For $k = 1$ to n
        For $i = 1$ to n
            For $j = 1$ to n
                Compute $\alpha_{ij}^{(k)} = \min\left(\alpha_{ij}^{(k-1)}, \alpha_{ik}^{(k-1)} + \alpha_{kj}^{(k-1)}\right)$
            Endfor
        Endfor
    Endfor
Output: $D = SP(v_i, v_j) = \alpha_{ij}^{(k)}$ for $i, j = 1, 2,..., n$.

The final matrix of $D = SP(v_i, v_j) = \alpha_{ij}^{(k)}$ stores the shortest paths of all the pairs of nodes. Floyd-Warshall's algorithm differs from Dijkstra's algorithm in that the shortest paths between all pairs of nodes in the former are computed using a single nested loop. Dijkstra's algorithm can also be used to compute the shortest paths of all pairs of nodes but this is done by calling Dijkstra() in *Code4A* n times, which may be quite tedious and not practical.

One practical application of Floyd-Warshall's algorithm is finding the *transitive closure* of a directed graph. In this problem, the idea is to construct a matrix for the reachability of the paths from v_i to v_j where the solution is already described in Floyd-Warshall's algorithm. The matrix is obtained by replacing the weights between the pairs of nodes with the binary values of 1 if the directed edge exists and 0 otherwise.

4.3.1 *Code4B*: Implementing Floyd-Warshall's Algorithm

Code4B is the implementation of Floyd-Warshall's algorithm for finding the shortest paths between all pairs of nodes in the weighted graph $G(V, E)$ for $n = |V| = 20$. Figure 4.5 shows the output of *Code4B*, which displays the shortest paths between all the nodes as a matrix in the top list view table and its weight matrix at the bottom. The application allows the user to select a pair of source-destination nodes by left-clicking the mouse on the respective nodes, which draws the shortest path between them. The graph is generated randomly and is refreshed with the click of the *New Graph* button.

Code4B has a single class called *CCode4B* with source files *Code4B.h* and *Code4B.cpp*. Table 4.2 lists some important objects and variables in the program. One particular variable of interest is the three-dimensional array called *alpha* that computes the shortest path between the source and destination nodes represented by its first two elements. The path goes through several intermediate nodes that are represented as its third element in the array.

Figure 4.6 is the organization of *Code4B* showing all the functions. The application starts with the constructor *CCode4B()* that creates the main window and the *New Graph* button.

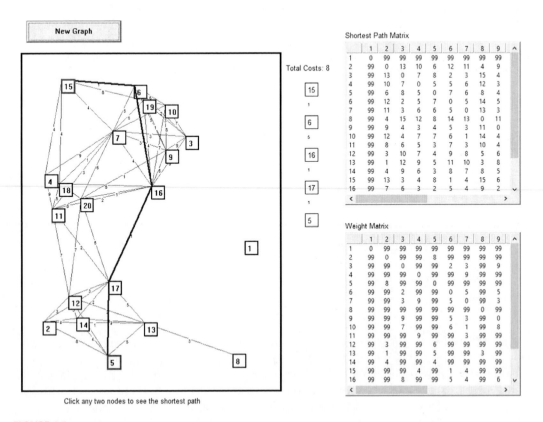

FIGURE 4.5
Output from *Code4B*.

TABLE 4.2

Important Objects/Variables in *Code4B* and Their Descriptions

Variable	Type	Description
n	constant	Number of nodes in the graph
bNGraph	CButton	Button for generating a new graph
bFlag	int	Event flag for the left button of the mouse with *bFlag* = 0 indicates no event, *bFlag* = 1 indicates that the source node has been selected, and *bFlag* = 2 denotes that the destination node has been selected
LinkRange	constant	Range for adjacency between two nodes in the graph
Sv, Dv	int	Source, destination nodes
v[i].wt[j]	int	Weight between (v_i, v_j)
v[i].sp[j]	int	Shortest path between (v_i, v_j)
e[i][j].via[k]	int	Path between (v_i, v_j) through the intermediate node v_k
alpha[i][j][k]	int	Distance between (v_i, v_j) through the intermediate node v_k
*table*1	CListCtrl	Top table displaying the shortest path matrix between (v_i, v_j)
*table*2	CListCtrl	Bottom table displaying the edge weight matrix between (v_i, v_j)
home, end	CPoint	Top-left and bottom-right corners of the drawing area

In activating the whole engine, the constructor calls *Initialize*() to construct the graph by assigning random coordinates to the nodes and random weights on the edges of the graph. After *Initialize*(), the graph is updated and displayed through *OnPaint*().

The shortest paths are computed offline through *Initialize*(). In computing the shortest paths, the function calls *FloydWarshall*(), which is the engine of Floyd-Warshall's algorithm. The results from the computation are displayed in the shortest path table through *ShowTable*().

The single shortest path between a pair of two nodes in the graph is displayed in the drawing area by first selecting the two nodes through the left clicks of the mouse. The event *ON_WM_LBUTTONDOWN*() detects the left clicks of the mouse for the two nodes, and this event is handled by *OnLButtonDown*(). A Boolean variable called *bFlag* monitors the selected nodes: *bFlag* = 0 denotes no event, *bFlag* = 1 denotes that the source node has been selected, and *bFlag* = 2 denotes that the destination node has been selected. An MFC function called *PtInRect*() in *OnLButtonDown*() detects the location of the click in Windows called *pt*, and the value returned by *pt* is compared to the rectangle *v[i].rct* that belongs to v_i. The click at the destination node produces *bFlag* = 2, and this activates *DrawNode*() for redrawing the node.

A button called *Compute* calls *OnCompute*() to produce the visual display of the single shortest path between two nodes in the graph. In displaying the single shortest path between two nodes, *OnCompute*() calls *DrawVCPath*() to trace the path from the source to its destination nodes. The path is drawn in a thick red line showing all the nodes along its path.

The full listing of the codes are given below.

```
//Code4B.h
#include <afxwin.h>
#include <afxcmn.h>
#include <math.h>
#define N 20
#define LinkRange 200
#define IDC_NGRAPH 800
```

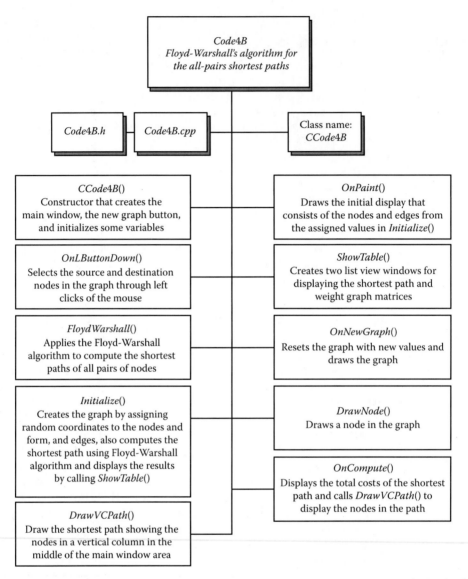

FIGURE 4.6
Design organization of *Code4B*.

```
class CCode4B: public CFrameWnd
{
private:
        int Sv,Dv,bFlag;
        CButton bNGraph;
        CListCtrl table1,table2;
        int alpha[N+1][N+1][N+1];
        typedef struct
        {
                int via[N+1];
        } LINK;
        LINK e[N+1][N+1];
```

```
        typedef struct
        {
                CPoint home;
                CRect rct;
                int wt[N+1],sp[N+1];
        } NODE;
        NODE v[N+1];
        CPoint home,end;
        CFont fArial,fHelvetica;
public:
        CCode4B();
        ~CCode4B()      {}
        void FloydWarshall(),Compute(),DrawNode(int);
        void Initialize(),DrawVCPath(int,int,int);
        void OnNewGraph(),ShowTable();
        afx_msg void OnPaint();
        afx_msg void OnLButtonDown (UINT, CPoint);
        DECLARE_MESSAGE_MAP()
};

class CMyWinApp:public CWinApp
{
public:
        virtual BOOL InitInstance();
};
CMyWinApp MyApplication;

BOOL CMyWinApp::InitInstance()
{
        m_pMainWnd=new CCode4B;
        m_pMainWnd->ShowWindow(m_nCmdShow);
        return TRUE;
}


//Code4B: Floyd-Warshall's Method
#include "Code4B.h"

BEGIN_MESSAGE_MAP(CCode4B, CFrameWnd)
ON_WM_PAINT()
        ON_WM_LBUTTONDOWN()
        ON_BN_CLICKED(IDC_NGRAPH,OnNewGraph)
END_MESSAGE_MAP()
CCode4B::CCode4B()
{
        Create(NULL,L"Floyd-Warshall's Shortest Path Algorithm",
                WS_OVERLAPPEDWINDOW,CRect(0,0,1000,800));
        bNGraph.Create(L"New Graph",WS_CHILD | WS_VISIBLE |
BS_DEFPUSHBUTTON,CRect(CPoint(30,30),CSize(180,40)),this,IDC_NGRAPH);
        home=CPoint(30,100); end=CPoint(500,700);
        fArial.CreatePointFont(60,L"Arial");
        fHelvetica.CreatePointFont(100,L"Helvetica");
        Initialize();
}
```

```cpp
void CCode4B::OnLButtonDown(UINT nFlags,CPoint pt)
{
        for (int i=1;i<=N;i++)
                if (v[i].rct.PtInRect(pt))
                {
                        bFlag++;
                        if (bFlag==1)
                        {
                                Sv=i;
                                DrawNode(Sv);
                        }
                        if (bFlag==2)
                        {
                                Dv=i;
                                DrawNode(Dv);
                                Compute();
                                bFlag=0;
                        }
                }
}

void CCode4B::ShowTable()
{
        CString s;
        int i,j;
        CRect rct;
        table1.DestroyWindow(); table2.DestroyWindow();
        //creates the list view window
        rct=CRect(CPoint(end.x+130,home.y-30),CSize(330,300));
        table1.Create(WS_VISIBLE | WS_CHILD | WS_DLGFRAME | LVS_REPORT
                | LVS_NOSORTHEADER,rct,this,802);
        rct=CRect(CPoint(end.x+130,home.y+320),CSize(330,300));
        table2.Create(WS_VISIBLE | WS_CHILD | WS_DLGFRAME | LVS_REPORT
                | LVS_NOSORTHEADER,rct,this,803);

        for (j=0;j<=N;j++)
        {
                s.Format(L"%d",j);
                table1.InsertColumn(j,((j==0)?L"SP":s),LVCFMT_CENTER,30);
                table2.InsertColumn(j,((j==0)?L"wt":s),LVCFMT_CENTER,30);
        }
        for (i=1;i<=N;i++)
        {
                s.Format(L"%d",i);
                table1.InsertItem(i-1,s,0);
                table2.InsertItem(i-1,s,0);
                for (j=1;j<=N;j++)
                {
                        s.Format(L"%d",v[i].sp[j]); table1.
SetItemText(i-1,j,s);
                        s.Format(L"%d",v[i].wt[j]); table2.
SetItemText(i-1,j,s);
                }
        }
}
```

```
void CCode4B::Initialize()
{
        int i,j,k;
        double distance;
        bFlag=0;
        srand(time(0));
        for (i=1;i<=N;i++)
        {
                v[i].home.x=home.x+20+rand()%(end.x-home.x-50);
                v[i].home.y=home.y+20+rand()%(end.y-home.y-50);
                v[i].rct=CRect(v[i].home.x,v[i].home.y,v[i].home.x+25,v[i].
                home.y+25);
        }
        for (i=1;i<=N;i++)
        {
                v[i].wt[i]=0;
                for (j=i+1;j<=N;j++)
                {
                        distance=sqrt(pow((double)(v[i].home.x-v[j].
home.x),2)+pow((double)(v[i].home.y-v[j].home.y),2));
                        if (distance<=LinkRange)
                                v[i].wt[j]=1+rand()%9;
                        else
                                v[i].wt[j]=99;
                        v[j].wt[i]=v[i].wt[j];
                }
        }
        for (i=1;i<=N;i++)
                for (j=1;j<=N;j++)
                {
                        v[i].sp[j]=v[i].wt[j];
                        for (k=1;k<=N;k++)
                                alpha[i][j][k]=99;
                }
        FloydWarshall();
        ShowTable();
}

void CCode4B::OnPaint()
{
        CPaintDC dc(this);
        CString s;
        CRect rct;
        CPen rPen(PS_SOLID,3,RGB(0,0,0));
        CPen mPen(PS_SOLID,2,RGB(0,0,0));
        CPen qPen(PS_SOLID,1,RGB(100,100,100));
        CPoint mPoint,pt;
        int i,j;
        dc.SelectObject(rPen);
        dc.Rectangle(home.x-10,home.y-10,end.x+10,end.y+10);
        dc.SelectObject(qPen); dc.SelectObject(fArial);
        for (i=1;i<=N;i++)
                for (j=1;j<=N;j++)
                        if (v[i].wt[j]!=99)
```

```
                              {
                                  dc.MoveTo(CPoint(v[i].home));
                                  dc.LineTo(CPoint(v[j].home));
                                  mPoint=CPoint((v[i].home.x+v[j].home.x)/2,
                                          (v[i].home.y+v[j].home.y)/2);
                                  s.Format(L"%d",v[i].wt[j]);
                                  dc.TextOut(mPoint.x,mPoint.y,s);
                              }
          dc.SelectObject(fHelvetica); dc.SelectObject(mPen);
          for (i=1;i<=N;i++)
                  DrawNode(i);
          dc.TextOutW(home.x+70,end.y+20,L"Click any two nodes to see the
    shortest path");
    }

    void CCode4B::FloydWarshall()
    {
        int i,j,k,m,r;
        for (i=1;i<=N;i++)
          for (j=1;j<=N;j++)
              for (k=1;k<=N;k++)
                  if (v[j].sp[i]!=99 || v[i].sp[k]!=99 || v[j].sp[k]!=99)
                      if (v[j].sp[i]+v[i].sp[k]<v[j].sp[k])
                      {
                              r=1;
                              v[j].sp[k]=v[j].sp[i]+v[i].sp[k];
                              for (m=1;m<=N;m++)
                                      alpha[j][k][m]=99;
                              for (m=1; m<=N; m++)
                                      if (alpha[j][i][m]!=99)
                                              alpha[j][k][r++]=alpha[j][i][m];
                              alpha[j][k][r++]=i;
                              for (m=1; m<=N; m++)
                                      if (alpha[i][k][m]!=99)
                                              alpha[j][k][r++]=alpha[i][k][m];

                      }
    }

    void CCode4B::Compute()
    {
          CClientDC dc(this);
          CString s;
          CPen pBlue(PS_SOLID,3,RGB(0,0,200));
          int k,u,w,r;
          dc.SelectObject(fHelvetica);
          s.Format(L"Total Costs:%d",v[Sv].sp[Dv]);
          dc.TextOut(end.x+20,home.y+10,s);

          r=1;
          DrawVCPath(r,Sv,0);
          dc.SelectObject(&pBlue);
          w=Sv; dc.MoveTo(v[w].home);
          for (k=1; k<=N; k++)                    //display the vias
          {
                  u=alpha[Sv][Dv][k];
```

```
                    if (u!=99)
                    {
                            r++; e[Sv][Dv].via[r]=u;
                            dc.LineTo(v[u].home);
                            DrawVCPath(r,u,w);
                            w=u;
                    }
            }
            if (v[Dv].sp[w]!=99)                //display the destination
                    dc.LineTo(v[Dv].home);
            e[Sv][Dv].via[r]=Dv;
            DrawVCPath(r+1,Dv,w);
            DrawNode(Sv); DrawNode(Dv);
}

void CCode4B::DrawVCPath(int r,int u,int w)
{
        CClientDC dc(this);
        CString s;
        CPoint pt;
        CRect rct;
        pt=CPoint(end.x+60,home.y+50+60*(r-1));
        rct=CRect(pt.x-5,pt.y-5,pt.x+20,pt.y+20); dc.Rectangle(rct);
        dc.SelectObject(fHelvetica);
        s.Format(L"%d",u);
        dc.TextOut(pt.x,pt.y,s);
        dc.SelectObject(fArial);
        if (r>1)
        {
                s.Format(L"%d",v[u].wt[w]);
                dc.TextOut(end.x+60,home.y+20+60*(r-1),s);
        }
}

void CCode4B::DrawNode(int u)
{
        CClientDC dc(this);
        CString s;
        CPoint pt;
        CPen pBlack(PS_SOLID,2,RGB(0,0,0));
        CPen pRed(PS_SOLID,3,RGB(200,0,0));
        if (bFlag==0)
                dc.SelectObject(pBlack);
        else
                dc.SelectObject(pRed);
        pt=v[u].home;
        dc.Rectangle(v[u].rct);
        s.Format(L"%d",u); dc.TextOut(pt.x+5,pt.y+5,s);
}

void CCode4B::OnNewGraph()
{
        Initialize(); Invalidate();
}
```

4.4 Mini-GPS

GPS is a navigation system in the form of a small device attached inside a car, aircraft, or ship. With today's technology, GPS is also available in other electronic devices such as cameras, cell phones, and computers. GPS receives useful real-time navigation information in the form of current coordinates from satellites every few minutes. A GPS device also displays a detailed map of a region with nearby cities and the network of roads linking the cities. In a car, GPS helps the driver to follow the shortest path from the source to the destination. Today's GPS has some extra features such as providing alternative paths to the shortest path in order to avoid traffic or road construction on that path. This is useful because the shortest path does not always guarantee arrival in the minimum amount of time. The presence of too many cars in that path due to traffic lights, road construction, or an accident may delay driving time. Therefore, one or two alternative paths that are the second and third shortest paths are provided and are readily available in the device.

A GPS device inside a car has digital maps that are activated when the device is turned on. A map is displayed relative to the current location that is determined in real time by a satellite. The current position is updated every few seconds as the GPS device has a receiver that picks up signals from the satellite. The maps inside the device are stored as a database, and they have the capability to zoom in or out, rotate, and translate according to the user's settings. The current position is the source node and it interacts with the road map in the device database for calculating the shortest path to the desired destination.

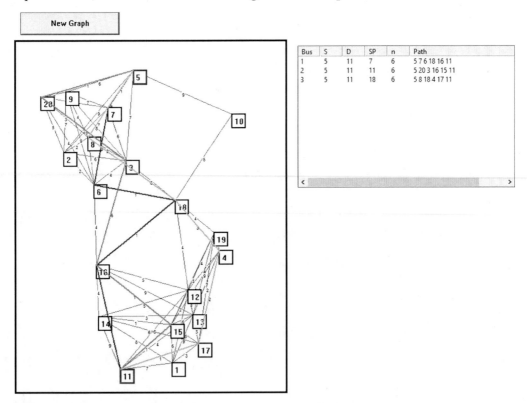

Bus	S	D	SP	n	Path
1	5	11	7	6	5 7 6 18 16 11
2	5	11	11	6	5 20 3 16 15 11
3	5	11	18	6	5 8 18 4 17 11

FIGURE 4.7
Output from *Code4C* showing three buses from the source to its destination.

Once the path is determined, the device then instructs the driver to follow along the path street by street until the destination is reached.

4.4.1 *Code4C*: Implementing the Mini-GPS

Code4C is the implementation of Dijkstra's algorithm for the *k* shortest path problem, which is about finding *k* different shortest paths between a pair of nodes. The project illustrates a simple GPS model for displaying three different routes between the source and its destination nodes. The project produces the offline version of GPS where no real-time data with regard to the current coordinates of the driver is included.

Figure 4.7 displays a sample output for *Code4C*. The output consists of a drawing area for the graph and a list view window for displaying the routing information. In the drawing area are 20 nodes with v_5 and the source node and v_{11} as its destination node. The program produces three different shortest paths from v_5 to v_{11}. Each path in the graph is called a *bus*, which is defined as a unique path from the source to its destination node in the graph. The output also displays the path information of the three buses in the list view window.

Code4C implements Dijkstra's algorithm for finding all three shortest paths. The first bus is obtained from the original graph *G* with 20 nodes. Once the path has been drawn the edges along the path are removed to reduce *G* to *G'*. Dijkstra's algorithm is applied again to *G'* to produce the second bus, which has a set of edges different from the first path. Next, the edges along the second bus are removed to reduce *G'* to *G"*. Again, the shortest path for *G"* is computed using Dijkstra's algorithm to produce the third bus.

The program has a single class called *CCode4C* with *Code4C.h* and *Code4C.cpp* as its source files. Most of the codes in *Code4C* are derived from *Code4A* although *Code4C* supports multiple shortest paths between a pair of nodes, whereas *Code4A* has only one. This is the mechanism of a GPS device where alternative routes are needed just in case the first shortest path becomes unavailable.

Table 4.3 lists some common variables and objects in *Code4C*. A structure called *BUS* that include different buses in an array called *Bus* to represent the successfully drawn paths

TABLE 4.3

Important Objects/Variables in *Code4C* and Their Descriptions

Variable	Type	Description
bNGraph	*CButton*	Button for generating a new graph
bCompute	*CButton*	Compute the shortest path
Bus[i].path[k]	*int*	Path *k* in Bus *i*
Bus[i].nNodes	*int*	Number of nodes in Bus *i*
Bus[i].sp	*int*	Total cost of Bus *i*
home	*CPoint*	Top left corner of the rectangular grid area
v[i].wt[j]	*int*	Weight between (v_i, v_j)
v[i].sp[j]	*int*	Shortest path between (v_i, v_j)
fv[i]	*bool*	Flag status of v_i
LinkRange	constant	Threshold value of the range for adjacency between two nodes in the graph
N	constant	Number of nodes in the graph
nBus	*int*	Number of successful buses
Sv, Dv	*int*	Source, destination nodes
Pv	*int*	Predecessor node to the current node
pBus[i]	*CPen*	Pen color of Bus *i*
table	*CListCtrl*	Table for displaying the successful buses

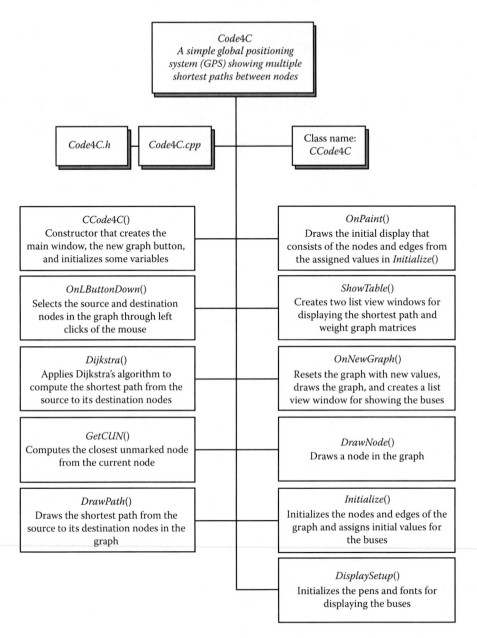

FIGURE 4.8
Organization of *Code4C*.

between the source and destination nodes. In our application, *Bus* is linked to the members of *BUS*, namely, the array *Path*, which is the number of nodes *nNodes* along its path and *sp*, which is its total cost.

Figure 4.8 is the organization chart of *Code4C*. The mini-GPS system starts with the constructor *CCode4C()*, which creates the main window and the *New Graph* button. The function also calls *DisplaySetup()* and *OnNewGraph()*, which sets the common display variables and creates the graph. *OnNewGraph()* calls *Initialize()*, which creates the graph by assigning random coordinates to the nodes and random weights to the edges of the graph. The initial graph is displayed and updated through *OnPaint()*.

The same method for selecting the source-destination nodes as in *Code4B* is deployed here. The event *ON_WM_LBUTTONDOWN()* detects the left click of the mouse whose position in Windows is returned by the *CPoint* object *pt*. The object value is checked with *v[i].rct* using *PtInRect()*. The source node is identified through *bFlag* = 1 while the destination has *bFlag* = 2.

Dijkstra() computes the shortest path using Dijkstra's algorithm. This function is called once the second node has been selected while the bus from the source to its destination node is drawn using *DrawPath()*. With the completion of the first bus, its information is updated and displayed in the list view window. Also, the edges of the first bus are removed from the original graph G. The removal of edges is performed by *Initialize()* by assigning a value of 99 to the edges.

The second bus is a repeat of the first bus by referring to the reduced graph G'. Therefore, the calculation for the new shortest path between the nodes will consider the fact that the first path has nonadjacent nodes along the way. The program then computes the new shortest path that clearly avoids the first shortest path. Similarly, the third shortest path is computed and drawn by blanking the weights of the edges along the first and second shortest paths.

The full listing of the codes are given below.

```
//Code4C.h
#include <afxwin.h>
#include <afxcmn.h>
#include <fstream>
#define IDC_TABLE 500
#define IDC_NGRAPH 501
#define N 20
#define LinkRange 200

class CCode4C:public CFrameWnd
{
private:
        int Sv,Dv,nBus,bFlag;
        int cColor;
        int Pv[N+1];
        bool fv[N+1];
        CButton bNGraph;
        CPen pBus[N+1];
        CPoint home,end;
        CFont fHelvetica,fArial;
        CListCtrl table;
        typedef struct
        {
                CPoint home;
                CRect rct;
                int wt[N+1];
                int sp[N+1];
        } NODE;
        NODE v[N+1];
```

```
        typedef   struct
        {
                int Path[N+1];
                int nNodes;
                int sp;
        } BUS;
        BUS Bus[N+1];
public:
        CCode4C();
        ~CCode4C()                {}
        int GetCUN();
        void Dijkstra(),DrawPath(),DrawNode(int);
        void Initialize(),ShowTable(),DisplaySetup();
        afx_msg void OnPaint();
        afx_msg void OnLButtonDown(UINT,CPoint);
        afx_msg void OnNewGraph();
        DECLARE_MESSAGE_MAP()
};

class CMyWinApp:public CWinApp
{
public:
        virtual BOOL InitInstance();
};
CMyWinApp MyApplication;

BOOL CMyWinApp::InitInstance()
{
        m_pMainWnd=new CCode4C;
        m_pMainWnd->ShowWindow(m_nCmdShow);
        return TRUE;
}

//Code4C.h
#include "Code4C.h"

BEGIN_MESSAGE_MAP(CCode4C, CFrameWnd)
ON_WM_PAINT()
        ON_WM_LBUTTONDOWN()
        ON_BN_CLICKED(IDC_NGRAPH,OnNewGraph)
END_MESSAGE_MAP()

CCode4C::CCode4C()
{
        home=CPoint(30,100); end=CPoint(500,700);
        Create(NULL, L"Mini GPS",WS_OVERLAPPEDWINDOW,CRect(0,0,1100,800));
        bNGraph.Create(L"New Graph",WS_CHILD | WS_VISIBLE |
BS_DEFPUSHBUTTON,CRect(CPoint(home.x,home.y-60),CSize(180,40)),this,
IDC_NGRAPH);
        DisplaySetup(); OnNewGraph();
}

void CCode4C::OnNewGraph()
{
        CString s[]={"Bus","S","D","SP","N","Path"};
```

```
        nBus=0; Initialize(); Invalidate();
        table.DestroyWindow();
        table.Create(WS_VISIBLE | WS_CHILD | WS_BORDER | LVS_REPORT
                | LVS_NOSORTHEADER,CRect(CPoint(end.x+30,home.y),CSize(390,2
                50)),this,IDC_TABLE);
        for (int i=0;i<=5;i++)
                table.InsertColumn(i,s[i],LVCFMT_LEFT,((i==5)?250:40));
}


void CCode4C::DisplaySetup()
{
        cColor=1;
        int Color[]={RGB(150,150,150),
                RGB(0,0,200),RGB(200,0,0),
                RGB(0,200,0),RGB(200,200,0),
                RGB(0,200,200),RGB(200,0,200),
                RGB(200,50,150),RGB(100,100,255),
                RGB(250,50,255),RGB(100,200,50)};
        for (int i=0;i<=10;i++)
                pBus[i].CreatePen(PS_SOLID,2,Color[i]);
        fHelvetica.CreatePointFont (100,L"Helvetica");
        fArial.CreatePointFont (60,L"Arial");
}


void CCode4C::Initialize()
{
        int i,j,u,w;
        double distance;
        srand(time(0));
        bFlag=0;
        if (nBus==0)
        {
                for (i=1;i<=N;i++)
                {
                        v[i].home.x=home.x+20+rand()%(end.x-home.x-50);
                        v[i].home.y=home.y+20+rand()%(end.y-home.y-50);
                        v[i].rct=CRect(v[i].home.x,v[i].home.y,v[i].home.
x+25,v[i].home.y+25);
                }
                for (i=1;i<=N;i++)
                        for (j=i;j<=N;j++)
                        {
                                v[i].wt[j]=99;
                                distance=sqrt(pow((double)(v[i].home.x-v[j].
                        home.x),2)+pow((double)(v[i].home.y-v[j].home.y),2));
                                if (distance<LinkRange)
                                        v[i].wt[j]=1+rand()%9;
                                v[j].wt[i]=v[i].wt[j];
                        }
        }
        else
        {
                u=Bus[nBus].Path[1];
                for (i=1;i<=Bus[nBus].nNodes-1;i++)
```

```
                {
                        w=Bus[nBus].Path[i+1];
                        v[u].wt[w]=99;  v[w].wt[u]=99;
                        u=w;
                }
        }
}

void CCode4C::DrawNode(int u)
{
        CClientDC dc(this);
        CString s;
        CPoint pt;
        CPen pBlack(PS_SOLID,2,RGB(0,0,0));
        CPen pRed(PS_SOLID,3,RGB(200,0,0));
        if (u==Sv || u==Dv)
                dc.SelectObject(pRed);
        else
                dc.SelectObject(pBlack);
        pt=v[u].home;
        dc.Rectangle(v[u].rct);
        s.Format(L"%d",u);  dc.TextOut(pt.x+5,pt.y+5,s);
}

void CCode4C::OnPaint()
{
        CPaintDC dc(this);
        CString s;
        CRect rct;
        CPen rPen(PS_SOLID,3,RGB(0,0,0));
        CPen mPen(PS_SOLID,2,RGB(0,0,0));
        CPen qPen(PS_SOLID,1,RGB(100,100,100));
        CPoint mPoint,pt;
        int i,j;
        dc.SelectObject(rPen);
        dc.Rectangle(home.x-10,home.y-10,end.x+10,end.y+10);
        dc.SelectObject(qPen);  dc.SelectObject(fArial);
        for (i=1;i<=N;i++)
                for (j=1;j<=N;j++)
                        if (v[i].wt[j]!=99)
                        {
                                dc.MoveTo(CPoint(v[i].home));
                                dc.LineTo(CPoint(v[j].home));
                                mPoint=CPoint((v[i].home.x+v[j].home.x)/2,
                                        (v[i].home.y+v[j].home.y)/2);
                                s.Format(L"%d",v[i].wt[j]);
                                dc.TextOut(mPoint.x,mPoint.y,s);
                        }
        dc.SelectObject(fHelvetica);  dc.SelectObject(mPen);
        for (i=1;i<=N;i++)
                DrawNode(i);
}
int CCode4C::GetCUN()
{
        int minDistance=99;
```

```
        int cun;
        for (int i=1;i<=N;i++)
              if ((!fv[i]) && (minDistance>=v[Sv].sp[i]))
              {
                    minDistance=v[Sv].sp[i];
                    cun=i;
              }
        return cun;
}

void CCode4C::Dijkstra()
{
        int minDistance=99;
        int cun;

        //initialize
        for (int i=1;i<=N;i++)
        {
              fv[i]=false;
              Pv[i]=-1;
              v[Sv].sp[i]=99;
        }
        v[Sv].sp[Sv]=0;

        //compute
        int k=0;
        while (k<N)
        {
              cun=GetCUN();
              fv[cun]=true;
              for (int i=1;i<=N;i++)
                    if ((!fv[i]) && (v[cun].wt[i]>0))
                          if (v[Sv].sp[i]>v[Sv].sp[cun]+v[cun].wt[i])
                          {
                                v[Sv].sp[i]=v[Sv].sp[cun]+v[cun].wt[i];
                                Pv[i]=cun;
                          }
              k++;
        }
        nBus++;
        Bus[nBus].sp=v[Sv].sp[Dv];
}

void CCode4C::DrawPath()
{
        CClientDC dc(this);
        int w,i,u,r,dummy[N+1];

        //display the source
        dc.SelectObject(pBus[1+cColor%10]); cColor++;
        w=Dv; Bus[nBus].nNodes=0; Bus[nBus].Path[1]=Sv;
        for (i=1;i<=N;i++)
        {
              u=Pv[w];
              if (u!=-1)
```

```
                {
                        r=++Bus[nBus].nNodes; dummy[r]=w;
                        w=u;
                }
                dummy[r+1]=Sv;
        }
        r=++Bus[nBus].nNodes; Bus[nBus].Path[r]=Dv;
        u=Sv; Bus[nBus].Path[1]=u;
        dc.MoveTo(v[u].home);
        for (i=1;i<=Bus[nBus].nNodes;i++)
        {
                w=dummy[Bus[nBus].nNodes-i+1]; Bus[nBus].Path[i]=w;
                dc.LineTo(v[w].home);
                u=w;
        }
        w=Dv; Bus[nBus].Path[i]=w;
        dc.LineTo(v[w].home);
}

void CCode4C::OnLButtonDown(UINT nFlags,CPoint pt)
{
        CClientDC dc(this);
        CString s;
        int i,j;
        dc.SelectObject(pBus[1]); dc.SelectObject(fArial);
        for (int i=1;i<=N;i++)
                if (v[i].rct.PtInRect(pt))
                {
                        bFlag++;
                        if (bFlag==1) //bFlag=1 is the source
                        {
                                Sv=i;
                                DrawNode(Sv);
                        }
                        if (bFlag==2) //bFlag=2 is the destination
                        {
                                Dv=i; DrawNode(Dv);
                                for (j=1;j<=3;j++)
                                {
                                        Dijkstra();
                                        if (v[Sv].sp[Dv]!=99)
                                        {
                                                DrawPath(); ShowTable();
Initialize();
                                        }
                                        else
                                        {
                                                bFlag=0;
                                                dc.SelectObject(pBus[0]);
                                                DrawNode(Sv); DrawNode(Dv);
                                        }
                                }
                        }
                }
}
```

```
void CCode4C::ShowTable()
{
        int i,k,w,u;
        CString s,S;
        for (i=1;i<=nBus;i++)
        {
                table.DeleteItem(i-1);
                s.Format(L"%d",i); table.InsertItem(i-1,s,0);
                s.Format(L"%d",Bus[i].Path[1]); table.SetItemText(i-1,1,s);
                w=Bus[i].nNodes;
                s.Format(L"%d",Bus[i].Path[w]); table.SetItemText(i-1,2,s);
                u=Bus[i].Path[1]; w=Bus[i].Path[w];
                s.Format(L"%d",Bus[i].sp); table.SetItemText(i-1,3,s);
                s.Format(L"%d",Bus[i].nNodes); table.SetItemText(i-1,4,s);
                S="";
                for (k=1;k<=Bus[i].nNodes;k++)
                {
                        s.Format(L"%d ",Bus[i].Path[k]); S +=s;
                }
                table.SetItemText(i-1,5,S);
        }
}
```

4.5 Multicolumn Interconnection Network

A muliticolumn interconnection network (MIN) has a set of n source nodes that connects to another set of n destination nodes through one or more columns (stages) of switches. The nodes in a MIN could be processing elements, memory modules, and network resources such as printers, network hubs, servers, and getaways. A MIN forms the backbone of many high-performance multiprocessor, supercomputer, and telecommunications systems due to its efficient topology for communication [3]. In telecommunication networks, for example, a MIN provides low latency with regard to node communication to support efficient routing for high-speed processing capabilities. These networks normally have regular topologies for efficient management and administration that incur high overhead in terms of data management. However, the arrangement is a trade-off for high bandwidth in its output.

Packet switching and *circuit switching* are two types of message and data transmission in a MIN. In packet switching, a message or data from the source node is broken into several smaller packets before they are delivered to the destination node. In this network, switches are configured dynamically to allow fast connections between the nodes. In most cases, the packets use different paths before they merge to the destination node. Each packet has the same header address for the destination node in the form of the IP address and the sequence for assembly once it merges with other packets. If a packet fails to arrive due to some problem along its path, the destination node will notify the source node ask for the packet to be re-sent. E-mails and other message-sending mechanisms through the Internet are the most common form of packet switching.

Circuit switching applies to a telephone call connecting the caller (source) to the receiver (destination) on a real-time basis. The caller and receiver are allocated with one channel each and the path between them is wholly reserved for them until the call is terminated. If another call enters the network and all the paths between the source and destination nodes are occupied, then the call is placed in a queue. The call is connected once a pair of source–destination paths becomes available. The connection between the source–destination pairs may go through several switches in the network.

The topology of an interconnection network can be regular or irregular. In a *regular* topology, the nodes have the same degrees to provide the same number of input and output. The networks are normally arranged as rectangles, squares, triangles, or hexagons of the same sizes. The nodes in the network are located at fixed positions, and because of this, extension and network scaling is rather difficult. *Irregular networks*, such as the Internet, have nodes with varying degrees that are scattered at nonfixed locations. Due to the nonfixed locations, these networks are more flexible in terms of network extension and scaling.

The jobs carried out in a MIN include tedious and laborious tasks such as real-time communication, data sorting, bitonic sorting, permutation, and cyclic shifting. One common application is data sorting, which is the rearrangement of data from the given raw data according to alphanumeric or numeric arrangement in ascending or descending order. In a MIN, the set of raw data to be sorted is placed on the source nodes. Sorting is then performed using some common techniques such as a bubble sort. This technique is performed by running a series of iterations in a loop that produces new arrangements of the data each time. At each iteration, the resulting data in the new arrangement is placed in the

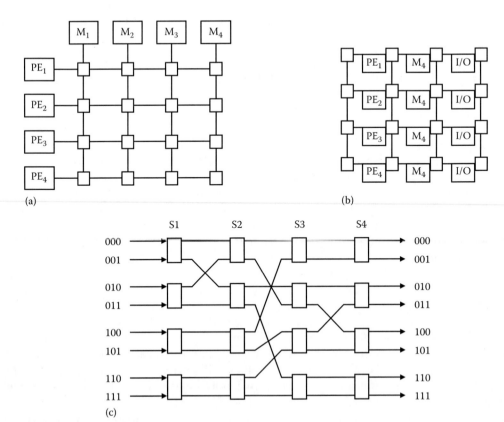

FIGURE 4.9
Some common MINs: (a) crossbar network, (b) network-on-chip, and (c) omega network.

destination nodes. The final iteration has the sorted data as the final result in the destination nodes.

The architecture of MINs extends from fundamental network topology, namely, the ring, star, bus, and hypercube designs. Figure 4.9 shows some common types of MINs: crossbar (a), network-on-chip (NOC) (b), and omega networks (c). The crossbar network connects any processing element (PE) with a memory module (M) through switches (clear rectangles) using a matrix layout. In the same manner, the network-on-chip has processing elements, memory modules, and input/output (I/O) drivers that communicate through switches and are compactly designed within a single chip. In the omega network, the switches (S) are arranged in four vertical columns to allow communications between the left and right processing elements.

A crossbar network has nodes called switches that provide connections between the processing modules in the network. The modules could be PEs with individual memory units or PEs with separate memory units. The switches are normally arranged in a rectangular grid as inner nodes in the network while the processing modules are positioned at the boundaries. For a maximum bandwidth, the crossbar network can deliver n messages from n source nodes to n destination nodes without blocking at an instance of time. With n nodes the maximum overhead is wn^2, where w is the data width.

An NOC is a network consisting of several PEs and memory modules embedded in a single chip. An NOC has a tremendous scalability and parallelism capabilities for supporting complex multicore processing requirements. An omega network is another common switching network that has switches arranged in several columns for connecting the left and right PEs. The switch complexity is $O(n \log n)$ and the network is powerful for tasks involving both packet and circuit switchings.

4.5.1 *Code4D*: Multicolumn Interconnection Network

Code4D is a program that maps a set of m source nodes to a set of m destination nodes through layers of switches in a rectangular interconnection network. The main objective in this program is to illustrate circuit switching by having the source–destination pairs connected on a real-time basis. In this network, the nodes and switches are arranged with the source and destination nodes forming the left and right boundaries, respectively. The source and destination nodes have the processing capabilities while the switches only function to relay messages from their neighbors with no processing capability. The source nodes serve as the input while the destination nodes are the output. The switches are arranged as vertical columns (stages) in a rectangular grid. The minimum number of columns in this network is one but connectivity between the two end nodes improves with a higher number of columns.

A switch in the rectangular interconnection network has four ports marked as north, south, east, and west. The internal switches have the same four ports, which allow them to communicate with the neighboring left, right, up, and down nodes. The switches at the top boundary have east, west, and south ports only, while the nodes at the bottom have east, west, and north ports only.

Circuit switching can be achieved through effective pairings between the source–destination nodes. Each successful pair is a bus whose concept is similar to the one described earlier in *Code4C*. The shortest path between the source and destination nodes is drawn using Dijkstra's algorithm by assuming every edge between two adjacent nodes has a weight of one unit. Circuit switching requires well-connected buses between the source and destination nodes. With m source–destination pairs it may not be possible to get 100%

pairings unless all the pairings involve horizontal paths from the sources to destinations. A *blocking* happens when a segment (or more) of the path from the source node to its destination has been taken by another path. An alternative shortest path(s) is sought and if a blocking happens in that path too, then the path for the source–destination pair is said to be blocked.

Code4D provides an idea for circuit switching between pairs of nodes in a rectangular interconnection network. It does not provide the optimal solution for drawing the buses between the source–destination pairs. Due to the topology, it will not be possible to have all the source–destination nodes connected because at least one pair is blocked. One idea for the optimal solution is to replace the rectangular grid with a toroid (rectangular torus) where all the switches have regular degrees of four. This is an improvement from our current model where the north switches on the top row are connected to the south switches of the bottom row. We leave it to the reader to do this by continuing the work from this program.

Figure 4.10 shows the output from *Code4D* with an interconnection network of size 6 × 8 in the order according to the sequence of {243561}. Matchings between the source nodes and destination nodes are performed by left-clicking the mouse on the respective pairs of nodes. The sequence {243561} is read as

$$(v_2, v_{11}) \rightarrow (v_4, v_7) \rightarrow (v_3, v_{12}) \rightarrow (v_5, v_{10}) \rightarrow (v_6, v_9) \rightarrow (v_1, v_8).$$

With the above sequence, five successful buses (first five pairs) and one unsuccessful bus (last pair) are produced. The successful buses are drawn as different colors between the source-destination nodes with their paths listed in the table on the right. The pair (v_1, v_8) that is placed last in the sequence has been blocked.

Table 4.4 lists and describes some important variables and objects in *Code4D*. *Bus* is the main variable and the array is created from a structure called *BUS*. Its contents and purposes are similar to the one in *Code4C*. The buttons for activating the new graph and computing the shortest paths are *bNGraph* and *bCompute*, respectively. The nodes of the graph are represented by the array *v* whose linkages with its members are defined in *NODE*.

The source node in the graph is denoted *Sv* while its destination is *Dv*. The nodes in the graph are marked with *fv* to denote their current status: if it is marked for the path the value is 1; otherwise, the value is 0.

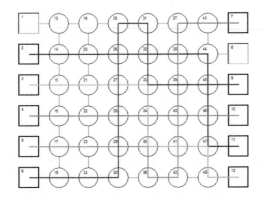

Bus	S	D	SP	n	Path
1	2	11	10	11	2 14 20 26 32 38 44 45 46 47 11
2	4	7	10	11	4 16 22 28 34 40 39 38 37 43 7
3	3	12	10	11	3 15 21 27 33 34 35 41 47 48 12
4	5	10	10	11	5 17 23 29 35 36 42 41 40 46 10
5	6	9	14	15	6 18 24 30 29 28 27 26 25 31 32 33 3!

FIGURE 4.10
An interconnection network of size 6 × 8.

TABLE 4.4

Important Objects/Variables in *Code4D* and Their Descriptions

Variable	Type	Description
bNGraph	*CButton*	Button for generating a new graph
bCompute	*CButton*	Compute the shortest path
Bus[i].Path[k]	*int*	Path k in Bus i
Bus[i].nNodes	*int*	Number of nodes in Bus i
Bus[i].sp	*int*	Total cost of Bus i
home	*CPoint*	Top left corner of the rectangular grid area
v[i].wt[j]	*int*	Weight between (v_i, v_j)
v[i].sp[j]	*int*	Shortest path between (v_i, v_j)
fv[i]	*bool*	Flag status of v_i
LinkRange	constant	Threshold value of the range for adjacency between two nodes in the graph
n	constant	Number of nodes in the graph
nBus	*int*	Number of successful buses
Sv, Dv	*int*	Source, destination nodes
pV	*int*	Predecessor node to the current node
pBus[i]	*CPen*	Pen color of Bus i
table	*CListCtrl*	Table for displaying the successful buses

Figure 4.11 lists and describes the functions in the program. The functions are further described here in their sequence. First, the constructor creates the main window and two edit boxes, *eNx* and *eNy*, for the row and column sizes of the rectangular grid, respectively. *OnStart()* reads these row and column input values, then calls *DisplaySetup()* and *OnReset()* to make up the display that consists of the drawing area and table to display the routing results. *OnPaint()*, which is invoked from *Invalidate()*, draws the rectangular grid showing the left (source) and right (destination) nodes and the switches.

The source and its destination nodes are selected by the user by left-clicking the mouse on the respective nodes. The function *OnLButtonDown()* reads the nodes, then calls *Dijkstra()* to compute the shortest path between these two nodes. It follows that *DrawPath()* draws a new bus if the source–destination path is not blocked and the routing information is updated in the table through *ShowTable()*. If the path is blocked, no new bus is drawn and the information is also updated in the table. *Dijkstra()* calls *GetCUN()* repeatedly to mark the closest unmarked node in computing the shortest path from the source to the destination nodes.

The event *ON_WM_LBUTTONDOWN()* detects the left click of the mouse and it is responded to by *OnLButtonDown()*. A Boolean variable called *LineFlag* is used to differentiate the nodes clicked by the user: *LineFlag* = 0 is the initial value denoting that the event is not triggered yet, *LineFlag* = 1 denotes that the source node has been selected, and *LineFlag* = 2 denotes that the destination node has been selected. The location of the click is read and represented by the *CPoint()* object, *pt*. The value returned by *pt* is checked if it belongs to the rectangle *v[k].rct* using *PtInRect()*. If it does, then *LineFlag* is updated according to the current event. Obviously, *LineFlag* = 2 creates a new bus by updating *nBus* value and activates *Dijkstra()* and *DrawPath* for computing and drawing the shortest path. Once the new bus has been created, the *LineFlag* value is reset to 0 to mark the end of the current event.

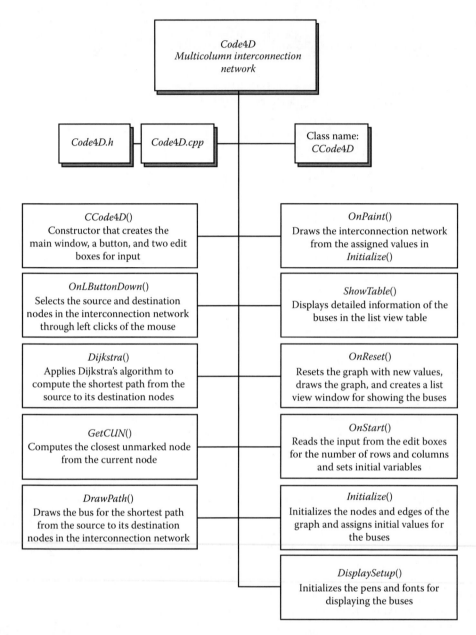

FIGURE 4.11
Organization of *Code4D*.

Code listing for *Code4D* are given as follows:

```
//Code4D.h
#include <afxwin.h>
#include <afxcmn.h>
#define IDC_TABLE 500
#define IDC_RESET 501
#define IDC_START 502
#define NMax 100
```

```
class CCode4D:public CFrameWnd
{
private:
        int Nx,Ny,n;
        int Sv,Dv,nBus,idc,LineFlag;
        int cColor;
        int Pv[NMax+1];
        bool fv[NMax+1],flag;
        CButton Reset,Generate,Start;
        CPen pBus[NMax+1];
        CPoint home;
        CFont fCourier,fArial,fTimes;
        CListCtrl table;
        CEdit eNx,eNy;
        CStatic sNx,sNy;

        typedef struct
        {
                CPoint home;
                CRect rct;
                int wt[NMax+1];
                int sp[NMax+1];
        } NODE;
        NODE v[NMax+1];

        typedef struct
        {
                int Path[NMax+1];
                int nNodes;
        } BUS;
        BUS Bus[NMax+1];
public:
        CCode4D();
        ~CCode4D()              {}
        int GetCUN();
        void Dijkstra(), DrawPath();
        void Initialize(),ShowTable(),DisplaySetup();
        afx_msg void OnPaint();
        afx_msg void OnLButtonDown(UINT,CPoint);
        afx_msg void OnReset();
        afx_msg void OnStart();
        DECLARE_MESSAGE_MAP()
};

class CMyWinApp:public CWinApp
{
public:
        virtual BOOL InitInstance();
};
CMyWinApp MyApplication;

BOOL CMyWinApp::InitInstance()
{
        m_pMainWnd=new CCode4D;
```

```
        m_pMainWnd->ShowWindow(m_nCmdShow);
        return TRUE;
}

//Code4D.cpp: Multi-stage Interconnection network
#include "Code4D.h"

BEGIN_MESSAGE_MAP(CCode4D, CFrameWnd)
ON_WM_PAINT()
        ON_WM_LBUTTONDOWN()
        ON_BN_CLICKED (IDC_RESET,OnReset)
        ON_BN_CLICKED (IDC_START,OnStart)
END_MESSAGE_MAP()

CCode4D::CCode4D()
{
        idc=550; flag=0; home=CPoint(10,20);
        Create(NULL, L"Multi-column Interconnection Network",
                WS_OVERLAPPEDWINDOW,CRect(0,0,1100,800));
        sNy.Create(L"Ny (1-10)",WS_VISIBLE | SS_CENTERIMAGE | SS_CENTER,
                CRect(CPoint(670,100),CSize(100,30)),this,IDC_STATIC);
        sNx.Create(L"Nx (1-10)", WS_VISIBLE | SS_CENTERIMAGE | SS_CENTER,
                CRect(CPoint(670,140),CSize(100,30)),this,IDC_STATIC);
        eNy.Create(WS_CHILD | WS_VISIBLE | WS_BORDER | SS_CENTER,
                CRect(CPoint(780,100),CSize(70,30)),this,idc++);
        eNx.Create(WS_CHILD | WS_VISIBLE | WS_BORDER | SS_CENTER,
                CRect(CPoint(780,140),CSize(70,30)),this,idc++);
        Start.Create(L"Start",WS_CHILD | WS_VISIBLE | BS_DEFPUSHBUTTON,
                CRect(CPoint(670,180),CSize(180,30)),this,IDC_START);
}

void CCode4D::OnStart()
{
        CString s;
        eNy.GetWindowText(s); Ny=_ttoi(s);
        eNx.GetWindowText(s); Nx=_ttoi(s);
        n=Nx*Ny;
        if (n<=NMax)
        {
                sNy.DestroyWindow(); sNx.DestroyWindow();
                eNy.DestroyWindow(); eNx.DestroyWindow();
                Reset.Create(L"Reset",WS_CHILD | WS_VISIBLE | BS_DEFPUSHBUTT
ON,CRect(CPoint(670,10),CSize(190,30)),this,IDC_RESET);
                flag=1; DisplaySetup(); OnReset();
        }
}

void CCode4D::OnReset()
{
        nBus=0;         Initialize(); Invalidate();

        //create the table
        CString s[]={"Bus","S","D","SP","n","Path"};
        table.DestroyWindow();
```

```
        table.Create(WS_VISIBLE | WS_CHILD | WS_BORDER | LVS_REPORT
                | LVS_NOSORTHEADER,CRect(CPoint(670,50),CSize(390,250)),
                this,IDC_TABLE);
        for (int i=0;i<=5;i++)
                table.InsertColumn(i,s[i],LVCFMT_LEFT,((i==5)?250:40));
}

void CCode4D::DisplaySetup()
{
        cColor=1;
        int Color[]={RGB(150,150,150),
                RGB(0,0,200),RGB(200,0,0),
                RGB(0,200,0),RGB(200,200,0),
                RGB(0,200,200),RGB(200,0,200),
                RGB(200,100,150),RGB(100,100,255),
                RGB(250,50,255),RGB(255,255,255)};
        pBus[0].CreatePen(PS_SOLID,1,Color[0]);
        for (int i=1;i<=9;i++)
                pBus[i].CreatePen(PS_SOLID,2,Color[i]);
        fTimes.CreatePointFont (80,L"Times New Roman");
        fCourier.CreatePointFont (60,L"Courier");
        fArial.CreatePointFont (60,L"Arial");
}

void CCode4D::Initialize()
{
        int i,j,k,u,w;
        LineFlag=0;
        if (nBus==0)
        {
                //label the nodes
                k=0;
                for (j=1;j<=Ny;j++)
                {
                        k++;
                        v[k].home=CPoint(80+home.x,20+home.y+60*(j-1));
                        v[k].rct=CRect(CPoint(v[k].home.x-20,
                                v[k].home.y-20),CSize(40,40));
                        v[k+Ny].home=CPoint(80+home.x+60*(Nx-1),
                                20+home.y+60*(j-1));
                        v[k+Ny].rct=CRect(CPoint(v[k+Ny].home.x-20,
                                v[k+Ny].home.y-20),CSize(40,40));
                }
                k=2*Ny;
                for (i=2;i<=Nx-1;i++)
                        for (j=1;j<=Ny;j++)
                        {
                                k++;
                                v[k].home=CPoint(80+home.x+60*(i-1),
                                        20+home.y+60*(j-1));
                                v[k].rct=CRect(CPoint(v[k].home.x-20,
                                        v[k].home.y-20),CSize(40,40));
                        }
```

```
            //draw edges between nodes
            for (i=1;i<=n;i++)
            {
                    v[i].wt[i]=99;
                    for (j=i+1;j<=n;j++)
                    {
                            v[i].wt[j]=99; v[j].wt[i]=99;
                            if (i>2*Ny)              //middle columns
                            {
                                    if (j-i==1 && i%Ny!=0)
                                    {
                                            v[i].wt[j]=1; v[j].wt[i]=1;
                                    }
                                    if (j-i==Ny)
                                    {
                                            v[i].wt[j]=1; v[j].wt[i]=1;
                                    }
                            }
                            if (i<=Ny && j-i==2*Ny)
                            {
                                    v[i].wt[j]=1; v[j].wt[i]=1;
                            }
                            if ((i>Ny && i<=2*Ny) && j-i==(Nx-2)*Ny)
                            {
                                    v[i].wt[j]=1; v[j].wt[i]=1;
                            }
                    }
            }
    }
    else
    {
            u=Bus[nBus].Path[1];
            for (i=1;i<=Bus[nBus].nNodes-1;i++)
            {
                    w=Bus[nBus].Path[i+1];
                    v[u].wt[w]=99; v[w].wt[u]=99;
                    u=w;
            }
    }
}

void CCode4D::OnPaint()
{
    int i,j,k;
    CPaintDC dc(this);
    CString s;
    CPen gPen(PS_SOLID,1,RGB(150,150,150));
    CPen hPen(PS_SOLID,1,RGB(0,0,200));
    if (flag)
    {
            dc.SelectObject(gPen); dc.SelectObject(fArial);
            for (i=1;i<=n;i++)
                    for (j=i;j<=n;j++)
                            if (v[i].wt[j]!=99)
```

```
                                    {
                                            dc.MoveTo(v[i].home);
                                            dc.LineTo(v[j].home);
                                    }
                    dc.SelectObject(&fArial);
                    dc.SetTextColor(RGB(0,0,255));
                    for (k=1;k<=n;k++)
                            if (k<=2*Ny)
                            {
                                    dc.SelectObject(gPen);
                                    dc.Rectangle(v[k].rct);
                                    s.Format(L"%d",k);
                                    dc.TextOut(v[k].home.x-15,v[k].home.y-18,s);
                            }
                            else
                            {
                                    dc.SelectObject(hPen);
                                    dc.Ellipse(v[k].rct);
                                    s.Format(L"%d",k);
                                    dc.TextOut(v[k].home.x-10,v[k].home.y-15,s);
                            }
            }
}

int CCode4D::GetCUN()
{
        int minDistance=99;
        int cun;
        for (int i=1;i<=n;i++)
                if ((!fv[i]) && (minDistance>=v[Sv].sp[i]))
                {
                        minDistance=v[Sv].sp[i];
                        cun=i;
                }
        return cun;
}

void CCode4D::Dijkstra()
{
        int minDistance=99;
        int cun;

        //initialize
        for (int i=1;i<=n;i++)
        {
                fv[i]=false;
                Pv[i]=-1;
                v[Sv].sp[i]=99;
        }
        v[Sv].sp[Sv]=0;

        //compute
        int k=0;
        while (k<n)
```

```
            {
                    cun=GetCUN();
                    fv[cun]=true;
                    for (int i=1;i<=n;i++)
                            if ((!fv[i]) && (v[cun].wt[i]>0))
                                    if (v[Sv].sp[i]>v[Sv].sp[cun]
                                                    +v[cun].wt[i])
                                    {
                                            v[Sv].sp[i]=v[Sv].sp[cun]
                                                    +v[cun].wt[i];
                                            Pv[i]=cun;
                                    }
                    k++;
            }
}

void CCode4D::DrawPath()
{
        CClientDC dc(this);
        int w,i,u,r,dummy[NMax+1];

        //display the source
        dc.SelectObject(pBus[1+cColor%10]); cColor++;
        w=Dv;  Bus[nBus].nNodes=0;  Bus[nBus].Path[1]=Sv;
        for (i=1;i<=n;i++)
        {
                u=Pv[w];
                if (u!=-1)
                {
                        r=++Bus[nBus].nNodes; dummy[r]=w;
                        w=u;
                }
                dummy[r+1]=Sv;
        }
        r=++Bus[nBus].nNodes; Bus[nBus].Path[r]=Dv;
        u=Sv; Bus[nBus].Path[1]=u;
        dc.MoveTo(v[u].home);
        for (i=1;i<=Bus[nBus].nNodes;i++)
        {
                w=dummy[Bus[nBus].nNodes-i+1]; Bus[nBus].Path[i]=w;
                dc.LineTo(v[w].home);
                u=w;
        }
        w=Dv; Bus[nBus].Path[i]=w;
        dc.LineTo(v[w].home);
}

void CCode4D::OnLButtonDown(UINT nFlags,CPoint pt)
{
        CClientDC dc(this);
        CString s;
        dc.SelectObject(pBus[1]); dc.SelectObject(fArial);
        for (int k=1;k<=n;k++)
                if (v[k].rct.PtInRect(pt))
```

```
                        {
                                LineFlag++;
                                if (LineFlag==1 && k<=Ny)
                                {
                                        Sv=k; dc.Rectangle(v[Sv].rct);
                                        s.Format(L"%d",Sv);
                                        dc.TextOut(v[Sv].home.x-15,
                                                v[Sv].home.y-18,s);
                                }
                                if (LineFlag==2 && (k>Ny && k<=2*Ny))
                                {
                                        Dv=k;
                                        dc.Rectangle(v[Dv].rct);
                                        Dijkstra();
                                        if (v[Sv].sp[Dv]!=99)
                                        {
                                                nBus++; DrawPath(); ShowTable();
                                                Initialize();
                                                s.Format(L"%d",Dv);
                                                dc.TextOut(v[Dv].home.x-15,
                                                        v[Dv].home.y-18,s);
                                        }
                                        else
                                        {
                                                LineFlag=0;
                                                dc.SelectObject(pBus[0]);
                                                dc.Rectangle(v[Sv].rct);
                                                dc.Rectangle(v[Dv].rct);
                                                s.Format(L"%d",Sv);
                                                dc.TextOut(v[Sv].home.x-15,
                                                        v[Sv].home.y-18,s);
                                                s.Format(L"%d",Dv);
                                                dc.TextOut(v[Dv].home.x-15,
                                                        v[Dv].home.y-18,s);
                                        }
                                }
                                if (LineFlag==2)
                                        LineFlag=0;
                        }
}

void CCode4D::ShowTable()
{
        int i,k,w,u;
        CString s,S;
        for (i=1;i<=nBus;i++)
        {
                table.DeleteItem(i-1);
                s.Format(L"%d",i); table.InsertItem(i-1,s,0);
                s.Format(L"%d",Bus[i].Path[1]); table.SetItemText(i-1,1,s);
                w=Bus[i].nNodes;
                s.Format(L"%d",Bus[i].Path[w]); table.SetItemText(i-1,2,s);
                u=Bus[i].Path[1]; w=Bus[i].Path[w];
                s.Format(L"%d",v[u].sp[w]); table.SetItemText(i-1,3,s);
```

```
            s.Format(L"%d",Bus[i].nNodes); table.SetItemText(i-1,4,s);
            S="";
            for (k=1;k<=Bus[i].nNodes;k++)
            {
                    s.Format(L"%d ",Bus[i].Path[k]); S +=s;
            }
            table.SetItemText(i-1,5,S);
    }
}
```

5

Computing the Minimum Spanning Tree

5.1 Problem Description

Every connected graph $G(V, E)$ may have several spanning trees. A spanning tree $T(V, E')$ is a subgraph, with the criteria that it includes all the nodes of the original graph, $V = \{v_i\}$, $i = 1, 2,...n$, and all these nodes are connected by only selected edges $E' \subseteq E\,(G)$. Another important criterion of a spanning tree is that no cycle is allowed when selecting the edges. This is in line with the definition of a tree as a connected graph without any cycles [1]. As a rule of thumb, a tree with n nodes will form a spanning tree with only $n - 1$ edges [2,3]. For example, a cyclic graph with four nodes, as illustrated in Figure 5.1a, has a total of four spanning trees with each formed by three edges, as illustrated in Figure 5.1b. Another example, in Figure 5.2a, is a diamond graph with four nodes, and it has a total of eight spanning trees with each also formed by three edges, as illustrated in Figure 5.2b.

Suppose a weight or length is assigned to each of the edges of the graph. The total weight of the resulting spanning tree can be computed by having the summation of all the weights assigned to the selected edges; hence, different total weights could be obtained from these different spanning trees. The problem is to find the minimum spanning tree of a graph. The minimum spanning tree is the spanning tree where the sum of weights on its edges is the minimum given by

$$MST = Min \sum_{ij} x_{ij} w_{ij} \tag{5.1}$$

$$\text{Subject to } x_{ij} = \begin{cases} 1 & \text{if } e_{ij} \text{ is selected} \\ 0 & \text{otherwise} \end{cases}$$

for $i, j = 1, 2...n$. In the above equation, MST is the minimum spanning tree and x_{ij} denotes the decision variable, e_{ij} is the edge, w_{ij} is the weight assigned to e_{ij}, and n is the total number of nodes in the graph.

Prim's algorithm and Kruskal's algorithm are among the prominent methods used to find the minimum spanning tree. Finding the minimum spanning tree is a very common problem as it has many real-world applications. Among these applications are large-scale planning of distribution networks, transportation and communication, data storage reduction, and in data mining. This chapter presents three different projects related to the minimum spanning tree problems: *Code5A*, *Code5B*, and *Code5C*. *Code5A* is the greedy algorithm used to compute the minimum spanning tree using Kruskal's algorithm, while *Code5B* and *Code5C* are applications of Kruskal's algorithm for a transportation problem and a broadcasting in ad hoc wireless network problem, respectively.

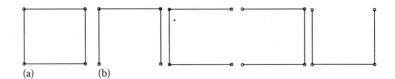

FIGURE 5.1
(a) A cyclic graph and (b) spanning trees for a cyclic graph.

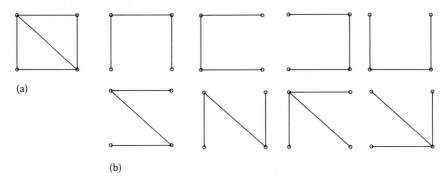

FIGURE 5.2
(a) A diamond graph and (b) spanning trees for a diamond graph.

5.2 Algorithms for Computing Minimum Spanning Tree

The minimum spanning tree problem has its solutions in the form of a greedy algorithm; the most common are Kruskal's algorithm and Prim's algorithm. Kruskal's algorithm focuses on the selection of edges while Prim's algorithm is based on the selection of nodes.

5.2.1 Kruskal's Algorithm

Kruskal's algorithm is a greedy algorithm, which is the easiest method for finding the minimum spanning tree of a graph. The idea of Kruskal's algorithm is to find the minimum spanning tree by initially having each node as its own component with no edges. An edge with minimum weight will be added to merge the components or the nodes. This procedure is repeated until all the nodes are connected and become a tree. The selection of the edges continues until the total number of edges becomes $n - 1$. Note that the selection of the edges during the procedure must not create a cycle.

Figure 5.3a through f illustrates the execution of Kruskal's algorithm. In (a) weight w_{ij} is assigned to the graph $G(V, E)$ with $V = \{v_i\}$ and $E = \{e_{ij}\}$ where $i, j = 1, 2, \ldots 5$. In order to find the minimum spanning tree, $T(V, E')$ using Kruskal's algorithm, it is necessary to sort the edges in ascending order based on the assigned weight. Therefore, the list of edges in ascending order for this example is

$$e_{34} = 2$$
$$e_{15} = 3$$
$$e_{14} = 5$$
$$e_{54} = 7$$

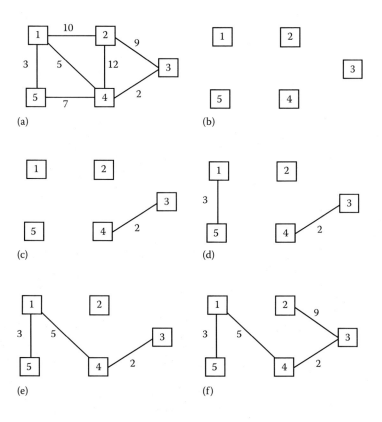

FIGURE 5.3
Execution of Kruskal's algorithm. (a) The graph, (b) initial minimum spanning tree, (c) e_{34} is selected, (d) e_{15} is selected, (e) e_{14} is selected, and (f) final minimum spanning tree.

$$e_{23} = 9$$
$$e_{12} = 10$$
$$e_{24} = 12$$

The initial minimum spanning tree has only the nodes with no edge, $E' = \{\phi\}$, as illustrated in (b). Based on the list of edges, which has been sorted in ascending order, e_{34} has the minimum weight; therefore, it is selected to be added to the minimum spanning tree, as illustrated in (c). The updated set of edges becomes

$$E' = \{e_{34}\}, \ |E'| = 1.$$

The same procedure of selecting the next smallest weight that does not create a cycle is repeated; therefore, e_{15} is selected and added to the minimum spanning tree, as illustrated in (d). The updated set of edges becomes

$$E' = \{e_{34}, e_{15}\}, \ |E'| = 2.$$

The next edge with the smallest weight to be added to the minimum spanning tree is e_{14}, as illustrated in (e). The updated set of edges becomes

$$E' = \{e_{34}, e_{15}, e_{14}\}, \ |E'| = 3.$$

It can be seen that the next edge with minimum weight is e_{54}; however, it could not be selected because it will create a cycle if added to the minimum spanning tree.

The next smallest weight is e_{23}; therefore, it is selected to be added to the minimum spanning tree, as illustrated in (f). The updated set of edges becomes

$$E' = \{e_{34}, e_{15}, e_{14}, e_{23}\}, \ |E'| = 4.$$

Since the original tree has five nodes, the minimum spanning tree should therefore have only four edges. Hence the final minimum spanning tree is already obtained, as illustrated in (f). The total weight of the minimum spanning tree, MST_{cost} is

$$MST_{cost} = w_{34} + w_{15} + w_{14} + w_{23} = 2 + 3 + 5 + 9 = 19.$$

Kruskal's algorithm can be summarized as follows:

Given: The graph $G(V, E)$ for $V = \{v_i\}$, $E = \{e_{ij}\}$ with weight matrix $W = \{w_{ij}\}$ and $|V| = n$ for $i, j = 1, 2,\ldots, n$.
Initialization:
Let $MST_{cost} = 0$ // MST_{cost} is the total cost of the minimum spanning tree;
$E' = \{\phi\}$;
$m = 0$ //m is the number of edges in the minimum spanning tree;
Process:
Sort edges, $E = \{e_{ij}\}$ in ascending order of weight, $W = \{w_{ij}\}$;
While ($m < n - 1$)
   Select the next edge e_{ij} of minimum weight;
   If $i \neq j$
      Add e_{ij} in E';
      $m++$;
      $MST_{cost} = MST_{cost} + w_{ij}$;
   Endif
Endwhile
Output: The minimum spanning tree, $T = (V, E')$ with its MST_{cost}.

5.2.2 Prim's Algorithm

Prim's algorithm is a greedy algorithm used to obtain the minimum spanning tree. With this algorithm, the size of the minimum spanning tree, which initially consists of only a single node, is increasing continuously by adding one edge at a time until it spans all the nodes. In Prim's algorithm, the initial set of node selection for the minimum spanning tree is set as $V_{new} = \{v_i\}$, where v_i is an arbitrary node selection of the original tree. On the other hand, the initial set of the edges will be an empty set, $E_{new} = \{\phi\}$. The number of elements of V_{new} and E_{new} will keep on increasing as a node and an edge will be added in each iteration.

At every iteration, the selection of the edge $E = \{e_{ij}\}$ must follow the condition that the node v_i (the initial node of the edge) is in V_{new} but v_j (the terminal node of the edge) is not in the V_{new}. One logical reason for considering this condition is to avoid creating a cycle. Among all the considered edges, the one with the least minimum weight is selected to be added into E_{new}. Automatically, the node v_j during edge selection for E_{new} will be added to V_{new}. Suppose the original tree has n nodes; this procedure is repeated until all the nodes are in V_{new} and a total of $n - 1$ edges are selected in E_{new}. Figure 5.4a through g illustrates the execution of Prim's algorithm.

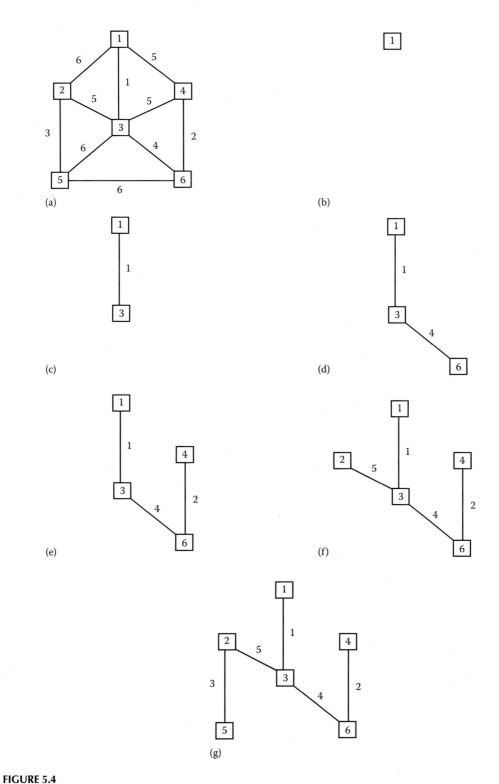

FIGURE 5.4
Execution of Prim's algorithm. (a) The graph, (b) v_1 is selected, (c) v_3 and e_3 are selected, (d) v_6 and e_{36} are selected, (e) v_4 and e_{64} are selected, (f) v_2 and e_{32} are selected, and (g) the minimum spanning tree.

The original graph is illustrated in (a), where there are 6 nodes and 10 edges with the weight w_{ij}. As explained previously, any node can be selected at first. Hence, v_1 is selected, as illustrated in (b), and therefore

$$V_{new} = \{v_1\} \text{ and } E_{new} = \{ \}.$$

The next procedure is to find the edge to be added to the minimum spanning tree. It can be seen that, from the original tree, v_1 is connected to the neighboring nodes through the edges with the assigned weight as the following:

$$L = \{e_{12}, e_{13}, e_{14}\}.$$

Among all the edges listed above, the edge with the minimum weight is obviously e_{13}. Therefore, the terminal nodes, which are v_3 and e_{13}, are selected and added to the minimum spanning tree, as illustrated in (c). Hence, the updated set of new nodes and set of new edges become

$$V_{new} = \{v_1, v_3\} \text{ and } E_{new} = \{e_{13}\}.$$

The next step is to consider all the edges with the condition that the initial nodes are among v_1 and v_3 and the terminal node is other than these two nodes. Hence, the edges to be considered are

$$L = \{e_{12}, e_{14}, e_{32}, e_{35}, e_{36}, e_{34}\}.$$

Among all the six edges, the one with minimum weight is e_{36}. Therefore, v_6 and e_{36} are selected and added to the minimum spanning tree, as illustrated in (d). Hence, the updated set of new nodes and set of new edges become

$$V_{new} = \{v_1, v_3, v_6\} \text{ and } E_{new} = \{e_{13}, e_{36}\}.$$

The same procedure is repeated where all the edges with the condition of the initial nodes are among v_1, v_3, and v_6 and the terminal nodes other than these three nodes are being considered. Hence, the edges to be considered are

$$L = \{e_{12}, e_{14}, e_{32}, e_{35}, e_{34}, e_{65}, e_{64}\}.$$

Among all the seven edges, the edge with the minimum weight is e_{64}. Therefore, v_4 and e_{64} are selected and added to the minimum spanning tree, as illustrated in (e). Hence, the updated set of new nodes and set of new edges become

$$V_{new} = \{v_1, v_3, v_6, v_4\} \text{ and } E_{new} = \{e_{13}, e_{36}, e_{64}\}.$$

Again, the same procedure of considering all the edges when the condition of the initial nodes is among v_1, v_3, v_6, and v_4 and the terminal node other than these four nodes is applied. Hence, the edges to be considered are

$$L = \{e_{12}, e_{32}, e_{35}, e_{65}\}.$$

Among all the four edges, the one with the minimum weight is e_{32}. Therefore, v_2 and e_{32} are selected and added to the minimum spanning tree, as illustrated in (f). Hence, the updated set of new nodes and set of new edges become

$$V_{new} = \{v_1, v_3, v_6, v_4, v_2\} \text{ and } E_{new} = \{e_{13}, e_{36}, e_{64}, e_{32}\}.$$

The same procedure of considering the edges with the condition that the initial nodes are among $v_1, v_3, v_6, v_4,$ and v_2 and the terminal node is other than these five nodes is repeated. Hence, the edges to be considered are

$$L = \{e_{35}, e_{65}, e_{25}\}.$$

Among all the three edges, the edge with minimum weight is e_{25}. Therefore, v_5 and e_{25} are selected and added to the minimum spanning tree, as illustrated in (g). Hence, the updated set of new nodes and set of new edges become

$$V_{new} = \{v_1, v_3, v_6, v_4, v_2, v_5\} \text{ and } E_{new} = \{e_{13}, e_{36}, e_{64}, e_{32}, e_{25}\}.$$

It is noticeable that the total number of edges in the set of new edges is now five and all the nodes are in the set of new nodes. Hence, the complete minimum spanning tree is finally obtained, as illustrated in (g). The total weight of the minimum spanning tree, MST_{cost} is

$$MST_{cost} = w_{13} + w_{36} + w_{64} + w_{32} + w_{25} = 1 + 4 + 2 + 5 + 3 = 15.$$

Prim's algorithm is summarized as follows:

> **Given:** The graph $G(V, E)$ for $V = \{v_i\}$, $E = \{e_{ij}\}$ with weight matrix $W = \{w_{ij}\}$ and $|V| = n$
> for $i, j = 1, 2,...,n$.
> **Initialization:**
> Let $MST_{cost} = 0$ //MST_{cost} is the total cost of the minimum spanning tree;
> Let $E_{new} = \{\phi\}$;
> Let $V_{new} = \{v_i\}$ //v_i is the starting node selected randomly;
> Let $m = 0$ //m is the number of edges in the minimum spanning tree;
> **Process:**
> While ($m < n - 1$)
>   Select e_{ij} with least minimum w_{ij} such that $v_i \in V_{new}$ but $v_j \notin V_{new}$. If more than one
>   edge has the same value of weight, select only one;
>   Add e_{ij} to E_{new};
>   Add v_j to V_{new};
> $MST_{cost} = MST_{cost} + w_{ij}$;
> m++;
> Endwhile
> **Output:** The minimum spanning tree, $T = (V, E')$ with its total cost, MST_{cost}, $E' = E_{new}$.

5.2.3 Code5A: Kruskal's Algorithm

Code5A is the implementation of Kruskal's algorithm for computing the minimum spanning tree for a graph. The output of *Code5A* is shown in Figure 5.5, which consists of

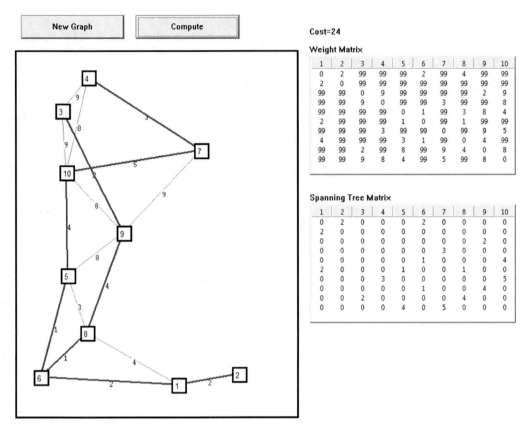

Cost=24

Weight Matrix

1	2	3	4	5	6	7	8	9	10
0	2	99	99	99	2	99	4	99	99
2	0	99	99	99	99	99	99	99	99
99	99	0	9	99	99	99	99	2	9
99	99	9	0	99	99	3	99	99	8
99	99	99	99	0	1	99	3	8	4
2	99	99	99	1	0	99	1	99	99
99	99	99	3	99	99	0	99	9	5
4	99	99	99	3	1	99	0	4	99
99	99	2	99	8	99	9	4	0	8
99	99	9	8	4	99	5	99	8	0

Spanning Tree Matrix

1	2	3	4	5	6	7	8	9	10
0	2	0	0	0	2	0	0	0	0
2	0	0	0	0	0	0	0	0	0
0	0	0	0	0	0	0	0	2	0
0	0	0	0	0	0	3	0	0	0
0	0	0	0	0	1	0	0	0	4
2	0	0	0	1	0	0	1	0	0
0	0	0	3	0	0	0	0	0	5
0	0	0	0	0	1	0	0	4	0
0	0	2	0	0	0	0	4	0	0
0	0	0	0	4	0	5	0	0	0

FIGURE 5.5
Output of *Code5A*.

10 randomly scattered nodes and 17 edges. The minimum spanning tree of the generated graph is computed once the push button *Compute* is clicked. The interface also allows a new graph to be generated at any time. In the output, a blue line is used to show the minimum spanning tree.

The weight matrix shows the weight of the edges connected by the pairing nodes. For example, the edge (v_9, v_{10}) has weight $w_{9,10} = 8$. It can be seen that some of the elements in the weight matrix are being assigned as 99 in order to denote that the pairing nodes are not adjacent to each other. In the spanning tree matrix, the element that denotes values other than 0 is the pairing nodes that form edges in the minimum spanning tree. Obviously, the minimum spanning tree $T(V, E')$ where: $E' = \{(v_5, v_6), (v_6, v_8), (v_1, v_6), (v_1, v_2), (v_3, v_9), (v_4, v_7), (v_5, v_{10}), (v_8, v_9), (v_7, v_{10})\}$.

The total cost of the minimum spanning tree is

$$MST_{cost} = w_{5,6} + w_{6,8} + w_{1,6} + w_{1,2} + w_{3,9} + w_{4,7} + w_{5,10} + w_{8,9} + w_{7,10}$$

$$= 1+1+2+2+2+3+4+4+5$$

$$= 24.$$

Code5A has two source files, *Code5A.h* and *Code5A.cpp*. The nodes and edges of the graph are represented by the structures *NODE* and *EDGE*, respectively. There are also two button

TABLE 5.1

Important Objects/Variables in *Code5A* and Their Descriptions

Variable	Type	Description
N	constant	Number of nodes in the graph
R	constant	Range for adjacency between two nodes in the graph
nEdges	*int*	Number of edges in the graph
graph_edge[i][l]	*int*	$l = 1$ denotes initial node of e_i $l = 2$ denotes terminal node of e_i $l = 3$ denotes weight assigned to e_i
tree[i][j]	*int*	Weight of the selected edges for the minimum spanning tree
sets[i][j]	*int*	Path in the list
top[i]	*int*	Node in the top priority
MST_{cost}	*int*	Total weight of the minimum spanning tree
e[i][j].*Wt*	*int*	Weight between v_i and v_j
Home, *End*	*CPoint*	Top-left and bottom-right corners of the drawing area
bNGraph	*CButton*	Button for generating a new graph
bCompute	*CButton*	Button for computing the minimum spanning tree

objects called *bNGraph* for *New Graph* and *bCompute* for *Compute*. Other important objects and variables are briefly described in Table 5.1.

The organization of *Code5A* is shown in Figure 5.6. The figure shows all the functions involved in the project. The processing starts with the constructor *CCode5A()*, which draws the main window and the *New Graph* and *Compute* button. The constructor calls *Initialize()*, which creates a new graph with N nodes by assigning random coordinates and weights to the edges. Each node is drawn using *UpdateNodes()* and the whole graph is drawn inside the drawing area through *OnPaint()*. Once the *Compute* button is pushed, the function *OnCompute()* will then call *Kruskal()* to compute the minimum spanning tree and *ShowResults()* to show the results.

The main engine is *Kruskal()*, which applies Kruskal's algorithm for computing the minimum spanning tree. The results of the minimum spanning tree in terms of the weight matrix, spanning tree matrix, and the cost of the minimum spanning tree generated from the graph are displayed through *ShowResults()*. In computing the minimum spanning tree, *Kruskal()* calls *SortEdges()* to sort all the edges in ascending order. The selected edge is drawn in blue and at the same time the value of its weight is added iteratively and stored in MST_{cost}. A new graph with randomly located positions in Windows and randomly determined weights on the edges is produced by *Initialize()*. This graph is later drawn in Windows through *OnPaint()*. Another function, *UpdateNodes()*, serves to draw the node specified in its argument.

```
//Code5A.h
#include <afxwin.h>
#include <afxcmn.h>
#include <math.h>
#include <time.h>
#define N 10
#define R 250
#define IDC_NGRAPH 501
#define IDC_tWEIGHT 502
#define IDC_tST 503
#define IDC_COMPUTE 600
```

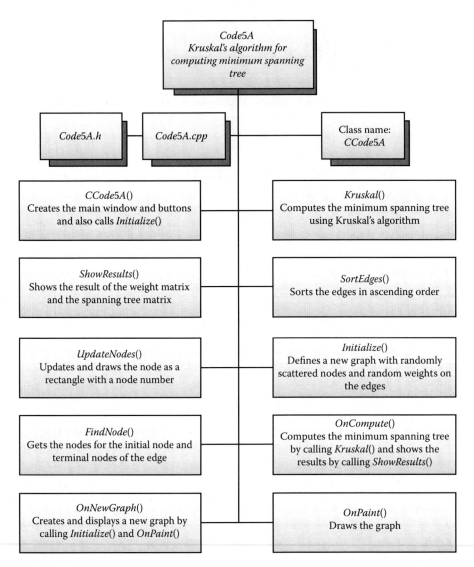

FIGURE 5.6
Organization of *Code5A*.

```
class CCode5A: public CFrameWnd
{
private:
        CButton bNGraph,bCompute;
        CPoint Home,End;
        CFont fCourier,fArial;
        CListCtrl tWeight,tST;
        int nEdges;
        int graph_edge[N+1][N+1];
        int tree[N+1][N+1];
        int sets[N+1][N+1];
        int top[N+1];
        int MSTcost;
```

```
        typedef struct
        {
                CPoint Home;
                CRect rct;
        } NODE;
        NODE v[N+1];

        typedef struct
        {
                int Wt;
                bool ST;
        } EDGE;
        EDGE e[N+1][N+1];

public:
        CCode5A();
        ~CCode5A()      {}
        afx_msg void OnPaint();
        afx_msg void OnCompute();
        afx_msg void OnNewGraph();
        void Kruskal(),ShowResults(),SortEdges(),Initialize(),UpdateNodes();
        int FindNode(int);
        DECLARE_MESSAGE_MAP();
};

class CMyWinApp: public CWinApp
{
public:
        virtual BOOL InitInstance();
};
CMyWinApp MyApplication;

BOOL CMyWinApp::InitInstance()
{
        m_pMainWnd=new CCode5A;
        m_pMainWnd->ShowWindow(m_nCmdShow);
        return TRUE;
}

//Code5A.cpp
#include "Code5A.h"
BEGIN_MESSAGE_MAP(CCode5A,CFrameWnd)
        ON_WM_PAINT()
        ON_BN_CLICKED(IDC_COMPUTE,OnCompute)
        ON_BN_CLICKED(IDC_NGRAPH,OnNewGraph)
END_MESSAGE_MAP()

CCode5A::CCode5A()
{
        Home=CPoint(30,100); End=CPoint(500,700);
        Create(NULL,L"Kruskal's Algorithm for Minimum Spanning Tree",
                WS_OVERLAPPEDWINDOW,CRect(0,0,1000,800));
        bNGraph.Create(L"New Graph",WS_CHILD | WS_VISIBLE | BS_DEFPUSHBUTTON,
                CRect(CPoint(30,30),CSize(180,40)),this,IDC_NGRAPH);
```

```
        bCompute.Create(L"Compute",WS_CHILD | WS_VISIBLE
              | BS_DEFPUSHBUTTON,CRect(CPoint(Home.x+200,30),CSize(180,40)),
              this,IDC_COMPUTE);
        fArial.CreatePointFont(100,L"Arial");
        fCourier.CreatePointFont(60,L"Courier");
        Initialize();
}

void CCode5A::Initialize()
{
        int i,j;
        double distance;
        tWeight.DestroyWindow(); tST.DestroyWindow();
        srand(time(0));
        for (i=1;i<=N;i++)
        {
                v[i].Home.x=Home.x+20+rand()%(End.x-Home.x-50);
                v[i].Home.y=Home.y+20+rand()%(End.y-Home.y-50);
                v[i].rct=CRect(CPoint(v[i].Home.x-10,v[i].
                      Home.y-10),CSize(25,25));
        }
        nEdges=0;
        for (i=1;i<=N;i++)
        {
                e[i][i].Wt=0; e[i][i].ST=0;
                for (j=i+1;j<=N;j++)
                {
                        distance=sqrt(pow(double(v[i].Home.x-v[j].Home.x),2)
                              +pow(double(v[i].Home.y-v[j].Home.y),2));
                        e[i][j].ST=0; e[j][i].ST=0;
                        if (distance<R)
                        {
                                e[i][j].Wt=1+rand()%9;
                                nEdges++;
                                graph_edge[nEdges][1]=i;
                                graph_edge[nEdges][2]=j;
                                graph_edge[nEdges][3]=e[i][j].Wt;
                        }
                        else
                                e[i][j].Wt=99;
                        e[j][i].Wt=e[i][j].Wt;
                }
        }
}

void CCode5A::OnNewGraph()
{
        Initialize(); Invalidate();
}
void CCode5A::OnPaint()
{
        CPaintDC dc(this);
        CString s;
        int i,j;
```

```
        CPoint mPoint;
        CPen mPen(PS_SOLID,3,RGB(0,0,0));
        CPen qPen(PS_SOLID,1,RGB(150,150,150));
        dc.SelectObject(mPen);
        dc.Rectangle(Home.x-10,Home.y-10,End.x+10,End.y+10);
        dc.SelectObject(qPen); dc.SelectObject(fCourier);
        for (i=1;i<=N;i++)
                for (j=1;j<=N;j++)
                        if (e[i][j].Wt!=99)
                        {
                                dc.MoveTo(CPoint(v[i].Home));
                                dc.LineTo(CPoint(v[j].Home));
                                s.Format(L"%d",e[i][j].Wt);
                                dc.TextOut((v[i].Home.x+v[j].Home.x)/2,
                                        (v[i].Home.y+v[j].Home.y)/2,s);
                                mPoint=CPoint((v[i].Home.x+v[j].Home.x)/2,
                                        (v[i].Home.y+v[j].Home.y)/2);
                        }
        UpdateNodes();
}

void CCode5A::UpdateNodes()
{
        CClientDC dc(this);
        CString s;
        CPen mPen(PS_SOLID,3,RGB(0,0,0));
        dc.SelectObject(mPen); dc.SelectObject(fArial);
        for (int i=1;i<=N;i++)
        {
                dc.Rectangle(v[i].rct);
                s.Format(L"%d",i);
                dc.TextOutW(v[i].Home.x-5,v[i].Home.y-5,s);
        }
}

void CCode5A::ShowResults()
{
        CClientDC dc(this);
        CString s;
        int i,j;
        tWeight.Create(WS_VISIBLE|WS_CHILD|WS_DLGFRAME|LVS_REPORT
                |LVS_NOSORTHEADER,CRect(CPoint(End.
                    x+30,Home.y),CSize(355,200)),
                this,IDC_tWEIGHT);
        tWeight.InsertColumn(0,L"",LVCFMT_CENTER,1);
        tST.Create(WS_VISIBLE|WS_CHILD|WS_DLGFRAME|LVS_REPORT
                |LVS_NOSORTHEADER,CRect(CPoint(End.x+30,Home.
                    y+250),CSize(355,200)),
                this,IDC_tST);
        tST.InsertColumn(0,L"",LVCFMT_CENTER,1);
        for (j=1;j<=N;j++)
        {
                s.Format(L"%d",j);
                tWeight.InsertColumn(j,s,LVCFMT_CENTER,35);
```

```
              s.Format(L"%d",j);
              tST.InsertColumn(j,s,LVCFMT_CENTER,35);
       }
       for (i=1;i<=N;i++)
              for (j=0;j<=N;j++)
              {
                     if (j==0)
                     {
                            s.Format(L"%d",i);
                            tWeight.InsertItem(i-1,s,0);
                            s.Format(L"%d",i);
                            tST.InsertItem(i-1,s,0);
                     }
                     if (j>0)
                     {
                            s.Format(L"%d",e[i][j].Wt);
                            tWeight.SetItemText(i-1,j,s);
                            s.Format(L"%d",((e[i][j].ST)?e[i][j].Wt:0));
                            tST.SetItemText(i-1,j,s);
                     }
              }
       UpdateNodes();
       s.Format(L"Cost=%d",MSTcost);
       dc.TextOut(End.x+30,Home.y-50,s);
       dc.TextOutW(End.x+30,Home.y-20,L"Weight Matrix");
       dc.TextOutW(End.x+30,Home.y+230,L"Spanning Tree matrix");
}

void CCode5A::OnCompute()
{
       Kruskal();
       ShowResults();
}

void CCode5A::SortEdges()
{
       int i,j,t;
       for (i=1;i<=nEdges-1;i++)
              for (j=1;j<=nEdges-i;j++)
                     if (graph_edge[j][3]>graph_edge[j+1][3])
                     {
                            t=graph_edge[j][1];
                            graph_edge[j][1]=graph_edge[j+1][1];
                            graph_edge[j+1][1]=t;
                            t=graph_edge[j][2];
                            graph_edge[j][2]=graph_edge[j+1][2];
                            graph_edge[j+1][2]=t;
                            t=graph_edge[j][3];
                            graph_edge[j][3]=graph_edge[j+1][3];
                            graph_edge[j+1][3]=t;
                     }
}
```

```
void CCode5A::Kruskal()
{
        CClientDC dc(this);
        CString s;
        CPen pMSTpath(PS_SOLID,3,RGB(0,100,200));
        int p1,p2,i,j;
        int p,q;
        SortEdges();
        for (i=1;i<=N;i++)
        {
                sets[i][1]=i;
                top[i]=1;
        }
        dc.SelectObject(pMSTpath);
        MSTcost=0;
        for (i=1;i<=nEdges;i++)
        {
                p1=FindNode(graph_edge[i][1]);
                p2=FindNode(graph_edge[i][2]);
                if (p1!=p2)
                {
                        p=graph_edge[i][1]; q=graph_edge[i][2];
                        tree[graph_edge[i][1]][graph_edge[i][2]]=graph_
                                edge[i][3];
                        tree[graph_edge[i][2]][graph_edge[i][1]]=graph_
                                edge[i][3];
                        dc.MoveTo(v[p].Home); dc.LineTo(v[q].Home);
                        MSTcost +=graph_edge[i][3];
                        e[p][q].ST=1; e[q][p].ST=1;
                        for (j=1;j<=top[p2];j++)
                        {
                                top[p1]++;
                                sets[p1][top[p1]]=sets[p2][j];
                        }
                        top[p2]=0;
                }
        }
}


int CCode5A::FindNode(int n)
{
        int i,j;
        for (i=1;i<=nEdges;i++)
                for (j=1;j<=top[i];j++)
                        if (n==sets[i][j])
                                return i;
        return -1;
}
```

5.3 Case Study of the Pavement Construction Problem

An owner of a huge piece of private land has multiple businesses running on the land. One of his businesses is operating a campsite with 10 huts. The huts are available for renting. In order to increase the profits from the campsite business, he plans to upgrade his campsite by developing a pavement. The pavement needs to be able to link all his huts.

The owner has hired a contractor to develop the pavement. Currently, there are multiple routes that connect all the huts. In order to minimize the cost, the contractor needs to have only one selected route for developing the pavement in order to link all the huts. This problem can be solved by using the concept of a minimum spanning tree. The cost can be minimized if the selected route is the route with minimum distance. The minimum spanning tree for computing the minimum cost of developing the pavement can be represented by the following equation:

$$\text{Pavement cost} = Min \sum_{ij} x_{ij} w_{ij} \qquad (5.2)$$

$$\text{Subject to } x_{ij} = \begin{bmatrix} 1 & \text{if } e_{ij} \text{ is selected} \\ 0 & \text{otherwise} \end{bmatrix}$$

$$i, j = 1, 2 \ldots n.$$

In the above equation, x_{ij} denotes decision variable, e_{ij} denotes the routes connecting huts i and j, and w_{ij} is the distance for route v_i to v_j. The pavement cost is in kilometer units.

5.3.1 *Code5B*: Pavement Construction

Code5B is a construction plan using a pavement development model that applies Kruskal's algorithm. The output of *Code5B* is shown in Figure 5.7, which consists of 10 randomly scattered nodes representing the location of the huts. The minimum spanning tree for developing the pavement in order to link all the huts using Kruskal's algorithm is computed once the push button is clicked. The interface also allows a new graph to be generated at any time. The minimum cost (in terms of distance) is 23 km. The selected edges or route for the minimum spanning tree are drawn in brown.

Code5B has a single class called *CCode5B* with two source files, *Code5B.h* and *Code5B.cpp*. The nodes and edges of the graph are represented by the structures *NODE* and *EDGE*, respectively. There are also two button objects called *bNGraph* for *The Huts* and *bCompute* for *Compute Minimum Cost*. Other important objects and variables are briefly described in Table 5.2.

The organization of *Code5B* is shown in Figure 5.8. The figure shows all the functions involved in the project. The processing starts with the constructor *CCode5B()*, which draws the main window and the *The Huts* and *Compute Minimum Cost* buttons. The constructor calls *Initialize()*, which creates a new graph with *N* nodes by assigning random coordinates and weights to the edges. Each node is drawn using *UpdateNodes()*, and the whole graph is drawn inside the drawing area through *OnPaint()*. Once the *Compute Minimum Cost* button is pushed, the function *OnCompute()* will then call *Kruskal()* to compute the minimum spanning tree to determine the minimum cost path and *ShowResults()* to show the results.

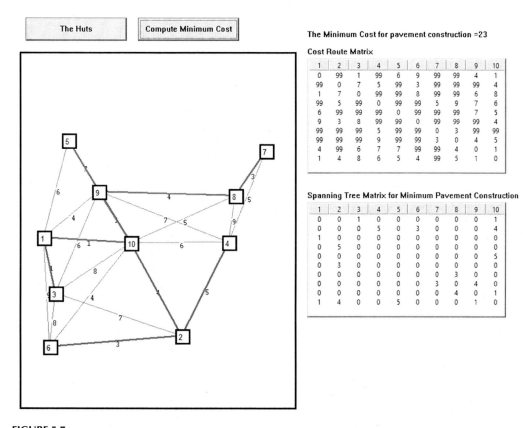

FIGURE 5.7
Output of *Code5B*.

TABLE 5.2

Important Objects/Variables in *Code5B* and Their Descriptions

Variable	Type	Description
N	constant	Number of nodes in the graph
R	constant	Range for adjacency between two nodes in the graph
nEdges	int	The number of edges in the graph
graph_edge[i][l]	int	If l = 1; the initial node of edge i
		If l = 2; the terminal node of edge i
		If l = 3; the weight assigned to edge i
tree[i][j]	int	The weight of the selected edges for minimum spanning tree
sets[i][j]	int	Path in the list
top[i]	int	Node with the top priority
MST_{cost}	int	The total weight of the minimum spanning tree
e[i][j].Wt	int	The weight on edge (v_i, v_j)
Home, End	CPoint	Top-left and bottom-right corners of the drawing area
bNGraph	CButton	Button for generating a new graph
bCompute	CButton	Button for computing the minimum spanning tree

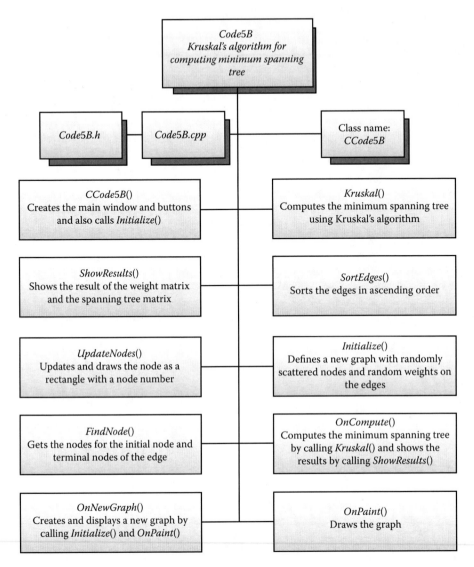

FIGURE 5.8
Organization of *Code5B*.

The main engine is *Kruskal()*, which applies Kruskal's algorithm for computing the minimum spanning tree to determine the minimum cost of pavement construction. The results of the minimum cost of pavement construction with the cost route matrix and spanning tree matrix for minimum pavement construction are displayed through *ShowResults()*. In computing the minimum cost construction, *Kruskal()* calls *SortEdges()* to sort all the edges in ascending order. The selected edge is drawn in brown and at the same time the value of its weight is added iteratively and stored in *MSTcost*. A new graph with randomly located positions in Windows and randomly determined weights on the edges is produced by *Initialize()*. This graph is later drawn in Windows through *OnPaint()*. Another function, *UpdateNodes()*, serves to draw the node specified in its argument.

```
//Code5B.h
#include <afxwin.h>
#include <afxcmn.h>
#include <math.h>
#include <time.h>
#define N 10
#define R 250
#define IDC_NGRAPH 501
#define IDC_tWEIGHT 502
#define IDC_tST 503
#define IDC_COMPUTE 600

class CCode5B: public CFrameWnd
{
private:
        CButton bNGraph,bCompute;
        CPoint Home,End;
        CFont fCourier,fArial;
        CListCtrl tWeight,tST;
        int nEdges;
        int graph_edge[N+1][N+1];
        int tree[N+1][N+1];
        int sets[N+1][N+1];
        int top[N+1];
        int MSTcost;

        typedef struct
        {
                CPoint Home;
                CRect rct;
        } NODE;
        NODE v[N+1];

        typedef struct
        {
                int Wt;
                bool ST;
        } EDGE;
        EDGE e[N+1][N+1];
public:
        CCode5B();
        ~CCode5B()      {}
        afx_msg void OnPaint();
        afx_msg void OnCompute();
        afx_msg void OnNewGraph();
        void Kruskal(),ShowResults(),SortEdges(),Initialize(),UpdateNodes();
        int FindNode(int);
        DECLARE_MESSAGE_MAP();
};

class CMyWinApp: public CWinApp
{
public:
        virtual BOOL InitInstance();
```

```cpp
};
CMyWinApp MyApplication;

BOOL CMyWinApp::InitInstance()
{
        m_pMainWnd=new CCode5B;
        m_pMainWnd->ShowWindow(m_nCmdShow);
        return TRUE;
}


//Code5B.cpp
#include "Code5B.h"
BEGIN_MESSAGE_MAP(CCode5B,CFrameWnd)
        ON_WM_PAINT()
        ON_BN_CLICKED(IDC_COMPUTE,OnCompute)
        ON_BN_CLICKED(IDC_NGRAPH,OnNewGraph)
END_MESSAGE_MAP()

CCode5B::CCode5B()
{
        Home=CPoint(30,100); End=CPoint(500,700);
        Create(NULL,L"Kruskal's Algorithm for Construction Planning Problem",
                WS_OVERLAPPEDWINDOW,CRect(0,0,1000,800));
        bNGraph.Create(L"The Huts",WS_CHILD | WS_VISIBLE | BS_DEFPUSHBUTTON,
                CRect(CPoint(30,30),CSize(180,40)),this,IDC_NGRAPH);
        bCompute.Create(L"Compute Minimum Cost",WS_CHILD | WS_VISIBLE
                | BS_DEFPUSHBUTTON,CRect(CPoint(Home.x+200,30),CSize(180,40)),
                this,IDC_COMPUTE);
        fArial.CreatePointFont(100,L"Arial");
        fCourier.CreatePointFont(60,L"Courier");
        Initialize();
}

void CCode5B::Initialize()
{
        int i,j;
        double distance;
        tWeight.DestroyWindow(); tST.DestroyWindow();
        srand(time(0));
        for (i=1;i<=N;i++)
        {
                v[i].Home.x=Home.x+20+rand()%(End.x-Home.x-50);
                v[i].Home.y=Home.y+20+rand()%(End.y-Home.y-50);
                v[i].rct=CRect(CPoint(v[i].Home.x-10,v[i].
                        Home.y-10),CSize(25,25));
        }
        nEdges=0;
        for (i=1;i<=N;i++)
        {
                e[i][i].Wt=0; e[i][i].ST=0;
                for (j=i+1;j<=N;j++)
                {
                        distance=sqrt(pow(double(v[i].Home.x-v[j].Home.x),2)
```

```
                                  +pow(double(v[i].Home.y-v[j].Home.y),2));
                        e[i][j].ST=0; e[j][i].ST=0;
                        if (distance<R)
                        {
                                e[i][j].Wt=1+rand()%9;
                                nEdges++;
                                graph_edge[nEdges][1]=i;
                                graph_edge[nEdges][2]=j;
                                graph_edge[nEdges][3]=e[i][j].Wt;
                        }
                        else
                                e[i][j].Wt=99;
                        e[j][i].Wt=e[i][j].Wt;
                }
        }
}

void CCode5B::OnNewGraph()
{
        Initialize(); Invalidate();
}

void CCode5B::OnPaint()
{
        CPaintDC dc(this);
        CString s;
        int i,j;
        CPoint mPoint;
        CPen mPen(PS_SOLID,3,RGB(0,0,0));
        CPen qPen(PS_SOLID,1,RGB(150,150,150));
        dc.SelectObject(mPen);
        dc.Rectangle(Home.x-10,Home.y-10,End.x+10,End.y+10);
        dc.SelectObject(qPen); dc.SelectObject(fCourier);
        for (i=1;i<=N;i++)
                for (j=1;j<=N;j++)
                        if (e[i][j].Wt!=99)
                        {
                                dc.MoveTo(CPoint(v[i].Home));
                                dc.LineTo(CPoint(v[j].Home));
                                s.Format(L"%d",e[i][j].Wt);
                                dc.TextOut((v[i].Home.x+v[j].Home.x)/2,
                                        (v[i].Home.y+v[j].Home.y)/2,s);
                                mPoint=CPoint((v[i].Home.x+v[j].Home.x)/2,
                                        (v[i].Home.y+v[j].Home.y)/2);
                        }
        UpdateNodes();
}

void CCode5B::UpdateNodes()
{
        CClientDC dc(this);
        CString s;
        CPen mPen(PS_SOLID,3,RGB(0,0,0));
        dc.SelectObject(mPen); dc.SelectObject(fArial);
```

```
        for (int i=1;i<=N;i++)
        {
                dc.Rectangle(v[i].rct);
                s.Format(L"%d",i);
                dc.TextOutW(v[i].Home.x-5,v[i].Home.y-5,s);
        }
}


void CCode5B::ShowResults()
{
        CClientDC dc(this);
        CString s;
        int i,j;
        tWeight.Create(WS_VISIBLE|WS_CHILD|WS_DLGFRAME|LVS_REPORT
                |LVS_NOSORTHEADER,CRect(CPoint(End.x+30,Home.y),
                        CSize(355,200)),
                this,IDC_tWEIGHT);
        tWeight.InsertColumn(0,L"",LVCFMT_CENTER,1);
        tST.Create(WS_VISIBLE|WS_CHILD|WS_DLGFRAME|LVS_REPORT
                |LVS_NOSORTHEADER,CRect(CPoint(End.x+30,Home.y+250),
                        CSize(355,200)),
                this,IDC_tST);
        tST.InsertColumn(0,L"",LVCFMT_CENTER,1);

        for (j=1;j<=N;j++)
        {
                s.Format(L"%d",j);
                tWeight.InsertColumn(j,s,LVCFMT_CENTER,35);
                s.Format(L"%d",j);
                tST.InsertColumn(j,s,LVCFMT_CENTER,35);
        }

        for (i=1;i<=N;i++)
                for (j=0;j<=N;j++)
                {
                        if (j==0)
                        {
                                s.Format(L"%d",i);
                                tWeight.InsertItem(i-1,s,0);
                                s.Format(L"%d",i);
                                tST.InsertItem(i-1,s,0);
                        }
                        if (j>0)
                        {
                                s.Format(L"%d",e[i][j].Wt);
                                tWeight.SetItemText(i-1,j,s);
                                s.Format(L"%d",((e[i][j].ST)?e[i][j].Wt:0));
                                tST.SetItemText(i-1,j,s);
                        }
                }
        UpdateNodes();
        s.Format(L"The Minimum Cost for pavement construction=%d",MSTcost);
        dc.TextOut(End.x+30,Home.y-50,s);
        dc.TextOutW(End.x+30,Home.y-20,L"Cost Route Matrix");
```

```
        dc.TextOutW(End.x+30,Home.y+230,L"Spanning Tree Matrix for Minimum
            Pavement Construction");
}

void CCode5B::OnCompute()
{
        Kruskal();
        ShowResults();
}

void CCode5B::SortEdges()
{
        int i,j,t;
        for (i=1;i<=nEdges-1;i++)
                for (j=1;j<=nEdges-i;j++)
                        if (graph_edge[j][3]>graph_edge[j+1][3])
                        {
                                t=graph_edge[j][1];
                                graph_edge[j][1]=graph_edge[j+1][1];
                                graph_edge[j+1][1]=t;
                                t=graph_edge[j][2];
                                graph_edge[j][2]=graph_edge[j+1][2];
                                graph_edge[j+1][2]=t;
                                t=graph_edge[j][3];
                                graph_edge[j][3]=graph_edge[j+1][3];
                                graph_edge[j+1][3]=t;
                        }
}

void CCode5B::Kruskal()
{
        CClientDC dc(this);
        CString s;
        CPen pMSTpath(PS_SOLID,3,RGB(200,100,100));
        int p1,p2,i,j;
        int p,q;
        SortEdges();
        for (i=1;i<=N;i++)
        {
                sets[i][1]=i;
                top[i]=1;
        }
        dc.SelectObject(pMSTpath);
        MSTcost=0;
        for (i=1;i<=nEdges;i++)
        {
                p1=FindNode(graph_edge[i][1]);
                p2=FindNode(graph_edge[i][2]);
                if (p1!=p2)
                {
                        p=graph_edge[i][1]; q=graph_edge[i][2];
                        tree[graph_edge[i][1]][graph_edge[i][2]]=graph_
                            edge[i][3];
```

```
                    tree[graph_edge[i][2]][graph_edge[i][1]]=graph_
                        edge[i][3];
                    dc.MoveTo(v[p].Home); dc.LineTo(v[q].Home);
                    MSTcost +=graph_edge[i][3];
                    e[p][q].ST=1; e[q][p].ST=1;
                    for (j=1;j<=top[p2];j++)
                    {
                            top[p1]++;
                            sets[p1][top[p1]]=sets[p2][j];
                    }
                    top[p2]=0;
            }
        }
}

int CCode5B::FindNode(int n)
{
        int i,j;
        for (i=1;i<=nEdges;i++)
                for (j=1;j<=top[i];j++)
                        if (n==sets[i][j])
                                return i;
        return -1;
}
```

5.4 Case Study of a Broadcasting Problem

An ad hoc wireless network is a type of network that does not depend on any preexisting infrastructure. It is a group of two or more electronic devices where the devices have the capability for wireless communication and networking. This collection of devices forms a spontaneous structure, as well as causing deformation of the network. It is also simply known as a self-organizing and adaptive system. On the other hand, an infrastructure network such as a cellular system has fixed base stations in which the communication between the devices uses dedicated nonwireless lines and access points. Unlike an infrastructure network, all the devices in the ad hoc wireless network are able to communicate with each other directly because they can serve as routers and hosts [4].

Broadcasting is disseminating information or messages from one source to many receivers. It is a fundamental task with various applications ranging from mobile ad hoc wireless networks to satellite communication. Among the well-known issues of wireless network communication is the operating capability in limited energy environments. It must be noted that most of the devices are battery-driven. Therefore, one of the objectives of these limited energy issues in a wireless networking environment is to minimize the total power consumption during broadcasting sessions in the network. Hence, optimum energy efficiency is often used as the performance metric to evaluate broadcasting in an ad hoc wireless network.

In this section, the broadcast communication in an ad hoc wireless network will be a case study for discussion. In general, the whole wireless communication network can be represented as a graph where the nodes represent the electronic devices and the edges represent the link between the nodes or the wireless units. In other words, the communication model of the ad hoc wireless network can be modeled as $G(V, E)$, where V is the set of n nodes and E is the set of edges in which the communication between the devices is available. In this case, e_{ij} belongs to E if v_i is able to transmit to v_j with a certain value of transmitted power. It must be noted that the elements of E depend on the position of the nodes as well as the communication range of the given nodes.

The problem of how to minimize the total cost of power or energy consumption when broadcasting in an ad hoc wireless network originating from the source node to all the nodes in the network can be solved by applying the concept of a minimum spanning tree. Kruskal's algorithm is able to develop a minimum energy tree for broadcasting that is rooted at the source and reaches all the desired destinations of all nodes. Once the minimum spanning tree has been obtained, the minimum energy consumption in the network is evaluated by determining the transmission nodes followed by the summation of the transmission power at each transmitting node.

In the process of developing the minimum energy tree representing the network, note that the whole network connectivity depends on the transmission power. The edge can be viewed as the transmitted power strength required in order to support a link between two nodes. Figure 5.9 shows an example of a subset of a broadcasting tree that involves only three nodes.

In this example, v_8 transmits to its neighbors, v_2 and v_4, with power strength of 1 and 5 dB, respectively. With the property of a wireless network, a single transmission from a node to reach both nodes is actually sufficient by only evaluating the maximum power required to reach any of the nodes individually, as follows:

$$Q_{ijk} = \max\{P_{ij}, P_{ik}\}. \tag{5.3}$$

In the above equation, Q_{ijk} denotes a single value of the transmission power to reach both node j and node k from node i. P_{ij} denotes the transmission power from node i to node j while P_{ik} denotes the transmission power from node i to node k. Therefore, the single value of the transmission power to reach from the transmitted nodes to both nodes for Figure 5.9 is $Q_{8,2,4} = \max\{P_{82}, P_{84}\} = \max\{1, 5\} = 5$. Hence, the total power required to maintain the minimum energy tree of the broadcasting session in the ad hoc wireless network is as follows:

$$P = \sum_{i=1}^{m} \max(P_{ij}, \ldots, P_{ik}). \tag{5.4}$$

In the above equation, m is the number of transmission nodes and $j \ldots k$ is the receiver nodes. P is the total power required in the minimum energy tree of broadcasting. Obviously, the total cost is the sum of the transmitted power strength at each of the selected transmitting nodes.

Figure 5.10 shows the whole ad hoc wireless network with a total of 10 nodes. It is assumed that omnidirectional antennas are used in which every transmission from a node

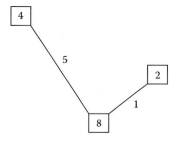

FIGURE 5.9
Subset of the broadcast tree.

could be received by all the nodes that specifically lie within its communication range. The energy required for the transmission between the nodes, or in other words, to support the link between the nodes in the network, is shown in Table 5.3.

The edges in red in Figure 5.11 are the selected edges forming a minimum spanning tree using Kruskal's algorithm for a minimum energy tree of broadcasting in the network.

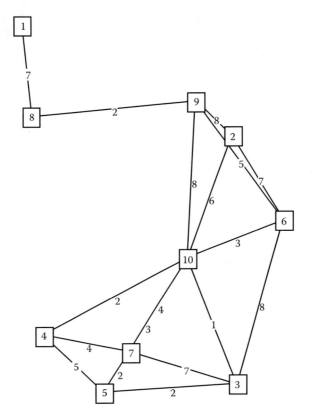

FIGURE 5.10
Ad hoc wireless network with 10 nodes.

TABLE 5.3

Energy Required to Maintain a Link between the Nodes in a Wireless Network

Node	v_1	v_2	v_3	v_4	v_5	v_6	v_7	v_8	v_9	v_{10}
v_1	0	0	0	0	0	0	0	7	0	0
v_2	0	0	0	0	0	7	0	0	8	6
v_3	0	0	0	0	2	8	7	0	0	1
v_4	0	0	0	0	5	0	4	0	0	2
v_5	0	0	2	5	0	0	2	0	0	3
v_6	0	7	8	0	0	0	0	0	5	3
v_7	0	0	7	4	2	0	0	0	0	4
v_8	7	0	0	0	0	0	0	0	2	0
v_9	0	8	0	0	0	5	0	2	0	8
v_{10}	0	6	1	2	3	3	4	0	8	0

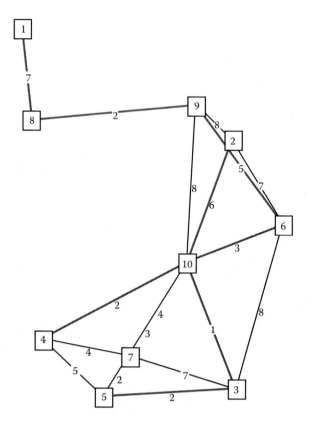

FIGURE 5.11
Minimum spanning tree (MST) for the minimum energy tree of broadcasting in the network.

The transmission nodes are nodes v_2, v_1, v_4, v_5, v_6, and v_7. The order in which the edges are added in this example is

$$e_{3,10} \rightarrow e_{3,5} \rightarrow e_{4,10} \rightarrow e_{5,7} \rightarrow e_{8,9} \rightarrow e_{6,10} \rightarrow e_{6,9} \rightarrow e_{2,10} \rightarrow e_{1,8}.$$

The energy cost for each selected edge in the minimum spanning tree can be seen in Table 5.4.

The cost or the overall transmitter power for the minimum energy tree of broadcasting in the network using Equation 5.4 is computed as follows:

$$P = Q_{3,10,5} + P_{4,10} + P_{5,7} + P_{8,9} + Q_{6,10,9} + P_{2,10} + P_{1,8}$$

$$= \max(P_{3,10}, P_{3,5}) + P_{4,10} + P_{5,7} + P_{8,9} + \max(P_{6,10}, P_{6,9}) + P_{2,10} + P_{1,8}$$

$$= 2 + 2 + 2 + 2 + 5 + 6 + 7$$

$$= 26.$$

5.4.1 *Code5C*: Broadcasting Problem

Code5C is a simple broadcasting problem in the ad hoc wireless communication network using Kruskal's algorithm. The output of *Code5C* is shown in Figure 5.12, which consists of

TABLE 5.4

Transmission Power for Each Selected Edge
of the Minimum Energy Tree

	v_i	v_j	Transmission Power, $P_{i,j}$
1	3	10	1
2	3	5	2
3	4	10	2
4	5	7	2
5	8	9	2
6	6	10	3
7	6	9	5
8	2	10	6
9	1	8	7

10 randomly scattered nodes representing the wireless device units. The minimum spanning tree for the broadcasting problem in the network using Kruskal's algorithm is computed once the push button is clicked.

The output of *Code5C* shows four tables. The first table is the transmission cost matrix, which displays the power strength needed to have a link between the nodes, and the second one is the bottom left table, which displays the matrix for selected edges forming the minimum spanning tree together with its weight on the edge representing the power strength or energy value. These edges are also displayed in the bottom-right table, where *p* is the transmission node and *q* is the receiver node associated with the cost for each link. The top-right table is the one that extracts only the transmission nodes and the maximum transmission power at each transmission node. It must be noted that this table uses Equation 5.3 to obtain each single maximum transmission power at each transmission node. The total cost of transmission power in the ad hoc wireless network which uses Equation 5.4 is displayed at the top of the interface. The interface also allows a new graph to be generated at any time.

Code5C has two source files, *Code5C.h* and *Code5C.cpp*. The nodes and edges of the graph are represented by the structures *NODE* and *EDGE*, respectively. There are also two button objects called *bNGraph* for *Ad Hoc Wireless Network* and *bCompute* for *Compute Minimum Cost*. Other important objects and variables are briefly described in Table 5.5.

The organization of *Code5C* is shown in Figure 5.13. The figure shows all the functions involved in the project. The processing starts with the constructor *CCode5C()*, which draws the main window and the *Ad Hoc Wireless Network* and *Compute Minimum Cost* buttons. The constructor calls *Initialize()*, which creates a new graph with *N* nodes by assigning random coordinates and weights to the edges. Each node is drawn using *UpdateNodes()* and the whole graph is drawn inside the drawing area through *OnPaint()*. Once the *Compute* button is pushed, the function *OnCompute()* will then call *Kruskal()* to compute the minimum spanning tree for broadcasting in the ad hoc wireless network and *ShowResults()* to show the results.

The main engine is *Kruskal()*, which applies Kruskal's algorithm for computing the minimum spanning tree in order to determine the minimum cost of broadcasting in the ad hoc wireless network. The result of the minimum spanning tree generated from the graph is displayed through *ShowResults()*. In computing the minimum spanning tree, *Kruskal()* calls *SortEdges()* to sort all the edges in ascending order. The selected edge for the minimum spanning tree is drawn in red. The determination of the transmission node as well as the maximum power required as in Equation 5.3 is computed. The evaluation of the minimum energy expenditure for broadcasting in the network using Equation 5.4 is stored in *total_cost*. A new

Total cost of overall transmission power in the Ad Hoc Wireless Network is 18

Transmission Cost Matrix

	1	2	3	4	5	6	7	8	9	10
1	0	99	99	99	99	1	4	99	99	99
2	99	0	4	5	99	99	99	99	7	6
3	99	4	0	9	9	99	99	99	5	2
4	99	5	9	0	99	99	5	3	1	8
5	99	99	9	99	0	99	99	9	3	99
6	1	99	99	99	99	0	8	1	99	8
7	4	99	99	5	99	8	0	9	99	99
8	99	99	99	3	9	1	9	0	2	99
9	99	7	5	1	3	99	99	2	0	99
10	99	6	2	8	99	8	99	99	99	0

Spanning Tree Matrix

	1	2	3	4	5	6	7	8	9	10
1	0	0	0	0	0	1	4	0	0	0
2	0	0	4	5	0	0	0	0	0	0
3	0	4	0	0	0	0	0	0	0	2
4	0	5	0	0	0	0	0	0	1	0
5	0	0	0	0	0	0	0	0	3	0
6	1	0	0	0	0	0	0	1	0	0
7	4	0	0	0	0	0	0	0	0	0
8	0	0	0	0	0	1	0	0	2	0
9	0	0	0	1	3	0	0	2	0	0
10	0	0	2	0	0	0	0	0	0	0

Transmission Node	Max Transmission Power
1	4
4	1
6	1
3	2
8	2
5	3
2	5

Edges selected for MST

k	p	q	cost
1	1	6	1
2	4	9	1
3	6	8	1
4	3	10	2
5	8	9	2
6	5	9	3
7	1	7	4
8	2	3	4
9	2	4	5

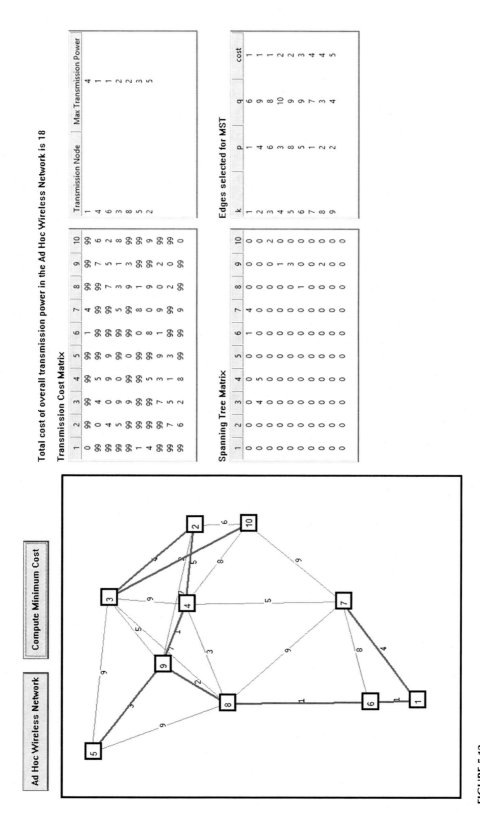

Ad Hoc Wireless Network | Compute Minimum Cost

FIGURE 5.12
Output of *Code5C*.

TABLE 5.5

Important Objects/Variables in *Code5C* and Their Descriptions

Variable	Type	Description
N	constant	Number of nodes in the graph
R	constant	Range for adjacency between two nodes in the graph
nEdges	int	Number of edges in the graph
graph_edge[i][l]	int	If l = 1; the initial node of edge i
		If l = 2; the terminal node of edge i
		If l = 3; the weight assigned to edge i
tree[i][j]	int	Weight of the selected edges for the minimum spanning tree
sets[i][j]	int	Path in the list
top[i]	int	Node with the top priority
total_cost	int	Total weight of the minimum spanning tree for broadcasting
e[i][j].Wt	int	The weight on edge (v_i, v_j)
Home, End	CPoint	Top-left and bottom-right corners of the drawing area
bNGraph	CButton	Button for generating a new graph
bCompute	CButton	Button for computing the minimum cost

graph with randomly located positions in Windows and randomly determined weights on the edges is produced by *Initialize*(). This graph is later drawn in Windows through *OnPaint*(). Another function, *UpdateNodes*(), serves to draw the node specified in its argument.

```
//Code5C.h
#include <afxwin.h>
#include <afxcmn.h>
#include <math.h>
#include <time.h>
#define N 10
#define R 250
#define IDC_NGRAPH 501
#define IDC_tWEIGHT 502
#define IDC_tST 503
#define IDC_COMPUTE 600

class CCode5C: public CFrameWnd
{
private:
        CButton bNGraph,bCompute;
        CPoint Home,End;
        CFont fCourier,fArial;
        CListCtrl tWeight,tST,table, tabletn, tableP;
        int nEdges;
        int graph_edge[N+1][N+1];
        int tree[N+1][N+1];
        int sets[N+1][N+1];
        int top[N+1];
        int total_cost;

        typedef struct
        {
                CPoint Home;
```

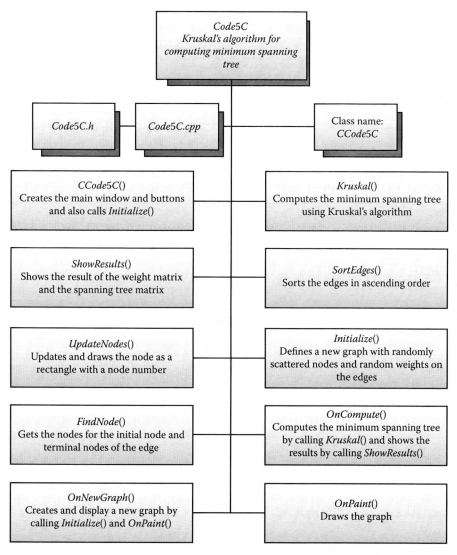

FIGURE 5.13
Organization of *Code5c*.

```
        CRect rct;
    } NODE;
    NODE v[N+1];
    typedef struct
    {
        int Wt;
        bool ST;
    } EDGE;
    EDGE e[N+1][N+1];

public:
    CCode5C();
    ~CCode5C()    {}
```

```
        afx_msg void OnPaint();
        afx_msg void OnCompute();
        afx_msg void OnNewGraph();
        void Kruskal(),ShowResults(),SortEdges(),Initialize(),UpdateNodes();
        int FindNode(int);
        DECLARE_MESSAGE_MAP();
};

class CMyWinApp: public CWinAp
//Code5C.cpp
#include "Code5C.h"
BEGIN_MESSAGE_MAP(CCode5C,CFrameWnd)
        ON_WM_PAINT()
        ON_BN_CLICKED(IDC_COMPUTE,OnCompute)
        ON_BN_CLICKED(IDC_NGRAPH,OnNewGraph)
END_MESSAGE_MAP()

CCode5C::CCode5C()
{
        Home=CPoint(30,100); End=CPoint(500,700);
        Create(NULL,L"Kruskal's Algorithm for broadcasting problems in a
            network",
              WS_OVERLAPPEDWINDOW,CRect(0,0,1000,800));
        bNGraph.Create(L"Ad Hoc Wireless Network",WS_CHILD | WS_VISIBLE |
            BS_DEFPUSHBUTTON,
              CRect(CPoint(30,30),CSize(180,40)),this,IDC_NGRAPH);
        bCompute.Create(L"Compute Minimum Cost",WS_CHILD | WS_VISIBLE
            | BS_DEFPUSHBUTTON,CRect(CPoint(Home.x+200,30),CSize(180,40)),
              this,IDC_COMPUTE);
        fArial.CreatePointFont(100,L"Arial");
        fCourier.CreatePointFont(60,L"Courier");
        Initialize();
}

void CCode5C::Initialize()
{
        int i,j;
        double distance;
        tWeight.DestroyWindow(); tST.DestroyWindow();
        srand(time(0));
        for (i=1;i<=N;i++)
        {
                v[i].Home.x=Home.x+20+rand()%(End.x-Home.x-50);
                v[i].Home.y=Home.y+20+rand()%(End.y-Home.y-50);
                v[i].rct=CRect(CPoint(v[i].Home.x-10,v[i].
                    Home.y-10),CSize(25,25));
        }
        nEdges=0;
        for (i=1;i<=N;i++)
        {
                e[i][i].Wt=0; e[i][i].ST=0;
                for (j=i+1;j<=N;j++)
                {
                        distance=sqrt(pow(double(v[i].Home.x-v[j].Home.x),2)
```

```
                                    +pow(double(v[i].Home.y-v[j].Home.y),2));
                        e[i][j].ST=0; e[j][i].ST=0;
                        if (distance<R)
                        {
                                e[i][j].Wt=1+rand()%9;
                                nEdges++;
                                graph_edge[nEdges][1]=i;
                                graph_edge[nEdges][2]=j;
                                graph_edge[nEdges][3]=e[i][j].Wt;
                        }
                        else
                                e[i][j].Wt=99;
                        e[j][i].Wt=e[i][j].Wt;
                }
        }
}

void CCode5C::OnNewGraph()
{
        Initialize(); Invalidate();
}

void CCode5C::OnPaint()
{
        CPaintDC dc(this);
        CString s;
        int i,j;
        CPoint mPoint;
        CPen mPen(PS_SOLID,3,RGB(0,0,0));
        CPen qPen(PS_SOLID,1,RGB(150,150,150));
        dc.SelectObject(mPen);
        dc.Rectangle(Home.x-10,Home.y-10,End.x+10,End.y+10);
        dc.SelectObject(qPen); dc.SelectObject(fCourier);
        for (i=1;i<=N;i++)
                for (j=1;j<=N;j++)
                        if (e[i][j].Wt!=99)
                        {
                                dc.MoveTo(CPoint(v[i].Home));
                                dc.LineTo(CPoint(v[j].Home));
                                s.Format(L"%d",e[i][j].Wt);
                                dc.TextOut((v[i].Home.x+v[j].Home.x)/2,
                                        (v[i].Home.y+v[j].Home.y)/2,s);
                                mPoint=CPoint((v[i].Home.x+v[j].Home.x)/2,
                                        (v[i].Home.y+v[j].Home.y)/2);
                        }
        UpdateNodes();
}

void CCode5C::UpdateNodes()
{
        CClientDC dc(this);
        CString s;
        CPen mPen(PS_SOLID,3,RGB(0,0,0));
        dc.SelectObject(mPen); dc.SelectObject(fArial);
```

```
          for (int i=1;i<=N;i++)
          {
                  dc.Rectangle(v[i].rct);
                  s.Format(L"%d",i);
                  dc.TextOutW(v[i].Home.x-5,v[i].Home.y-5,s);
          }
}


void CCode5C::ShowResults()
{
          CClientDC dc(this);
          CString s;
          int i,j;
          tWeight.Create(WS_VISIBLE|WS_CHILD|WS_DLGFRAME|LVS_REPORT
                  |LVS_NOSORTHEADER,CRect(CPoint(End.x+30,Home.y),
                          CSize(355,200)),
                  this,IDC_tWEIGHT);
          tWeight.InsertColumn(0,L"",LVCFMT_CENTER,1);
          tST.Create(WS_VISIBLE|WS_CHILD|WS_DLGFRAME|LVS_REPORT
                  |LVS_NOSORTHEADER,CRect(CPoint(End.x+30,Home.y+250),
                          CSize(355,200)),
                  this,IDC_tST);
          tST.InsertColumn(0,L"",LVCFMT_CENTER,1);

          for (j=1;j<=N;j++)
          {
                  s.Format(L"%d",j);
                  tWeight.InsertColumn(j,s,LVCFMT_CENTER,35);
                  s.Format(L"%d",j);
                  tST.InsertColumn(j,s,LVCFMT_CENTER,35);
          }

          for (i=1;i<=N;i++)
                  for (j=0;j<=N;j++)
                  {
                          if (j==0)
                          {
                                  s.Format(L"%d",i);
                                  tWeight.InsertItem(i-1,s,0);
                                  s.Format(L"%d",i);
                                  tST.InsertItem(i-1,s,0);
                          }
                          if (j>0)
                          {
                                  s.Format(L"%d",e[i][j].Wt);
                                  tWeight.SetItemText(i-1,j,s);
                                  s.Format(L"%d",((e[i][j].ST)?e[i][j].Wt:0));
                                  tST.SetItemText(i-1,j,s);
                          }
                  }
          UpdateNodes();
          dc.TextOutW(End.x+30,Home.y-20,L"Transmission Cost Matrix");
          dc.TextOutW(End.x+30,Home.y+230,L"Spanning Tree Matrix");
}
```

```
void CCode5C::OnCompute()
{
        Kruskal();
        ShowResults();
}

void CCode5C::SortEdges()
{
        int i,j,t;
        for (i=1;i<=nEdges-1;i++)
                for (j=1;j<=nEdges-i;j++)
                        if (graph_edge[j][3]>graph_edge[j+1][3])
                        {
                                t=graph_edge[j][1];
                                graph_edge[j][1]=graph_edge[j+1][1];
                                graph_edge[j+1][1]=t;
                                t=graph_edge[j][2];
                                graph_edge[j][2]=graph_edge[j+1][2];
                                graph_edge[j+1][2]=t;
                                t=graph_edge[j][3];
                                graph_edge[j][3]=graph_edge[j+1][3];
                                graph_edge[j+1][3]=t;
                        }
}

void CCode5C::Kruskal()
{
        CClientDC dc(this);
        CString s;
        CPen pMSTpath(PS_SOLID,3,RGB(255,0,0));
        int p1,p2,i,j;
        int p,q;
        int a[N+1],b[N+1],c[N+1];
        bool statustn[N+1];
        SortEdges();
        for (i=1;i<=N;i++)
        {
                sets[i][1]=i;
                top[i]=1;
        }

        CString str, strp;
        table.Create(WS_VISIBLE|WS_CHILD|WS_DLGFRAME|LVS_REPORT
        |LVS_NOSORTHEADER,CRect(CPoint(End.x+400,Home.y+250),CSize(285,200)),
        this,IDC_tST);
        table.InsertColumn(0,L"k",LVCFMT_CENTER,70);
        table.InsertColumn(1,L"p",LVCFMT_CENTER,70);
        table.InsertColumn(2,L"q",LVCFMT_CENTER,70);
        table.InsertColumn(3,L"cost",LVCFMT_CENTER,70);

        dc.SelectObject(pMSTpath);
        int k=1;
        for (i=1;i<=nEdges;i++)
        {
```

```
                  p1=FindNode(graph_edge[i][1]);
                  p2=FindNode(graph_edge[i][2]);
                  if (p1!=p2)
                  {
                          p=graph_edge[i][1]; q=graph_edge[i][2];
                          tree[graph_edge[i][1]][graph_edge[i][2]]=graph_
                                  edge[i][3];
                          tree[graph_edge[i][2]][graph_edge[i][1]]=graph_
                                  edge[i][3];
                          dc.MoveTo(v[p].Home); dc.LineTo(v[q].Home);

                          a[k]=p;
                          b[k]=q;
                          c[k]=graph_edge[i][3];
                          k++;

                          e[p][q].ST=1; e[p][q].ST=1;
                          for (j=1;j<=top[p2];j++)
                          {
                                  top[p1]++;
                                  sets[p1][top[p1]]=sets[p2][j];
                          }
                          top[p2]=0;
                  }

        }
        int sum=k;
        for (k=1;k<=sum-1;k++)
        {
                          str.Format(L"%d",k); table.InsertItem(k-1,str,0);
                          str.Format(L"%d",a[k]); table.SetItemText(k-1,1,str);
                          str.Format(L"%d",b[k]); table.SetItemText(k-1,2,str);
                          str.Format(L"%d",c[k]); table.SetItemText(k-1,3,str);

        }
        int    tn[N+1][N+1], temp_edge[N], costtn[N+1][N+1];
        int m, totaltn, max_cost_tn[N+1];
        int number_tn[N+1];
        for(i=1;i<=N-1;i++)
                statustn[i]=true;
        m=1;
        for (i=1;i<=N-1;i++)
        {
                if (statustn[i]==true)
                {
                        k=1;
                        for (j=i+1;j<=N-1;j++)
                        {
                                if (a[i]==a[j])
                                {
                                        k++;
                                        tn[m][k]=a[j];
                                        costtn[m][k]=c[j];
                                        statustn[j]=false;
```

```
                                      }
                                 }
                                 tn[m][1]=a[i];
                                 costtn[m][1]=c[i];
                                 max_cost_tn[m]=costtn[m][k];
                                 m++;
                            }
                  }
                  totaltn=m-1;

                  s.Format(L"number of transmission node is%d",totaltn);
                  dc.TextOut(900,500,s);

                  dc.TextOutW(End.x+400,Home.y+230,L"Edges selected for MST");
                  tableP.Create(WS_VISIBLE|WS_CHILD|WS_DLGFRAME|LVS_REPORT
                  |LVS_NOSORTHEADER,CRect(CPoint(End.x+400,Home.y),CSize(285,200)),
                  this,IDC_tST);
                  tableP.InsertColumn(0,L"Transmission Node",LVCFMT_CENTER,130);
                  tableP.InsertColumn(1,L"Max Transmission Power",LVCFMT_CENTER,150);

                  for (k=1;k<=totaltn;k++)
                  {
                            str.Format(L"%d",tn[k][1]); tableP.InsertItem
                                (k-1,str,0);
                            str.Format(L"%d",max_cost_tn[k]); tableP.SetItemText
                                (k-1,1,str);
                  }

                  total_cost=0;
                  for (i=1;i<=totaltn;i++)
                        {
                            total_cost +=max_cost_tn[i];
                        }
                  s.Format(L"Total cost of overall transmission power in the Ad Hoc
                        Wireless Network is%d",total_cost);
                  dc.TextOut(End.x+30, Home.y-50,s);
}

int CCode5C::FindNode(int n)
{
      int i,j;
      for (i=1;i<=nEdges;i++)
            for (j=1;j<=top[i];j++)
                  if (n==sets[i][j])
                        return i;
      return -1;
}
```

6

Computing the Maximum Clique

6.1 Problem Description

In graph theory, a clique is a very crucial concept when describing the structure of a group of elements that share the same common features or interest. This group of elements forms a subgraph where, in this particular concept, all of the elements are closely related to each other. For the graph $G(V, E)$, where V is a set of nodes and E is a set of edges, a *clique* is a subgraph, $G'(V', E')$ for $V' \subseteq V$ and $E' \subseteq E$ cohere in such a way that every node is adjacent to every other node in the subgraph. In general, the maximum clique problem deals with finding the subgraph with the largest clique. The maximum clique is also called the maximum cardinality of the graph. The number of nodes in the maximum clique in G is denoted as $\omega(G)$. Note that the maximum clique problem is an NP-hard problem, where it is not possible to find the polynomial solution.

The integer programming can be used to formulate the maximum clique problem as follows:

$$Maximum\ Clique = Max \sum_{i=1}^{n} x_i \tag{6.1}$$

subject to

$$x_i + x_j \leq 1, \quad \forall (i, j) \notin E$$

$$x_i \in \{0, 1\}, \quad i = 1, \ldots, n$$

where $x_i = \begin{cases} 1 & \text{if node is in the clique} \\ 0 & \text{otherwise.} \end{cases}$

The integer programming above is based on edge formulation. In the above equation, x_i is the decision variable and n is the total number of nodes in the graph.

There are various real applications that utilize the maximum clique problem. Among the domains are bioinformatics [1], economics [2], social network communication analysis [3], computer vision [4], and many more. For example, in social and information networks, each edge may represent a contact in the form of a call, an e-mail, or a physical proximity between two entities at a definite point of time. In the Facebook network, the members of a

TABLE 6.1

Summary of Athletes Who Share Common Sports

Athlete	Robert	Sally	Nicole	Larry	Kent
Robert	0	0	0	1	1
Sally	0	0	0	1	0
Nicole	1	0	0	0	0
Larry	1	1	0	0	0
Kent	1	0	0	1	0

group are represented as nodes while their social friends are the edges [5]. The maximum clique of Facebook can then be interpreted as follows:

Of all n number of friends one has, the maximum clique is the number of all the friends that are friends with each other, illustrating the strong connection between the group [6].

Table 6.1 illustrates the maximum clique problem and summarizes the top selected athletes who play common sports for one of the elementary schools in a state. The table only shows the names of the top five athletes. The entry of i, j is 1 if athletes i and j play a common sport, and 0 otherwise. The diagonal entries are 0 in this case. Referring to the table, Robert and Larry happen to play the same sport; therefore, the entry for $(i, j) = (1, 4)$ is 1. Meanwhile, Robert and Sally do not play any sport in common; therefore, the entry for $(i, j) = (1, 2)$ is 0.

The information from Table 6.1 can be represented conveniently in a graph. The nodes represent the athletes and they are represented as undirected edges if and only if the corresponding athletes play a common sport. The resulting undirected graph of Table 6.1 is shown in Figure 6.1, where the graph is $G(V, E)$ with $V = (v_1, v_2, v_3, v_4, v_5)$ and edge set $E \subseteq V \times V$. In this graph, the nodes represent Robert, Sally, Nicole, Larry, and Kent, respectively. For example, athletes v_2 and v_4 play the same sport; therefore, an undirected edge connects these two nodes.

The cliques of cardinality of two or higher in the graph in Figure 6.1 are

$\{v_1, v_4, v_5\}$

$\{v_1, v_4\}$

$\{v_1, v_5\}$

$\{v_4, v_5\}$

$\{v_1, v_3\}$

$\{v_2, v_4\}$

Obviously, the maximum clique of the graph in this example is $\{v_1, v_4, v_5\}$ with $\omega(G) = 3$. This represents Robert, Larry, and Kent as the maximum clique of the top athletes playing common sports in the school.

6.1.1 Greedy Algorithm for Finding the Maximum Clique

We discuss a greedy algorithm for finding the maximum clique. The graph $G(V, E)$ has nodes $V = \{v_1, v_2, v_3...v_n\}$. The algorithm starts with finding the nodes adjacent to the first node, v_1. The nodes are adjacent only if $e_{ij} \in E$ and $\{v_i, v_j\} \subseteq V$. Once all the adjacent nodes to v_1 have been determined, the total number of adjacent nodes is stored as variable k_1. At the same time, all the adjacent nodes to v_1 are stored in an array $p_1[k_1] = v_i$. Next, among all the

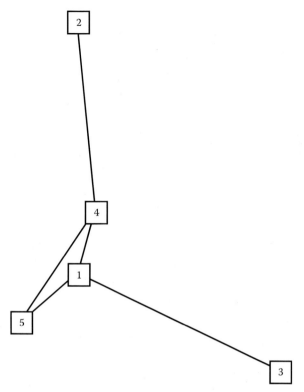

FIGURE 6.1
Graph representing the athletes who play common sports.

k_1 nodes stored in the array, the connectivity of every pair of each node is determined. All the adjacent nodes to v_1 that have been determined to be connected to each other are then stored as the first clique, C_1, and its total nodes in C_1 are determined. The same process of finding the clique of the graph originating from v_2 is applied. The process is repeated to all the nodes in order to find other cliques. Note that only isolated nodes do not have a clique originating from it. The maximum clique is then obtained from the clique with the highest number of nodes.

Figure 6.2 illustrates the execution of a greedy algorithm to find the maximum clique. The graph $G(V, E)$ with $V = \{v_1, v_2, v_3, v_4, v_5, v_6\}$ and $E = \{e_{ij}\}$ with $i, j = 1, 2...6$ is illustrated in (a). The procedure starts with finding the clique originating from v_1, which is circled as shown in (b). The adjacent nodes to v_1 are $p_1 = \{v_2, v_3, v_4, v_6\}$, as highlighted by the rectangle. It can be seen that the total number of adjacent nodes to v_1 is $k_1 = 4$. Therefore, the connectivity among every pair of these four nodes is determined as follows:

$$e_{23} \notin E$$
$$e_{24} \notin E$$
$$e_{26} \notin E$$
$$e_{34} \in E$$
$$e_{36} \in E$$
$$e_{46} \in E$$

174

Simulation for Applied Graph Theory Using Visual C++

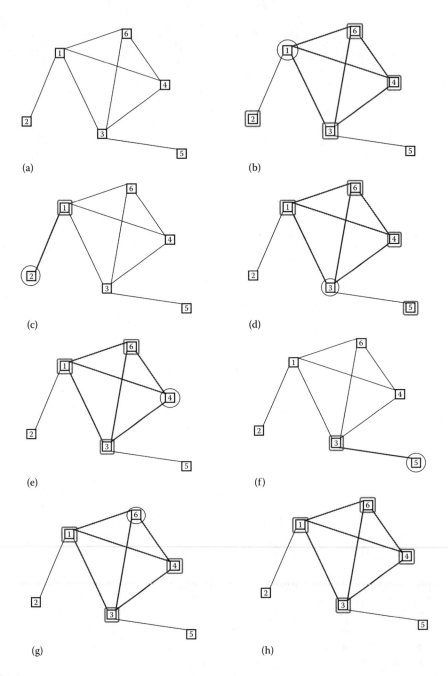

FIGURE 6.2
Execution of a greedy algorithm to find the maximum clique. (a) The graph, (b) finding clique originating from v_1, (c) finding clique originating from v_2, (d) finding clique originating from v_3, (e) finding clique originating from v_4, (f) finding clique originating from v_5, (g) finding clique originating from v_6, and (h) the maximum clique.

Based on the above determination of connectivity between all pairs of adjacent nodes, only the following can be seen: $\{e_{34}, e_{36}, e_{46}\} \in E$. Therefore, the first clique consists of nodes adjacent to each other, including v_1, as follows, where the edges connecting these nodes are highlighted in (b):

$$C_1 = \{v_1, v_3, v_4, v_6\} \text{ with } |C_1| = 4.$$

Next, the same procedure of finding the clique originating from v_2 is applied, in which the originating node is circled, as shown in (c). The adjacent node to v_2 is $p_2 = \{v_1\}$, as highlighted by the rectangle. Therefore, the total number of adjacent node to v_2 is $k_2 = 1$. Obviously, v_1 is connected to v_2 since

$$e_{12} \in E.$$

Since it is only $\{e_{12}\} \in E$, the second clique including v_2 is therefore as follows, where the edges connecting these nodes are highlighted in (c):

$$C_2 = \{v_1, v_2\} \text{ with } |C_2| = 2.$$

The same procedure of finding the clique originating from v_3 is repeated, in which the originated node is circled, as shown in (d). The adjacent nodes to v_3 are $p_3 = \{v_1, v_4, v_5, v_6\}$, as highlighted by the rectangle. It can be seen that the total number of adjacent nodes to v_3 is $k_3 = 4$. Therefore, the connectivity among every pair of these four nodes is determined as follows:

$e_{14} \in E$

$e_{15} \notin E$

$e_{16} \in E$

$e_{45} \notin E$

$e_{46} \in E$

$e_{56} \notin E$

Based on the above determination of connectivity between all pairs of adjacent nodes, only the following can be seen: $\{e_{14}, e_{16}, e_{46}\} \in E$. Therefore, the third clique consists of nodes adjacent to each, other including v_3, as follows, where the edges connecting these nodes are highlighted in (d):

$$C_3 = \{v_3, v_1, v_4, v_6\} \text{ with } |C_3| = 4.$$

The procedure of finding the clique originating from v_4 is repeated, in which the originating node is circled, as shown in (e). The adjacent nodes to v_4 are $p_4 = \{v_1, v_3, v_6\}$, as highlighted by the rectangle. It can be seen that the total number of adjacent nodes to v_4 is $k_4 = 3$. Therefore, the connectivity among every pair of these three nodes is determined as follows:

$e_{13} \in E$

$e_{16} \in E$

$e_{36} \in E$

Based on the above determination of connectivity between all pairs of adjacent nodes, all the edges are connected to each other. Therefore, the fourth clique consists of nodes adjacent to each other, including v_4, as follows, where the edges connecting these nodes are highlighted in (e):

$$C_4 = \{v_4, v_1, v_3, v_6\} \text{ with } |C_4| = 4.$$

The procedure of finding the clique originating from v_5 is repeated, in which the originating node is circled, as shown in (f). It can be seen that the adjacent node to v_5 is $p_5 = \{v_3\}$, as highlighted by the rectangle. Therefore, the total number of adjacent nodes to v_5 is $k_5 = 1$. Obviously, v_5 is connected to v_3 since

$$e_{35} \in E.$$

Since it is only $\{e_{35}\} \in E$, the fifth clique including v_5 is therefore as follows, where the edges connecting these nodes are highlighted in (f):

$$C_5 = \{v_5, v_3\} \text{ with } |C_5| = 2.$$

Finally, the procedure of finding the clique originating from v_6 is repeated, in which the originating node is circled, as shown in (g). The adjacent nodes to v_6 are $p_6 = \{v_1, v_3, v_4\}$, as highlighted by the rectangle. It can be seen that the total number of adjacent nodes to v_6 is $k_6 = 3$. Therefore, the connectivity among every pair of these three nodes is determined as follows:

$e_{13} \in E$

$e_{14} \in E$

$e_{34} \in E$

Based on the above determination of connectivity between all pairs of adjacent nodes, it can be seen that all of the nodes are connected to each other. Therefore, the final clique consists of all the nodes adjacent to each other, including v_6, as follows, where the edges connecting these nodes are highlighted in (g):

$$C_6 = \{v_6, v_1, v_3, v_4\} \text{ with } |C_6| = 4.$$

Obviously, the maximum clique of the graph in this example is $\{v_1, v_3, v_4, v_6\}$ with $\omega(G) = 4$, as shown in (h).

The algorithm to find the maximum clique can be summarized as follows:

Given: The graph $G(V, E)$ for $V = \{v_i\}$, $E = \{e_{ij}\}$ and $i, j = 1, 2, \ldots, n$
Problem: To find the maximum cliques $G'(V', E')$ where $V' \subseteq V$ and $E' \subseteq E$
Initialization:
Let $k = 0$;
$max[i] = 1$;
$a = 1$;

Process:
```
For i = 1 to n
    For j = 1 to n
        If (e_ij ∈ E)
            k++;
            p_i[k] = v_j;                //the array to store the adjacent nodes to v_i
    Endfor
    vCliq[1] = i;                        //vCliq[1] is the first node assigned to Clique i
    For r = 1 to k
        maxi=0;
        For l = r + 1 to k
            If (e_rl ∈ E)
                maxi++;
        Endfor
        If (maxi = k − r)
            a++;
            vCliq[a] = p[r];
            max[i] = 1;
    Endfor
    a++;
    vCliq[a] = p[k];
    max[i]++;
Endfor
```

Determine the clique with maximum number of nodes and update it as *maximum*;

```
For i = 1 to n
    If max[i] = maximum
        Maximum clique, nClique = max[i];
        Update all nodes of the Clique to vCliq[a] = p[r];
    Endif
Endfor
```
Output: Maximum Clique, $G'(V', E')$ with all the nodes listed $vCliq[a] = p[r]$.

6.1.2 *Code6A*: Implementing the Greedy Algorithm

Code6A is the implementation of the greedy algorithm for the maximum clique problem. The output of *Code6A* is shown in Figure 6.3, which displays 20 randomly scattered nodes inside a box. The maximum clique is computed once the button *Compute* is clicked. The interface also allows a new graph to be generated through the *New Graph* button. The maximum clique is displayed in the graph with both its nodes and edges highlighted in thick lines. In this example, the maximum clique originates from v_{13}. Referring to Figure 6.3, the adjacent nodes to v_{13} are $\{v_1, v_2, v_5, v_{10}, v_{12}, v_{16}\}$, but the only nodes that are connected to each other, including the originating node, are $\{v_{13}, v_1, v_2, v_{10}, v_{12}, v_{16}\}$. Therefore, the $\omega(G) = 6$ and the set of nodes for the maximum clique is $V' = \{v_{13}, v_1, v_2, v_{10}, v_{12}, v_{16}\}$.

Code6A has two source files, *Code6A.h* and *Code6A.cpp*. In *Code6A.h*, *nCliq* and *vCliq* represent the total number of nodes in the maximum clique and the array of nodes belonging to the maximum clique, respectively. The node and the edges of the graph are represented by the structures *EDGE* and *NODE*. There are also two button objects called *bNGraph* for

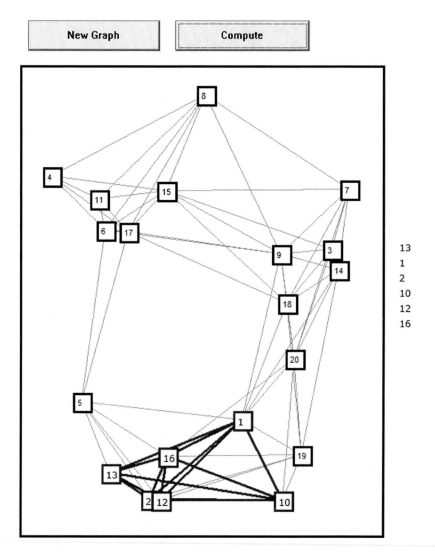

FIGURE 6.3
Output of *Code6A*.

New Graph and *bCompute* for *Compute*. Other important objects and variables are briefly described in Table 6.2.

The organization of *Code6A* is shown in Figure 6.4. The figure shows all the functions involved in the program. The processing starts with the constructor *CCode6A()*, which draws both the main window and the *New Graph* button. The constructor calls *Initialize()*, which creates a new graph with *N* nodes by assigning random coordinates and adjacency nodes to the edges. The whole graph is drawn inside the drawing area through *OnPaint()*. The computed nodes of the maximum clique from *MaxClique()* are read inside *OnCompute()*, from which the graph of the maximum clique is then drawn.

The main engine is *MaxClique()*, which applies the greedy algorithm for computing the maximum clique of the graph. The maximum clique from the whole graph is displayed through *OnCompute()*. In computing the maximum clique, *OnCompute()* calls *MaxClique()*

TABLE 6.2

Important Objects/Variables in *Code6A* and Their Descriptions

Variable	Type	Description
N	constant	Number of nodes in the graph
e[i][j].adj	int	Adjacency value between (v_i, v_j)
vCliq[i]	int	Nodes in the clique
max[i]	int	The number of nodes in each clique
nCliq	int	The number of nodes in the maximum clique
LinkRange	constant	Range for adjacency between two nodes in the graph
p[i]	int array	Predecessor node to the currently marked node
Home, End	CPoint	Top-left and bottom-right corners of the drawing area
bNGraph	CButton	Button for generating a new graph
bCompute	CButton	Button for computing the maximum clique

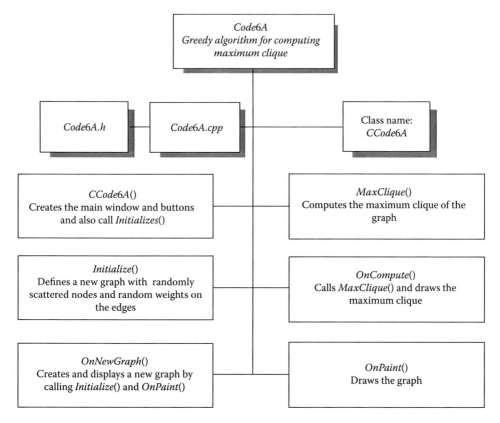

FIGURE 6.4
Organization of *Code6A*.

to obtain the nodes belonging to the maximum clique as well as the total number of nodes of the maximum clique. A new graph with randomly located positions in Windows and randomly determined weights on the edges is produced by *Initialize*(). This graph is later drawn in Windows through *OnPaint*(). This function also serves to draw the node specified in its argument.

The code is listed as follows:

```
//Code6A.h
#include <afxwin.h>
#include <fstream>
#define IDC_NGRAPH 500
#define IDC_COMPUTE 501
#define N 20
#define LinkRange 250

class CCode6A: public CFrameWnd
{
private:
        int vCliq[N+1],p[N+1];
        int max[N+1];
        int maxi,maximum,nCliq;
        CPoint Home,End;
        CFont fArial,fVerdana;
        CPen pColor[N+1];
        CButton bNGraph,bCompute;

        typedef struct
        {
                int adj;
        } EDGE;
        EDGE e[N+1][N+1];

        typedef struct
        {
                CPoint Home;
                CRect rct;
        } NODE;
        NODE v[N+1];
public:
        CCode6A();
        ~CCode6A()      {}
        void MaxClique(),Initialize();
        afx_msg void OnPaint(),OnNewGraph(),OnCompute();
        DECLARE_MESSAGE_MAP()
};

class CMyWinApp: public CWinApp
{
public:
        virtual BOOL InitInstance();
};
CMyWinApp MyApplication;

BOOL CMyWinApp::InitInstance()
{
        m_pMainWnd=new CCode6A;
        m_pMainWnd->ShowWindow(m_nCmdShow);
        return TRUE;
}
```

```cpp
//Code6A.cpp
#include "Code6A.h"

BEGIN_MESSAGE_MAP(CCode6A,CFrameWnd)
        ON_WM_PAINT()
        ON_BN_CLICKED(IDC_NGRAPH,OnNewGraph)
        ON_BN_CLICKED(IDC_COMPUTE,OnCompute)
END_MESSAGE_MAP()

CCode6A::CCode6A()
{
        Home=CPoint(30,100); End=CPoint(500,700);
        Create(NULL,L"Computing the Maximum Clique of a Graph",
                WS_OVERLAPPEDWINDOW,CRect(0,0,1000,800),NULL);
        bNGraph.Create(L"New Graph",WS_CHILD | WS_VISIBLE | BS_DEFPUSHBUTTON,
                CRect(CPoint(Home.x,30),CSize(180,40)),this,IDC_NGRAPH);
        bCompute.Create(L"Compute",WS_CHILD | WS_VISIBLE
                | BS_DEFPUSHBUTTON,CRect(CPoint(Home.x+200,30),CSize(180,40)),
                this,IDC_COMPUTE);
        fArial.CreatePointFont(80,L"Arial");
        fVerdana.CreatePointFont(100,L"Verdana");
        int Color[]={RGB(150,150,150),RGB(0,0,0),RGB(0,0,200)};
        for (int i=0;i<=2;i++)
                if (i==0)
                        pColor[i].CreatePen(PS_SOLID,1,Color[i]);
                else
                        pColor[i].CreatePen(PS_SOLID,3,Color[i]);
        Initialize();
}

void CCode6A::OnNewGraph()
{
        Initialize(); Invalidate();
}
void CCode6A::Initialize()
{
        int i,j;
        double distance;
        srand(time(0));
        for (i=1;i<=N;i++)
        {
                v[i].Home.x=Home.x+20+rand()%(End.x-Home.x-50);
                v[i].Home.y=Home.y+20+rand()%(End.y-Home.y-50);
                v[i].rct=CRect(CPoint(v[i].Home.x-10,v[i].Home.y-10),CSize
                        (25,25));
        }
        for (i=1;i<=N;i++)
        {
                e[i][i].adj=0;
                for (j=i+1;j<=N;j++)
                {
                        distance=sqrt(pow(double(v[i].Home.x-v[j].Home.x),2)
                                +pow(double(v[i].Home.y-v[j].Home.y),2));
                        e[i][j].adj=((distance<=LinkRange)?1:0);
```

```
                        e[j][i].adj=e[i][j].adj;
                }
        }
}

void CCode6A::OnPaint()
{
        CPaintDC dc(this);
        CString s;
        int i,j;
        dc.SelectObject(pColor[1]);
        dc.Rectangle(Home.x-10,Home.y-10,End.x+10,End.y+10);
        dc.SelectObject(pColor[0]);
        for (i=1;i<=N;i++)
                for (j=1;j<=N;j++)
                        if (e[i][j].adj)
                        {
                                dc.MoveTo(CPoint(v[i].Home));
                                dc.LineTo(CPoint(v[j].Home));
                        }
        dc.SelectObject(fArial); dc.SelectObject(pColor[1]);
        for (i=1;i<=N;i++)
        {
                dc.Rectangle(v[i].rct);
                s.Format(L"%d",i);
                dc.TextOutW(v[i].Home.x-5,v[i].Home.y-5,s);
        }
}

void CCode6A::MaxClique()
{
        int i,j,r,t,k,maxi,a;
        for (i=1; i<=N; i++)
        {
                k=0;
                for (j=1; j<=N; j++)
                        if (e[i][j].adj==1)
                        {
                                k++;
                                p[k]=j;
                        }
                vCliq[1]=i;
                max[i]=1;
                a=1;
                for (r=1; r<k; r++)
                {
                        maxi=0;
                        for (t=r+1; t<=k; t++)
                                if (e[p[r]][p[t]].adj==1)
                                        maxi++;
                        if (maxi==(k-r))
                        {
                                a++;
                                vCliq[a]=p[r];
                                max[i]++;
```

```
                        }
                }
                a++;
                vCliq[a]=p[k];
                max[i]++;
        }
        maximum=max[1];
        for (int c=2;c<=N;c++)
                if (max[c]>=maximum)
                        maximum=max[c];
        for (int i=1;i<=N;i++)
                if (max[i]==maximum)
                {
                        k=0;
                        for(j=1; j<=N; j++)
                                if(e[i][j].adj==1)
                                {
                                        k++;
                                        p[k]=j;
                                }
                        vCliq[1]=i;
                        max[i]=1;
                        a=1;
                        for (r=1; r<k; r++)
                        {
                                maxi=0;
                                for (t=r+1; t<=k; t++)
                                        if (e[p[r]][p[t]].adj==1)
                                                maxi++;
                                if (maxi==(k-r))
                                {
                                        a++;
                                        vCliq[a]=p[r];
                                        max[i]++;
                                }
                        }
                        a++;
                        vCliq[a]=p[k];
                        max[i]++; nCliq=max[i];
                }
}

void CCode6A::OnCompute()
{
        CClientDC dc(this);
        CString s;
        int i,j;
        MaxClique();
        dc.SelectObject(&pColor[2]);
        for (i=1;i<=nCliq;i++)      //draw the subgraph
                for (j=i+1;j<=nCliq;j++)
                {
                        dc.MoveTo(v[vCliq[i]].Home);
                        dc.LineTo(v[vCliq[j]].Home);
                }
```

```
dc.SetTextColor(RGB(0,0,0)); dc.SelectObject(fVerdana);
for (i=1;i<=nCliq;i++)
{
        dc.Rectangle(v[vCliq[i]].rct);
        s.Format(L"%d",vCliq[i]);
        dc.TextOutW(v[vCliq[i]].Home.x-5,v[vCliq[i]].Home.y-5,s);
        dc.TextOutW(End.x+30,Home.y+200+i*20,s);
}
}
```

6.2 Computing the Multiple Cliques of a Graph

A graph is capable of representing the behavior or the structure of relationship among its entities or nodes. In many situations, there is more than one group of complete interconnected nodes that can be seen in a graph. These groups of complete interconnected nodes form multiple groups of cliques. Therefore, a multiclique can be defined as a graph with multiple cliques.

Figure 6.5 illustrates a graph with multicliques. From a single graph with 20 nodes, the graph has five different cliques. In order to distinguish different cliques existing in the graph, each clique has been assigned a different color. The cliques in the graph are as follows:

$$C_1 = \{v_1, v_2, v_6, v_7, v_{12}, v_{13}, v_{16}, v_{18}\}$$

$$C_2 = \{v_5, v_9, v_{11}, v_{14}, v_{17}, v_{19}\}$$

$$C_3 = \{v_4, v_{10}, v_{16}, v_{20}\}$$

$$C_4 = \{v_{16}, v_{18}, v_{20}\}$$

$$C_5 = \{v_3, v_{15}, v_{19}\}.$$

There are various applications of multicliques, such as in bioinformatics, economics, and sociology. In economics, one of the applications of cliques is to provide a deeper insight into the internal structure of the stock market by using a market graph [2]. In the study, the nodes represent financial instruments while the edges represent the correlation coefficient between the financial instruments that exceed a specified threshold. A clique of the market graph represents a set of instruments that correlate to each other in such a way that the prices change similarly over time. The analysis of the market graph provides results that are able to predict the stock market behavior in the future.

6.2.1 Greedy Algorithm for Finding the Multicliques

A multiclique can be found using a greedy algorithm. The graph $G(V, E)$ has nodes $V = \{v_1, v_2, v_3...v_n\}$. The algorithm starts with finding the nodes adjacent to the first node, v_1. The nodes are adjacent only if $e_{ij} \in E$ and $\{v_i, v_j\} \subseteq V$. Once all the adjacent nodes to v_1 have been determined, the total number of the adjacent nodes is stored as variable k_1. All the adjacent nodes to v_1 are stored in an array $p_1[k_1] = v_i$. Next, among all the k_1 nodes, the connectivity of every pair of each node is determined. All the adjacent nodes to v_1 that have been determined to be connected to each other are then stored as the first clique, C_1, and the total

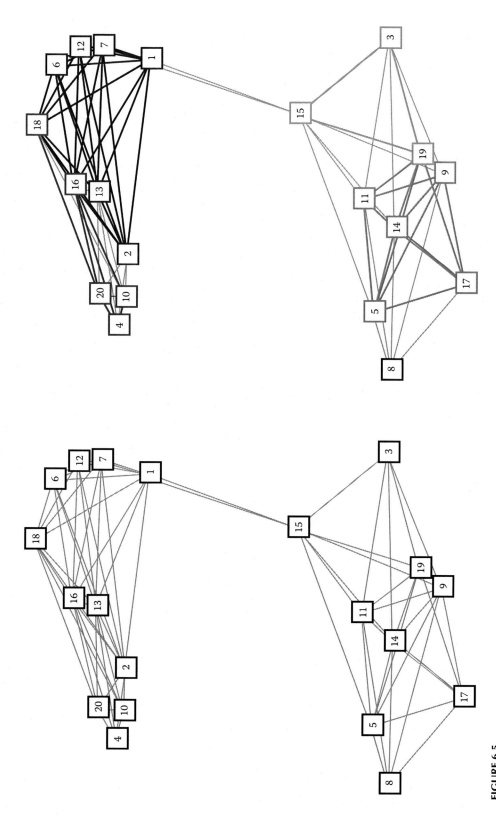

FIGURE 6.5
The graph (left) and its detected multicliques (right).

in C_1 is determined. The same process of finding the clique of the graph originating from v_2 is applied. If there is a clique originating from v_2, the second clique, C_2, is formed. The process is repeated for all the nodes in order to find other cliques. Note that only isolated nodes with no edges do not have a clique graph originating from them. All the cliques together with their total number of nodes are stored accordingly. Since this is a multiclique problem, the top m maximum cliques needed to be visualized are specified.

The algorithm to find multicliques can be summarized as follows:

Given: The graph $G(V, E)$ for $V = \{v_i\}$, $E = \{e_{ij}\}$ and $i, j = 1, 2, \ldots, n$
Problem: To find m maximum cliques $G'(V', E')$ where $V' \subseteq V$ and $E' \subseteq E$
Initialization:
Let $k = 0$;
max[i] = 1;
$a = 1$;
Process:
For $q = 1$ to m
  For $i = 1$ to n
    For $j = 1$ to n
      If $(e_{ij} \in E)$
        k++;
        $p_i[k] = v_j$;   //the array to store the adjacent nodes to v_i
    Endfor
    vCliq[1] = i;   //vCliq[1] is the first node assigned to Clique i
    For $r = 1$ to k
      maxi=0;
      For $l = r + 1$ to k
        If $(e_{rl} \in E)$
          maxi++;
      Endfor
      If (maxi = $k - r$)
        a++;
        vCliq[a] = p[r];
        max[i] = 1;
    Endfor
    a++;
    vCliq[a] = p[k];
    max[i]++;
Endfor

Determine the clique with maximum number of nodes and update it as *maximum*;

For $i = 1$ to n
    If max[i] = *maximum*
        Maximum clique, *nClique* = max[i];
        Update all nodes of the Clique to vCliq[a] = p[r];
    Endif
  Endfor
Endfor
Output: m maximum cliques, $G'(V', E')$ with all the nodes listed vCliq[a] = p[r].

6.2.2 *Code6B*: Implementing the Greedy Algorithm

Code6B is the implementation of the greedy algorithm for finding the multicliques in a graph. The output of *Code6B* is shown in Figure 6.6, which displays 20 randomly scattered nodes inside a box. The multicliques are computed and displayed in the graph once the *Compute* button is clicked. The interface also allows a new graph to be generated through the *New Graph* button.

Code6B has a single class called *CCode6B* with two source files, *Code6B.h* and *Code6B.cpp*. The nodes and edges of the graph are represented by the structures *NODE* and *EDGE*, respectively. There are also two button objects called *bNGraph* for *New Graph* and *bCompute* for *Compute*. Other important objects and variables are briefly described in Table 6.3.

The organization of *Code6B* is shown in Figure 6.7. The figure shows all the functions involved in the program. The processing starts with the constructor *CCode6B()*, which draws both the main window and the *New Graph* button. The constructor calls *Initialize()*, which creates a new graph with *N* nodes by assigning random coordinates and adjacency

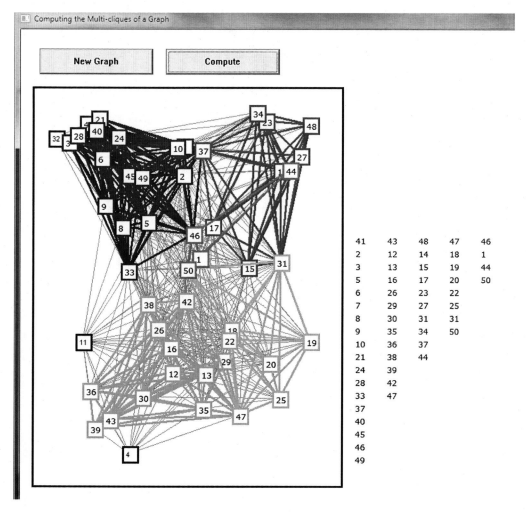

FIGURE 6.6
Output of *Code6B*.

TABLE 6.3

Important Objects/Variables in *Code6B* and Their Descriptions

Variable	Type	Description
N	constant	Number of nodes in the graph
e[i][j].adj	int	Adjacency value between (v_i, v_j)
vCliq[i]	int	Nodes in the clique
max[i]	int	The number of nodes in each clique
nCliq	int	The number of nodes in the top selected multicliques
LinkRange	constant	Range for adjacency between two nodes in the graph
p[i]	int array	Predecessor node to the currently marked node
Home, End	CPoint	Top-left and bottom-right corners of the drawing area
bNGraph	CButton	Button for generating a new graph
bCompute	CButton	Button for computing the maximum clique

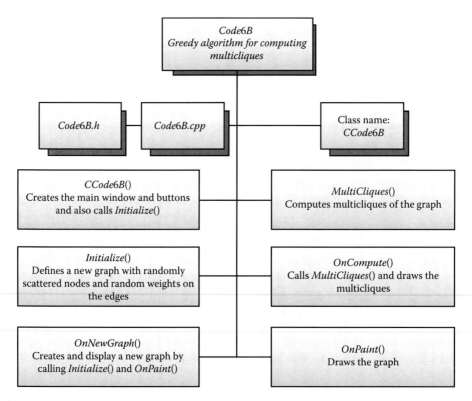

FIGURE 6.7
Organization of *Code6B*.

nodes to the edges. The whole graph is drawn inside the drawing area through *OnPaint()*. The computed nodes of the multicliques from *MultiCliques()* are read inside *OnCompute()*, and the multiclique graph is then drawn.

The main engine is *MultiCliques()*, which applies the greedy algorithm for computing the multicliques. The multicliques from the whole graph that represent the research clusters are displayed through *OnCompute()*. In computing the multiclique, *OnCompute()* calls *MultiCliques()* to obtain the nodes belonging to the five highest multicliques as well as the total number of nodes of the clique. A new graph with randomly located positions in Windows and randomly determined weights on the edges is produced by *Initialize*(). This graph is later drawn in Windows through *OnPaint*(). This function also serves to draw the node specified in its argument.

The codes for *Code6B* are listed as follows:

```
//Code6B.h
#include <afxwin.h>
#include <fstream>
#define IDC_NGRAPH 500
#define IDC_COMPUTE 501
#define N 50
#define R 250

class CCode6B: public CFrameWnd
{
private:
        int vCliq[N+1],p[N+1];
        int max[N+1];
        int maxi,maximum,nCliq;
        CPoint Home,End;
        CFont fArial,fVerdana;
        CPen pColor[N+1];
        CButton bNGraph,bCompute;

        typedef struct
        {
                int adj;
        } EDGE;
        EDGE e[N+1][N+1];

        typedef struct
        {
                CPoint Home;
                CRect rct;
        } NODE;
        NODE v[N+1];
public:
        CCode6B();
        ~CCode6B()        {}
        void MultiCliques(),Initialize();
        afx_msg void OnPaint(),OnNewGraph(),OnCompute();
        DECLARE_MESSAGE_MAP()
};

class CMyWinApp: public CWinApp
{
public:
        virtual BOOL InitInstance();
};
CMyWinApp MyApplication;

BOOL CMyWinApp::InitInstance()
{
```

```
        m_pMainWnd=new CCode6B;
        m_pMainWnd->ShowWindow(m_nCmdShow);
        return TRUE;
}
//Code6B.cpp
#include "Code6B.h"

BEGIN_MESSAGE_MAP(CCode6B,CFrameWnd)
        ON_WM_PAINT()
        ON_BN_CLICKED(IDC_NGRAPH,OnNewGraph)
        ON_BN_CLICKED(IDC_COMPUTE,OnCompute)
END_MESSAGE_MAP()

CCode6B::CCode6B()
{
        Home=CPoint(30,100); End=CPoint(500,700);
        Create(NULL,L"Computing the Multi-cliques of a Graph",
                WS_OVERLAPPEDWINDOW,CRect(0,0,1000,800),NULL);
        bNGraph.Create(L"New Graph",WS_CHILD | WS_VISIBLE |
BS_DEFPUSHBUTTON,
                CRect(CPoint(Home.x,30),CSize(180,40)),this,IDC_NGRAPH);
        bCompute.Create(L"Compute",WS_CHILD | WS_VISIBLE
                | BS_DEFPUSHBUTTON,CRect(CPoint(Home.x+200,30),
CSize(180,40)),
                this,IDC_COMPUTE);
        fArial.CreatePointFont(80,L"Arial");
        fVerdana.CreatePointFont(100,L"Verdana");
        int Color[]={RGB(150,150,150),RGB(0,0,0),RGB(0,0,200),
                RGB(0,200,0),RGB(200,0,0),RGB(0,200,200),RGB(255,0,50)};
        for (int i=0;i<=6;i++)
                if (i==0)
                        pColor[i].CreatePen(PS_SOLID,1,Color[i]);
                else
                        pColor[i].CreatePen(PS_SOLID,3,Color[i]);
        Initialize();
}

void CCode6B::OnNewGraph()
{
        Initialize(); Invalidate();
}

void CCode6B::Initialize()
{
        int i,j;
        double distance;
        srand(time(0));
        for (i=1;i<=N;i++)
        {
                v[i].Home.x=Home.x+20+rand()%(End.x-Home.x-50);
                v[i].Home.y=Home.y+20+rand()%(End.y-Home.y-50);
                v[i].rct=CRect(CPoint(v[i].Home.x-10,v[i].Home.y-10),CSize
                        (25,25));
        }
```

```
        for (i=1;i<=N;i++)
        {
                e[i][i].adj=0;
                for (j=i+1;j<=N;j++)
                {
                        distance=sqrt(pow(double(v[i].Home.x-v[j].Home.x),2)
                                +pow(double(v[i].Home.y-v[j].Home.y),2));
                        e[i][j].adj=((distance<=R)?1:0);
                        e[j][i].adj=e[i][j].adj;
                }
        }
}

void CCode6B::OnPaint()
{
        CPaintDC dc(this);
        CString s;
        int i,j;
        dc.SelectObject(pColor[1]);
        dc.Rectangle(Home.x-10,Home.y-10,End.x+10,End.y+10);
        dc.SelectObject(pColor[0]);
        for (i=1;i<=N;i++)
                for (j=1;j<=N;j++)
                        if (e[i][j].adj)
                        {
                                dc.MoveTo(CPoint(v[i].Home));
                                dc.LineTo(CPoint(v[j].Home));
                        }
        dc.SelectObject(fArial); dc.SelectObject(pColor[1]);
        for (i=1;i<=N;i++)
        {
                dc.Rectangle(v[i].rct);
                s.Format(L"%d",i);
                dc.TextOutW(v[i].Home.x-5,v[i].Home.y-5,s);
        }
}

void CCode6B::MultiCliques()
{
        int i,j,r,t,k,maxi,a;
        for (i=1; i<=N; i++)
        {
                k=0;
                for (j=1; j<=N; j++)
                        if (e[i][j].adj==1)
                        {
                                k++;
                                p[k]=j;
                        }
                vCliq[1]=i;
                max[i]=1;
                a=1;
                for (r=1; r<k; r++)
                {
```

```
                            maxi=0;
                            for (t=r+1; t<=k; t++)
                                    if (e[p[r]][p[t]].adj==1)
                                            maxi++;
                            if (maxi==(k-r))
                            {
                                    a++;
                                    vCliq[a]=p[r];
                                    max[i]++;
                            }
                    }
                    a++;
                    vCliq[a]=p[k];
                    max[i]++;
            }
            maximum=max[1];
            for (int c=2;c<=N;c++)
                    if (max[c]>=maximum)
                            maximum=max[c];
            for (int i=1;i<=N;i++)
                    if (max[i]==maximum)
                    {
                            k=0;
                            for(j=1; j<=N; j++)
                                    if(e[i][j].adj==1)
                                    {
                                            k++;
                                            p[k]=j;
                                    }
                            vCliq[1]=i;
                            max[i]=1;
                            a=1;
                            for (r=1; r<k; r++)
                            {
                                    maxi=0;
                                    for (t=r+1; t<=k; t++)
                                            if (e[p[r]][p[t]].adj==1)
                                                    maxi++;
                                    if (maxi==(k-r))
                                    {
                                            a++;
                                            vCliq[a]=p[r];
                                            max[i]++;
                                    }
                            }
                            a++;
                            vCliq[a]=p[k];
                            max[i]++; nCliq=max[i];
                    }
}

void CCode6B::OnCompute()
{
        CClientDC dc(this);
        CString s;
```

```
        int i,j,k;
        for (k=1;k<=5;k++)   //Compute and display 5 max cliques
        {
                MultiCliques();
                dc.SelectObject(&pColor[k+1]);
                for (i=1;i<=nCliq;i++)        //draw the subgraph
                        for (j=i+1;j<=nCliq;j++)
                        {
                                dc.MoveTo(v[vCliq[i]].Home);
                                dc.LineTo(v[vCliq[j]].Home);
                                e[vCliq[i]][vCliq[j]].adj=0; e[vCliq[j]]
[vCliq[i]].adj=0;
                        }
                dc.SetTextColor(RGB(0,0,0)); dc.SelectObject(fVerdana);

                for (i=1;i<=nCliq;i++)
                {
                        dc.Rectangle(v[vCliq[i]].rct);
                        s.Format(L"%d",vCliq[i]);
                        dc.TextOutW(v[vCliq[i]].Home.x-5,v[vCliq[i]].
Home.y-5,s);
                        dc.TextOutW(End.x+30+50*(k-1),Home.y+200+i*20,s);
                }
        }
}
```

6.3 Application of Clustering for Social Networking

Clustering is a process of identifying the underlying structure of nonuniform data followed by grouping the data elements based on some similarity measure in which the resulting group is called a *cluster* [7]. Graph clustering is a process of grouping the nodes of a graph into clusters. In general, there are two broad categories for graph clustering: the first one is exhaustive approaches that seek to partition the whole graph by assigning each node to a defined cluster, and the second one is approaches that seek only the cohesive subgraph [8], which is where each node coheres with the others.

The cohesive subgraph can be found by computing the clique, where each clique represents a cluster. An interesting application of clustering using a clique is the social network, which can represent human interaction. A social network is a type of graph in which the nodes are known as actors while the edges are known as ties. The social network graph can be directed or undirected and can have weighted or unweighted edges. The individual units in the society can be represented by the actor. The actor is the entity that socially interacts with others, such as people in teams or organizations. The social relationship is modeled by ties, which can be marriage, friendship, collaboration, authority, and so forth.

Constructing a social network and analyzing its structure leads to socially significant interpretations. These interpretations are mainly concerned with the role of individuals as well as the nature of the interaction between the individuals in the social network. One of the analyses is to determine the most significant actors or the ones who exhibit the most important role. In an undirected network, the most important actor, known as the *central actor*, is the one with many ties. The degree of a node is an obvious measure of centrality.

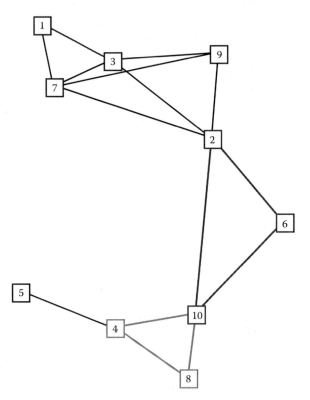

FIGURE 6.8
Interaction graph between undergraduate students in a discrete mathematics tutorial class.

For example, suppose that the actors in the social network shown in Figure 6.8 represent the undergraduate students in a discrete mathematics tutorial class. Two actors representing the undergraduate students have a tie between them if they are observed to be discussing the tutorial questions given to them during the tutorial session. The sets of actors $\{v_2, v_3, v_7, v_9\}$, $\{v_2, v_6, v_{10}\}$, and $\{v_4, v_8, v_{10}\}$ are cliques in this social network. Taking measure of centrality into account, the most central actor or the most important actor in the social network of discrete mathematics tutorial class in Figure 6.8 is v_2.

Another analysis in social networks is to determine the actor, who is also known as the *liaison*, whose removal would lead to disconnecting the social network graph. In Figure 6.8, v_2 is the liaison between $\{v_1, v_3, v_7, v_9\}$ and the rest of the actors while v_{10} is the liaison between $\{v_4, v_5, v_8\}$ and the rest of the actors. For example, if the ties in the same figure represent the lines of communication, in order for a message to reach v_5, v_4 or v_8 from one of the actors in $\{v_1, v_3, v_6, v_7, v_9\}$, the message must pass through v_2 and v_{10}.

Another interesting application is to study the interaction between academic researchers in a research center. A social network graph is used to represent this interaction, as shown in Figure 6.9a. The actors are the academic researchers who are involved in research. Instead of using names, the researchers are represented by numbers, where v_i represents researcher i. A tie between two actors exists if they are involved in the same research area. For example, v_{32} interact and share the same research area with v_{12}.

A group of academic researchers who share the same research domain and interact with each other based on the same research area form a cluster. These clusters, which are based

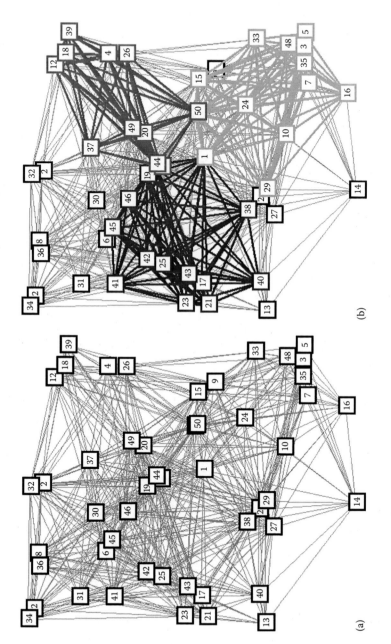

FIGURE 6.9
Interaction graph of academic researchers in the research center. (a) A social network graph and (b) research clusters formed by multicliques.

on a cohesive subgraph, are actually a concept of a clique in graph theory. As can be seen in Figure 6.9b, the social network graph has more than one clique. These multicliques form a group of research clusters in which Figure 6.9b highlights three research clusters. The highlighted research clusters are as follows:

$$\text{Research Cluster 1} = \{v_{40}, v_1, v_{17}, v_{19}, v_{21}, v_{23}, v_{25}, v_{38}, v_{41}, v_{42}, v_{43}, v_{44}, v_{45}, v_{46}\}$$

$$\text{Research Cluster 2} = \{v_5, v_1, v_3, v_7, v_{10}, v_{15}, v_{16}, v_{24}, v_{29}, v_{33}, v_{35}, v_{47}, v_{48}, v_{50}\}$$

$$\text{Research Cluster 3} = \{v_{39}, v_4, v_{11}, v_{12}, v_{18}, v_{20}, v_{26}, v_{37}, v_{44}, v_{49}, v_{50}\}$$

With this approach, the research center is able to determine the top research clusters based on the same research topic that the academic researchers are working on. At the same time, the center is able to analyze the roles of any researchers as well as the nature of interaction between them. For example, it can be seen that three researchers—1, 44, and 50—are involved in two different research clusters. This might be due to the nature of the multidisciplinary research that these three researchers are working on. The three researchers also can be regarded as a liaison, in that if any of them decided to exit the research center, the researcher social network would be disconnected. Therefore, an analysis of the researcher interaction in the form of a social network may provide useful insight for the management of the research center.

6.3.1 *Code6C*: Social Network Application

Code6C is the implementation of a greedy algorithm for computing the multicliques problem. The output of *Code6C* is shown in Figure 6.10, which displays 50 randomly scattered nodes inside a box. The multicliques are computed once the button *Compute* is clicked. The multicliques of the top three research clusters are displayed in the graph, with both its nodes and edges highlighted in different shades of grey. The total number of researchers that belong to the top three research clusters are displayed. At the same time, the academic researchers who are represented in numbers that belong to the cluster are displayed at the right side of the window. The interface also allows a new graph to be generated through the *Researcher Network* button.

Code6C has two source files, *Code6C.h* and *Code6C.cpp*. In *Code6C.h*, *nCliq* and *vCliq* represent the total number of nodes in the clique and the nodes belonging to the clique, respectively. The node and the edges of the graph are represented by the structures *EDGE* and *NODE*, respectively. There are also two button objects called *bNGraph* for *New Graph* and *bCompute* for *Compute*. Other important objects and variables are briefly described in Table 6.4.

The organization of *Code6C* is shown in Figure 6.11. The figure shows all the functions involved in the program. The processing starts with the constructor *Code6C()*, which draws the main window and the *New Graph* button. The constructor calls *Initialize()*, which creates a new graph with *N* nodes by assigning random coordinates and adjacency nodes to the edges. The whole graph is drawn inside the drawing area through *OnPaint()*. The computed nodes of the clique from *Research_Cluster()* are read inside *OnCompute()*, and the multicliques graph is then drawn.

The main engine is *Research_Cluster()*, which applies a greedy algorithm for computing the multicliques graph. The multicliques from the whole graph that represent the research clusters are displayed through *OnCompute()*. In computing the multicliques, *OnCompute()*

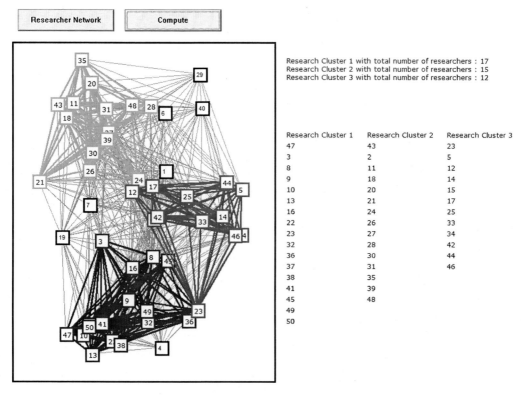

FIGURE 6.10
Output of *Code6C*.

calls *Research_Cluster*() to obtain the nodes belonging to the multicliques as well as the total number of nodes of the clique. A new graph with randomly located positions in Windows and randomly determined weights on the edges is produced by *Initialize*(). This graph is later drawn in Windows through *OnPaint*(). This function also serves to draw the node specified in its argument.

TABLE 6.4

Important Objects/Variables in *Code6C* and Their Descriptions

Variable	Type	Description
N	constant	Number of nodes in the graph
$e[i][j].adj$	*int*	Adjacency value between (v_i, v_j)
$vCliq[i]$	*int*	Nodes in the clique
$max[i]$	*int*	The number of nodes in each clique
$nCliq$	*int*	The number of nodes in the top selected multicliques
LinkeRange	constant	Range for adjacency between two nodes in the graph
$p[i]$	*int* array	Predecessor node to the currently marked node
Home, End	*CPoint*	Top-left and bottom-right corners of the drawing area
bNGraph	*CButton*	Button for generating a new graph
bCompute	*CButton*	Button for computing the maximum clique

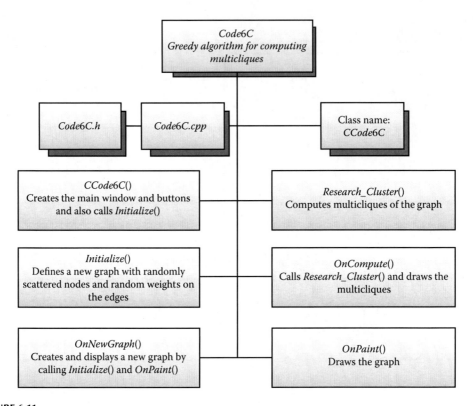

FIGURE 6.11
Organization of *Code6C*.

The codes in *Code6C* are given as follows:

```
//Code6C.h
#include <afxwin.h>
#include <fstream>
#define IDC_NGRAPH 500
#define IDC_COMPUTE 501
#define N 50
#define R 250

class CCode6C: public CFrameWnd
{
private:
        int vCliq[N+1],p[N+1];
        int max[N+1];
        int maxi,maximum,nCliq;
        CPoint Home,End;
        CFont fArial,fVerdana;
        CPen pColor[N+1];
        CButton bNGraph,bCompute;

        typedef struct
        {
                int adj;
        } EDGE;
        EDGE e[N+1][N+1];
```

```
            typedef struct
            {
                    CPoint Home;
                    CRect rct;
            } NODE;
            NODE v[N+1];
public:
            CCode6C();
            ~CCode6C()      {}
            void Research_Cluster(),Initialize();
            afx_msg void OnPaint(),OnNewGraph(),OnCompute();
            DECLARE_MESSAGE_MAP()
};

class CMyWinApp : public CWinApp
{
public:
            virtual BOOL InitInstance();
};
CMyWinApp MyApplication;

BOOL CMyWinApp::InitInstance()
{
            m_pMainWnd=new CCode6C;
            m_pMainWnd->ShowWindow(m_nCmdShow);
            return TRUE;
}

//Code6C.cpp
#include "Code6C.h"

BEGIN_MESSAGE_MAP(CCode6C,CFrameWnd)
        ON_WM_PAINT()
        ON_BN_CLICKED(IDC_NGRAPH,OnNewGraph)
        ON_BN_CLICKED(IDC_COMPUTE,OnCompute)
END_MESSAGE_MAP()

CCode6C::CCode6C()
{
        Home=CPoint(30,100); End=CPoint(500,700);
        Create(NULL,L"Computing the Top Research Clusters",
                WS_OVERLAPPEDWINDOW,CRect(0,0,1000,800),NULL);
        bNGraph.Create(L"Researcher Network",WS_CHILD | WS_VISIBLE |
BS_DEFPUSHBUTTON,
                CRect(CPoint(Home.x,30),CSize(180,40)),this,IDC_NGRAPH);
        bCompute.Create(L"Compute",WS_CHILD | WS_VISIBLE
                | BS_DEFPUSHBUTTON,CRect(CPoint(Home.x+200,30),CSize(180,40)),
                this,IDC_COMPUTE);
        fArial.CreatePointFont(80,L"Arial");
        fVerdana.CreatePointFont(100,L"Verdana");
        int Color[]={RGB(150,150,150),RGB(0,0,0),RGB(0,0,200),
                RGB(0,200,0),RGB(200,0,0),RGB(0,200,200),RGB(255,0,50)};
        for (int i=0;i<=6;i++)
                if (i==0)
                        pColor[i].CreatePen(PS_SOLID,1,Color[i]);
```

```
            else
                    pColor[i].CreatePen(PS_SOLID,3,Color[i]);
        Initialize();
}

void CCode6C::OnNewGraph()
{
        Initialize(); Invalidate();
}

void CCode6C::Initialize()
{
        int i,j;
        double distance;
        srand(time(0));
        for (i=1;i<=N;i++)
        {
                v[i].Home.x=Home.x+20+rand()%(End.x-Home.x-50);
                v[i].Home.y=Home.y+20+rand()%(End.y-Home.y-50);
                v[i].rct=CRect(CPoint(v[i].Home.x-10,v[i].Home.y-10),CSize
                    (25,25));
        }
        for (i=1;i<=N;i++)
        {
                e[i][i].adj=0;
                for (j=i+1;j<=N;j++)
                {
                        distance=sqrt(pow(double(v[i].Home.x-v[j].Home.x),2)
                                +pow(double(v[i].Home.y-v[j].Home.y),2));
                        e[i][j].adj=((distance<=R)?1:0);
                        e[j][i].adj=e[i][j].adj;
                }
        }
}

void CCode6C::OnPaint()
{
        CPaintDC dc(this);
        CString s;
        int i,j;
        dc.SelectObject(pColor[1]);
        dc.Rectangle(Home.x-10,Home.y-10,End.x+10,End.y+10);
        dc.SelectObject(pColor[0]);
        for (i=1;i<=N;i++)
                for (j=1;j<=N;j++)
                        if (e[i][j].adj)
                        {
                                dc.MoveTo(CPoint(v[i].Home));
                                dc.LineTo(CPoint(v[j].Home));
                        }
        dc.SelectObject(fArial); dc.SelectObject(pColor[1]);
        for (i=1;i<=N;i++)
        {
                dc.Rectangle(v[i].rct);
                s.Format(L"%d",i);
```

```
                    dc.TextOutW(v[i].Home.x-5,v[i].Home.y-5,s);
        }
}

void CCode6C::Research_Cluster()
{
        CClientDC dc(this);
        CString s;
        int i,j,r,t,k,maxi,a;
        for (i=1; i<=N; i++)
        {
                k=0;
                for (j=1; j<=N; j++)
                        if (e[i][j].adj==1)
                                {
                                        k++;
                                        p[k]=j;
                                }
                vCliq[1]=i;
                max[i]=1;
                a=1;
                for (r=1; r<k; r++)
                {
                        maxi=0;
                        for (t=r+1; t<=k; t++)
                                if (e[p[r]][p[t]].adj==1)
                                        maxi++;
                        if (maxi==(k-r))
                        {
                                a++;
                                vCliq[a]=p[r];
                                max[i]++;
                        }
                }
                a++;
                vCliq[a]=p[k];
                max[i]++;
        }

        maximum=max[1];
        for (int c=2;c<=N;c++)
                if (max[c]>=maximum)
                        maximum=max[c];
        for (int i=1;i<=N;i++)
                if (max[i]==maximum)
                {
                        k=0;
                        for(j=1; j<=N; j++)
                                if(e[i][j].adj==1)
                                {
                                        k++;
                                        p[k]=j;
                                }
                        vCliq[1]=i;
                        max[i]=1;
```

```
                        a=1;
                        for (r=1; r<k; r++)
                        {
                                maxi=0;
                                for (t=r+1; t<=k; t++)
                                        if (e[p[r]][p[t]].adj==1)
                                                maxi++;
                                if (maxi==(k-r))
                                {
                                        a++;
                                        vCliq[a]=p[r];
                                        max[i]++;
                                }
                        }
                        a++;
                        vCliq[a]=p[k];
                        max[i]++; nCliq=max[i];
                }
}
void CCode6C::OnCompute()
{
        CClientDC dc(this);
        CString s;
        int i,j,k;
        for (k=1;k<=3;k++)  //Compute and display 3 max cliques
        {

                Research_Cluster();
                dc.SelectObject(&pColor[k+1]);
                for (i=1;i<=nCliq;i++)       //draw the subgraph
                        for (j=i+1;j<=nCliq;j++)
                        {
                                dc.MoveTo(v[vCliq[i]].Home);
                                dc.LineTo(v[vCliq[j]].Home);
                                e[vCliq[i]][vCliq[j]].adj=0; e[vCliq[j]]
[vCliq[i]].adj=0;
                        }
                dc.SetTextColor(RGB(0,0,0)); dc.SelectObject(fVerdana);

                for (i=1;i<=nCliq;i++)
                {
                        s.Format(L"Research Cluster%d with total number of
researchers:%d",k, nCliq);
                        dc.TextOutW(Home.x+500,100+15*k,s);

                        s.Format(L"Research Cluster%d",k);
                        dc.TextOutW(End.x+30+150*(k-1),250,s);

                        dc.Rectangle(v[vCliq[i]].rct);
                        s.Format(L"%d",vCliq[i]);
                        dc.TextOutW(v[vCliq[i]].Home.x-5,v[vCliq[i]].
Home.y-5,s);
                        dc.TextOutW(End.x+30+150*(k-1),Home.y+150+i*20,s);
                }
        }
}
```

7

Triangulation Application

7.1 Convex Hull

The convex hull and Delaunay triangulation [1–5] are among the essential concepts in computational geometry and also in various scientific fields. The concepts certainly have an important relationship in which they are closely related to each other. This chapter will explain these essential concepts and discuss some algorithms to solve the problems of convex hull and Delaunay triangulation.

Polygons can be categorized into either convex or nonconvex (concave) polygons. A convex polygon is one where every point that lies in it can form a line segment in such a way that the line segment also must lie entirely within the polygon or on its boundary. Before studying the convex hull concept, it is useful to know what makes a polygon convex or nonconvex. In general, a polygon is a two-dimensional shape with only straight sides. Examples of polygons are triangles, rectangles, and pentagons. Since a polygon must have straight sides, a circle is not considered to be a polygon since it has a curved side.

Given that P is a polygon, it is a convex polygon if any point $p, q \in P$ forms a line segment $pq \subset P$. Another criterion of a polygon is that each of its interior angles is equal to or less then 180°. Figure 7.1a displays examples of convex polygons and Figure 7.1b displays examples of nonconvex polygons. Referring to (a), it can be seen that when a line segment is made from any two points in the convex polygons, it lies entirely in the polygon. In contrast, it can be seen in (b) that there are some points in the nonconvex polygons where, when they are connected to become a line segment, the line segment does not lie entirely inside the polygon.

It is useful to be able to differentiate between convex and nonconvex polygons in order to solve the problem of how to compute a convex hull. Given a set of points $P = \{v_i\}$ for $i = 1, 2, \ldots, n$, as shown in Figure 7.2a, the convex hull of P is the smallest convex polygon that contains all the points of P, as illustrated in Figure 7.2b. In other words, a convex hull can be imagined as the shape formed by a rubber band stretched around the points, keeping in mind that all the points must be within the rubber band and the shape of the rubber band must obey the criteria of convex polygons.

In order to compute the convex hull for a set of points, Figure 7.3a and b are used to illustrate the input and output of the process. For example, the set of the points as an input is $\{v_1, v_2, v_3, v_4, v_5, v_6, v_7, v_8, v_9, v_{10}\}$, as illustrated in Figure 7.3a. The output will be the convex hull representation where it will be only points $\{v_2, v_3, v_6, v_7, v_9\}$. Based on these five points, there will be five line segments or edges forming a convex hull.

The convex hull is one of the fundamental concepts in computational geometry. There are various practical applications for computing the convex hull, such as in image processing, robotics, pattern recognition, game theory, and a geographic information system.

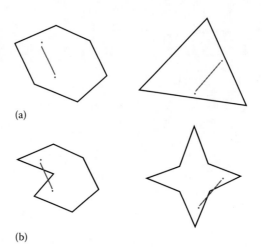

(a)

(b)

FIGURE 7.1
Examples of (a) convex and (b) nonconvex polygons.

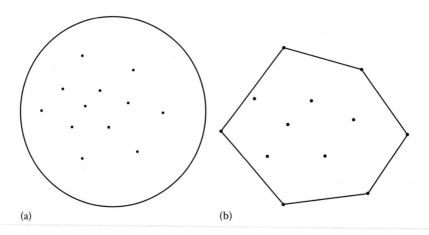

(a)   (b)

FIGURE 7.2
(a) A set of points P and (b) a convex hull for P.

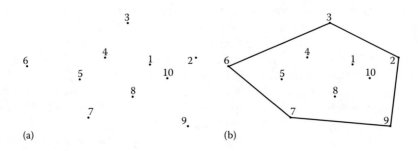

(a)   (b)

FIGURE 7.3
(a) Input and (b) output of the convex hull process.

Since this is a fundamental concept, many algorithms are being developed to compute the convex hull. Among these algorithms are brute force, gift wrapping, divide-and-conquer, and quickhull.

7.2 Algorithms for Computing the Convex Hull

7.2.1 Gift Wrapping Algorithm

The gift wrapping algorithm [2], also known as Jarvis March, is the simplest concept of a convex hull algorithm. It begins by identifying the leftmost point as the current point from which to start. In other words, the leftmost point is the point with the minimum value of the x-coordinate. The reason why the leftmost point is selected is because the point is absolutely one of the convex hull nodes. The leftmost point selected as the initial convex hull node is known as p. In order to find the next point, q, imagine that we are standing at the current point, p, and at the same time we are looking at the rest of the points simultaneously. The point that is the furthest to our right side is the point to be selected as the next convex hull node or as point q. Again, the same condition for finding the next point, r, or the next nodes for the convex hull when standing at point q is followed. Note that the selection point for the triplet p, q, and r must strictly follow a counterclockwise orientation.

Once the first triplet has been selected, then the next point p will be updated with the previous q while the next point q will be updated with the previous r. The next point r will be the point that is the furthest from the current point q, and again the triplet p, q, and r is following a counterclockwise orientation. This procedure must be repeated until the first convex hull node is reached. Figure 7.4 illustrates the execution of the gift wrapping algorithm to compute the convex hull.

The algorithm to find the convex hull using gift wrapping can be summarized as follows:

Given: Set P of all points in a two-dimensional domain
Problem: To find the convex hull of P
Initialization:
$p = \min x$;//the leftmost point
Select q;//the furthest point to the right side of current p
Select r;//the furthest point to the right side of current q
    //Orientation of (p, q, r) must be counterclockwise
Process:
While ($r \neq \min x$)
    Update $p = q$;
    Update $q = r$;
    Select r;
Endwhile
Output: The convex hull of P

7.2.2 Graham's Scan Algorithm

The bottom-most point among the set of points, which is the point with the minimum value of the y-coordinate needs to be determined. This can be achieved by sorting the

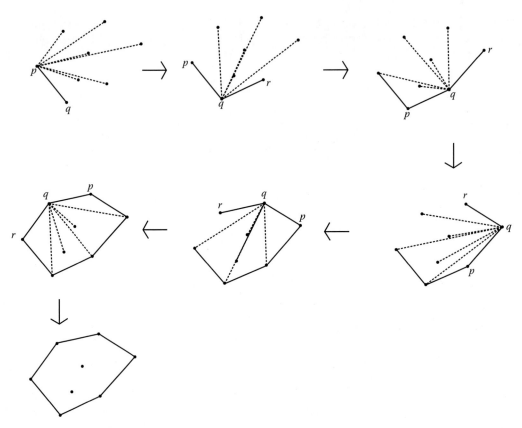

FIGURE 7.4
Execution of the gift wrapping algorithm to compute the convex hull.

points in ascending order of the y-coordinate. In a case where there are two points with the same value of y-coordinate, the one with the smaller value of x-coordinate will be selected. The selected bottom-most point will be the first point for the convex hull. The remaining points, $n-1$, are considered by sorting them in a polar angle following a counterclockwise direction around the chosen bottom-most point. Note that it is not efficient to compute the actual angle; therefore, another approach to compare the angle of the points around the bottom-most point is by using orientation.

Graham's scan algorithm [3] traverses the closed path and removes the concave points along the way. The first two nodes of the convex hull are the first two points in the sorted array. The algorithm determines the orientation of the recent combination of three points in order to reject or accept the third point. These points are denoted as previous point (p), current point (c), and next point (n). If the orientation of (p, c, n) is counterclockwise or making a left turn, then c is accepted. However, if the orientation is not counterclockwise or is making a right turn, then c needs to be rejected.

Figure 7.5a through f illustrates the execution of the Graham scan algorithm to compute a convex hull. A set of seven points is illustrated in (a). The bottom-most point as the anchor point is denoted as point p in (b). The blue line denotes the closed path formed by traversing all the points in increasing order of angle with respect to the chosen anchor point. In (c), the initial set of three points (p, c, n) is illustrated. The orientation of the points is counterclockwise or making a left turn; therefore, point c is accepted as the second vertex

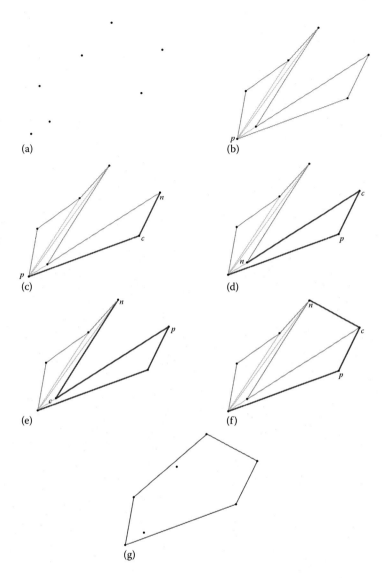

FIGURE 7.5
Execution of Graham's scan algorithm to find a convex hull. (a) The points, (b) p as anchor point, (c) initial (p, c, n), (d) update (p, c, n), (e) update (p, c, n) is not counterclockwise, (f) updated (p, c, n), and (g) the convex hull.

in the convex hull. The new set of (p, c, n) is updated and the orientation is determined, as illustrated in (d). Again, the orientation of the points is counteclockwise; therefore, point c is accepted as the third vertex in the convex hull.

The updated (p, c, n) is illustrated in (e) and it turns out that the orientation of the recent combination of points is not counterclockwise or making a right turn; therefore, point c is discarded. Hence, the previous points p and c are used but this time around, the next point following right after the discarded point, which is n, is used, as illustrated in (f). The orientation is counterclockwise; therefore, point c is accepted. The procedure is repeated until all the concave points are discarded, leaving only points that are selected as the convex hull. The resulting convex hull for these seven points is shown in (g).

Graham's scan algorithm [3] can be summarized as follows:

Given: Set $P = \{v_i\}$ of all points in a two-dimensional domain
Problem: To find the convex hull, S of P
Initialization:
Let $S = \{\phi\}$
Process:
$v_1 = \min y$ // the bottom-most point by comparing y-coordinates;
For $i = 1$ to n
   For $j = 1$ to $n - 1$
        Sort P in a polar angle around v_1;
   Endfor
Endfor
Add v_1, v_2 and v_3 into S
For $i = 4$ to n
   If add v_i is "left turn"
        Add v_i into S;
Endfor
Output: The convex hull S of P

7.2.3 Onion Peeling Algorithm [4]

Given a set of points $P = \{v_i\}$ for $i = 1,2,\ldots n$ on a plane, a convex hull of P is successfully computed using any available technique. The remaining points that are in the interior of the hull are denoted as P'. The procedure of computing the convex hull of P' is repeated again until there are no more points remaining for constructing convex hull in the interior. This is known as onion peeling of P, in which the procedure results in having a sequence or layers of nested convex hull. Figure 7.6a illustrates a set of 20 random points and Figure 7.6b illustrates the resulting onion peeling algorithm.

The algorithm of onion peeling with Graham's scan algorithm can be summarized as follows:

Given: Set $P = \{v_i\}$ for $i = 1,2,\ldots n$ of all points in a two-dimensional domain
Problem: To find the convex hull of P
Initialization:
Let $S_1 = \{\phi\}$
Process:
$v_1 = \min y$ // the bottom-most point by comparing y-coordinates
For $i = 1$ to n
   For $j = 1$ to $n - 1$
        Sort P in a polar angle around v_1;
   Endfor
Endfor
Add v_1, v_2 and v_3 into S_1
For $i = 4$ to n
   If add v_i is "left turn"
        Add v_i into S_1;
Endfor

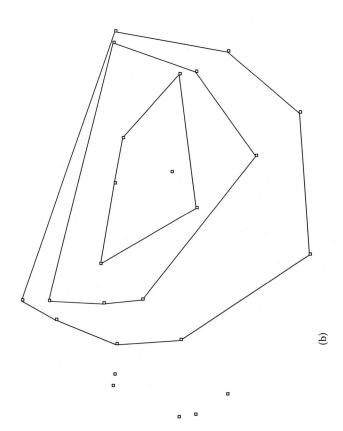

(b)

(a)

FIGURE 7.6
(a) A set of points *P* and (b) the resulting onion peeling algorithm.

$c = n - |S_1|$;
$k = 1$;
While $c > 2$
   k++;
   v_1 = min y // the bottom-most point among the remaining points by comparing
      y-coordinates;
   For $i = 1$ to n
        For $j = 1$ to $n - 1$
           Sort remaining P in a polar angle around v_1;
        Endfor
   Endfor
   Add v_1, v_2 and v_3 into S_k;
   For $i = 4$ to n
      If add v_i is "left turn"
        Add v_i into S_k; //v_i is the remaining point;
   Endfor
   Update $c = c - |S_k|$;
Endwhile
Output: The layers of convex hulls S_k of P

7.2.4 *Code7A*: Onion Peeling and Graham's Scan

Code7A is the implementation of Graham's scan and onion peeling algorithms for finding a convex hull. The output of *Code7A* is shown in Figure 7.7, which displays 30 randomly scattered nodes inside a box. It can be seen that there are three layers created with two remaining inner points in the figure. The convex hull is computed once the button *Compute* is clicked. The interface also allows a new graph to be generated through the *New Graph* button. *Code7A* first creates the outer layer as the first convex hull using Graham's scan algorithm. With the onion peeling algorithm, once a layer has been created, the next convex hull will be created for the remaining points. This procedure of creating the inner layers of the convex hull is repeated until the total number of final remaining inner points is equal to two or less.

    Code7A has two source files, *Code7A.h* and *Code7A.cpp*. In *Code7A.h*, n represents the number of points and the value is obtained from an edit box. The points are represented by the structure *Points*. They are also three button objects called *bCompute* for *Compute*, *bResult* for *Result*, and *bClear* for *Clear*. Other important objects and variables are briefly described in Table 7.1.

    The organization of *Code7A* is shown in Figure 7.8. The figure shows all the functions involved in the project. The processing starts with the constructor *CFP()*, which draws the main window and the *Compute*, *Result*, and *Clear* buttons. The constructor calls *OnPaint()*, which creates a set of random points and draws the points in specified small rectangles. The convex hull is computed and drawn through *OnResult()*. The points and the convex hull are cleared through *OnClear()*.

    The main engine is *OnResult()*, which applies the onion peeling algorithm and Graham's scan for computing the convex hull of a set of points. This function initially computes the outer layer of the convex hull. The anchor point is determined by means of the lowest y-coordinate point. The points are sorted in increasing polar angle and the first three

#Nodes (Please enter a number between 3 to 100)

| 30 | | Compute | | Result | | Clear |

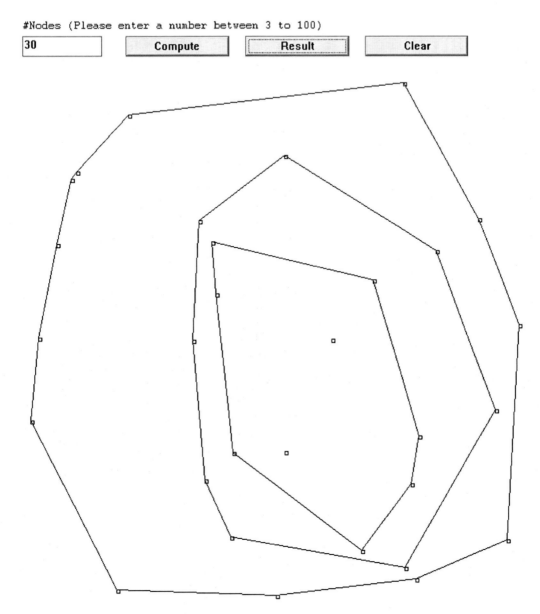

FIGURE 7.7
Output of *Code7A*.

points for the convex hull are determined. Among all the remaining points the concave points are disregarded if the orientation is turning right or is not counterclockwise. The points are accepted and stored as *que[i]* if the orientation is a left turn or following a counterclockwise direction. Another layer of the convex hull will be computed in case there are more than two remaining interior points. Every time a layer of convex hull is constructed, the number of remaining interior points will be updated. The procedure of creating the convex hull layers is repeated until the number of remaining points becomes two or less.

TABLE 7.1

Important Objects/Variables in *Code7A* and Their Descriptions

Variable	Type	Description
N	constant	Number of maximum points
n	int	Number of points
p[i].x	int	Coordinate x of the point
p[i].y	int	Coordinate y of the point
que[i]	int	Selected point for convex hull
bCompute	CButton	Button for generating random points
bResult	CButton	Button for computing convex hull
bClear	CButton	Button to clear the points

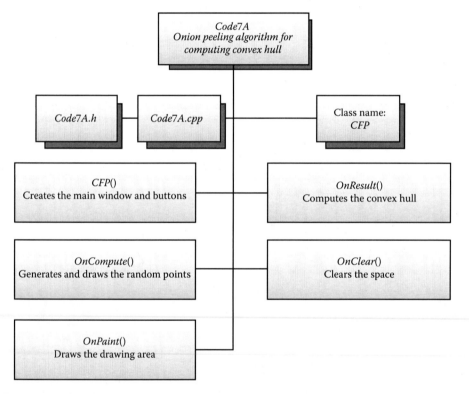

FIGURE 7.8
Organization of *Code7A*.

The following codes are the listings for *Code7A*:

```
//Code 7A.h
#include <afxwin.h>
#include <math.h>
#define N 100
#define R 100
```

```cpp
class CFP: public CFrameWnd
{
private:
       int n,top;
       typedef struct
       {
               int x;
               int y;
               CRect rct;
       }Point;
       Point p[N+1];
       Point que[N+1];

       CEdit eNodes;
       CButton bClear,bCompute,bResult;
public:
       CFP();
       ~CFP();
       afx_msg void OnPaint();
       afx_msg void OnClear();
       afx_msg void OnCompute();
       afx_msg void OnResult();
       DECLARE_MESSAGE_MAP()
};

class CMyWinApp:public CWinApp
{
       public:
               virtual BOOL InitInstance();
};
CMyWinApp MyApplication;

BOOL CMyWinApp::InitInstance(void)
{
       m_pMainWnd=new CFP;
       m_pMainWnd->ShowWindow(m_nCmdShow);
       return TRUE;
}

//Code 7A.cpp
#include "Code7A.h"
using namespace std;

BEGIN_MESSAGE_MAP(CFP,CFrameWnd)
       ON_WM_PAINT()
       ON_BN_CLICKED(20,OnCompute)
       ON_BN_CLICKED(21,OnResult)
       ON_BN_CLICKED(22,OnClear)
END_MESSAGE_MAP()
CFP::CFP()
{
       Create(NULL,L"Computation Geometry",WS_OVERLAPPEDWINDOW,CR
ect(0,0,900,900));
```

```
            eNodes.Create(WS_CHILD | WS_VISIBLE | WS_BORDER,
                    CRect(CPoint(120,40),CSize(100,25)),this,11);
            bCompute.Create(L"Compute",WS_CHILD | WS_VISIBLE |
BS_DEFPUSHBUTTON,
                    CRect(CPoint(250,40),CSize(130,25)),this,20);
            bResult.Create(L"Result",WS_CHILD | WS_VISIBLE | BS_DEFPUSHBUTTON,
                    CRect(CPoint(400,40),CSize(130,25)),this,21);
            bClear.Create(L"Clear",WS_CHILD | WS_VISIBLE | BS_DEFPUSHBUTTON,
                    CRect(CPoint(550,40),CSize(130,25)),this,22);
}
CFP::~CFP()
{
}
void CFP::OnPaint()
{
            CPaintDC dc(this);
            CFont F1;
            CRect rct1;
            F1.CreatePointFont(12,L"Courier");
            rct1=CRect(CPoint(30,10),CSize(800,800));
            dc.Rectangle(rct1);
            dc.SelectObject(F1);
            dc.TextOut(120,20,L"#Nodes (Please enter a number between 3 to
100)");
}
void CFP::OnCompute()
{
            CClientDC dc(this);
            int i;
            CFont F1;
            F1.CreatePointFont(50,L"Courier");
            dc.SelectObject(F1);
            CString s;
            eNodes.GetWindowText(s);
            n=_ttoi(s);
            srand(time(NULL));
            if(n>N || n<3)
            {
                        dc.TextOutW(100,100,L"invalid input");
            }
            else
            {
                    for(i=1;i<=n;i++)
                    {
                            p[i].x=90+rand()%660;
                            p[i].y=90+rand()%660;
                            p[i].rct=CRect(CPoint(p[i].x,p[i].y),CSize(5,5));
                            dc.Rectangle(p[i].rct);
                    }
            }
}
void CFP::OnResult()
```

```
{
      CClientDC dc(this);
      CString s;
      Point tmp;
      int i,j,k=0,a=1,b,c;
      for(j=1;j<=n;j++)
            for(i=1;i<=n-1;i++)
                  if((p[i].y < p[i+1].y) || ((p[i].y==p[i+1].y) &&
(p[i].x<p[i+1].x)))
                  {
                        tmp=p[i];
                        p[i]=p[i+1];
                        p[i+1]=tmp;
                  }
      for(j=1;j<=n;j++)
            for(i=1;i<=n-1;i++)
                  if((((p[i].x-p[1].x)*(-p[i+1].
y+p[1].y)-(p[i+1].x-p[1].x)
                        *(-p[i].y+p[1].y))<0) ||
                        (((p[i].x-p[1].x)*(-p[i+1].
y+p[1].y)-(p[i+1].x-p[1].x)
                        *(-p[i].y+p[1].y))<0) &&
                        (sqrt(double((p[1].x-p[i].x)*(p[1].x-p[i].x)+(p[1].y-
                        p[i].y)*(p[1].y-p[i].y)))<
                        sqrt(double((p[1].x-p[i+1].x)*(p[1].x-
p[i+1].x) + (p[1].y-
                        p[i+1].y)*(p[1].y-p[i+1].y)))))
                  {
                        tmp=p[i];
                        p[i]=p[i+1];
                        p[i+1]=tmp;
                  }
      top=0;
      que[++top]=p[1];
      que[++top]=p[2];
      que[++top]=p[3];//top=3
      for(i=4;i<=n;i++)
      {
            if((((p[i].x-que[top-1].x)*(-que[top].y+que[top-1].y)
            -(que[top].x-que[top-1].x)*(-p[i].y+que[top-1].y))>0))
            {
                  //turn right
                        que[top]=p[i];
                        top—;
                        i—;
            }
            else
            {
                        que[++top]=p[i];
            }
      }
      for(i=1; i<top;i++)
```

```
        {
                dc.MoveTo(que[i].x,que[i].y);
                dc.LineTo(que[i+1].x,que[i+1].y);
        }
dc.MoveTo(que[1].x,que[1].y);
dc.LineTo(que[top].x,que[top].y);
b=top;
c=n-top;
while(1)
{
        if(c>2)
        {
                for(i=1;i<=n;i++)
                        for(j=1;j<=b;j++)
                        {
                                if(p[i].x==que[j].x && p[i].y==
                        que[j].y)
                                {
                                        p[i].x=0;
                                        p[i].y=-1;
                                }
                        }
                for(j=1;j<=n;j++)
                        for(i=1;i<=n-1;i++)
                                if((p[i].y < p[i+1].y) || ((p[i].y==
                                p[i+1].y)
                                        && (p[i].x<p[i+1].x)))
                                {
                                        tmp=p[i];
                                        p[i]=p[i+1];
                                        p[i+1]=tmp;
                                }
                for(j=1;j<=c;j++)
                        for(i=1;i<=c-1;i++)
                                if((((p[i].x-p[1].x)*(-p[i+1].
                                        y+p[1].y)-(p[i+1].x-p[1].x)
                                                *(-p[i].y+p[1].y))<0) ||
                                                (((p[i].x-p[1].x)*(-p[i+1].
                                y+p[1].y)-(p[i+1].x-p[1].x)
                                                *(-p[i].y+p[1].y))<0) &&
                                                (sqrt(double((p[1].x-p[i].x)*
                                                (p[1].x-p[i].x)
                                                + (p[1].y-p[i].y)*(p[1].y-p
                                                [i].y)))<
                                                sqrt(double((p[1].x-p[i+1].x)*
                                                (p[1].x-p[i+1].x)
                                                + (p[1].y-p[i+1].y)*(p[1].y-p
                                                [i+1].y)))))
                                {
                                        tmp=p[i];
                                        p[i]=p[i+1];
                                        p[i+1]=tmp;
                                }
```

```
                              top=0;
                              que[++top]=p[1];
                              que[++top]=p[2];
                              que[++top]=p[3];//top=3
                              for(i=4;i<=c;i++)
                              {
                                      if((((p[i].x-que[top-1].x)*(-que[top].
                                             y+que[top-1].y)
                                         -(que[top].x-que[top-1].x)*(-p[i].
                                             y+que[top-1].y))>0))
                                      {
                                             //turn right
                                             que[top]=p[i];
                                             top-;
                                             i-;
                                      }
                                      else
                                      {
                                             que[++top]=p[i];
                                      }
                              }
                              for(i=1; i<top;i++)
                              {
                                      dc.MoveTo(que[i].x,que[i].y);
                                      dc.LineTo(que[i+1].x,que[i+1].y);
                              }
                              dc.MoveTo(que[1].x,que[1].y);
                              dc.LineTo(que[top].x,que[top].y);
                              b=top;
                              c=c-top;
                      }
                      else
                              break;
              }
}
void CFP::OnClear()
{
      CClientDC dc(this);
      CRect rc;
      InvalidateRect(&rc);
      Invalidate();
}
```

7.3 Delaunay Triangulation

In computational geometry, a tessellation or tiling of a domain into a set of connected polygons is an important procedure. This procedure is called mesh generation, which can be categorized according to the dimensionality and its elements. For example, triangular meshes and quadrilateral meshes are both two-dimensional meshes with triangle (three-sided polygon) and quadrilateral (four-sided polygon) elements, respectively. Tetrahedral meshes and hexahedral meshes are both three-dimensional meshes with four triangular faces and six quadrilateral faces, respectively. Mesh generation has been a significant procedure for various applications in various domains including graphics, modeling, and engineering. Among the famous applications of mesh generation are simulating various physical phenomena such as heat transfer, fluid flow, electromagnetic wave propagation, and mechanical deformation. Figure 7.9 illustrates an example of triangular mesh generation on a set of random points.

Meshes can also be categorized as structured or unstructured. Structured meshes often follow a uniform pattern and can be identified by a regular connectivity. In contrast, unstructured meshes do not follow a uniform pattern and can be identified by irregular connectivity. Unstructured meshes are becoming predominant because of their ability to model complex geometry. Triangular meshes are often chosen for two-dimensional modeling, since it has been proven that a triangle is the most flexible form for complex geometries. Therefore, this section only discusses two-dimensional unstructured triangulation mesh generation.

The requirement for a successful mesh application is that it must conform to the shape of the object or domain being modeled. Moreover, it is preferred that the triangular element itself is an optimal or right shape. In other words, it is desirable for the shape of the triangular element to be very close to an equilateral triangle. This can be achieved by

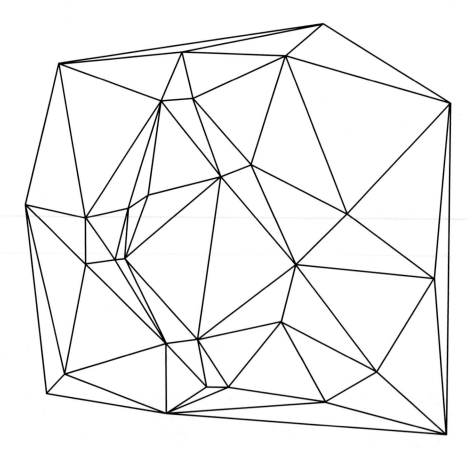

FIGURE 7.9
Mesh generation for a set of random points.

having the minimum angle in the triangle as large as possible. Note that elements with a long, stretched, or skinny shape must be avoided for mesh generation. However, in some special cases, such as in the application of anisotropic simulation, elements with length and stretch are needed at some predefined area of computational domain.

7.3.1 Triangulating a Set of Points

The triangulation of a set of nodes in a plane can be defined as follows:

Let V become a finite set of nodes in a plane. A simplicial complex T is a triangulation of V such that

1. V is the set of nodes in T
2. The union of all the simplices in T is the convex hull of V

Figure 7.10 illustrates an example of triangulation of a set of nodes in a plane. The figure shows a total of 10 points where these points are the set of nodes in T such as $V = \{v_i\}$ for $i = 1, 2, ..10$. The total number of triangular elements in this example is 11 and the total number of edges is 20. Each triangular element has three nodes connected to each other as follows:

Element 1 = $\{v_1, v_3, v_{10}\}$	Element 2 = $\{v_1, v_4, v_{10}\}$	Element 3 = $\{v_1, v_4, v_8\}$
Element 4 = $\{v_4, v_8, v_9\}$	Element 5 = $\{v_2, v_4, v_9\}$	Element 6 = $\{v_2, v_4, v_{10}\}$
Element 7 = $\{v_2, v_5, v_{10}\}$	Element 8 = $\{v_3, v_5, v_{10}\}$	Element 9 = $\{v_2, v_5, v_6\}$
Element 10 = $\{v_2, v_6, v_7\}$	Element 11 = $\{v_2, v_7, v_9\}$	

One of the well-known approaches to having an optimal mesh triangulation is Delaunay triangulation [1], created by Nikolaevich Delone, a Russian mathematician. He discovered the regular partitioning of the domain and the theory of Dirichlet partitioning or Voronoi diagram. One important criterion of Delaunay triangulation is the circumcircle properties in which it strictly sets that the nodes of a triangular element lie exactly in a circle and at the same time there will be no other nodes or points that lie in the same circle.

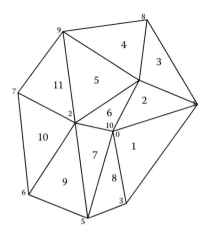

FIGURE 7.10
Triangulation of 10 elements for 11 random nodes.

These criteria ensure the optimal mesh triangulation as it maximizes the minimum angle of the triangular element. The Delaunay triangulation of P, $DT(P)$ can be defined as follows:

1. Three points, v_i, v_j, $v_k \in P$, are the set of nodes for a Delaunay triangulation element if the three nodes lie exactly at the circumcircle and the circle does not contain any other nodes. This circle is called the circumcircle of the triangle and can be defined by (v_i, v_j, v_k).
2. The edge of the Delaunay triangulation is formed by the two points, v_i, $v_j \in P$, if the circle contains the two points on its boundary and does not have any other points inside it.

Figure 7.11a and b illustrates four points. In (a), each of the triangles has a unique circumcircle in which only three nodes lie exactly on the circumcircle. The circumcircle can be defined as (v_1, v_2, v_3) and (v_1, v_2, v_4). It can be seen that no other nodes lie inside the circumcircle. Therefore, the triangles satisfy the circumcircle property; hence, Figure 7.11a shows the Delaunay triangulation. In contrast, if the connectivity of the same points is changed,

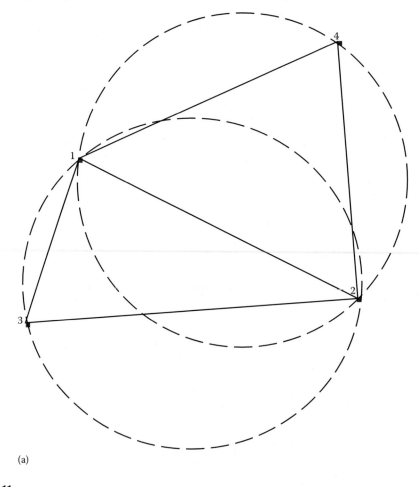

(a)

FIGURE 7.11
(a) Triangular elements that satisfy the circumcircle property.   *(Continued)*

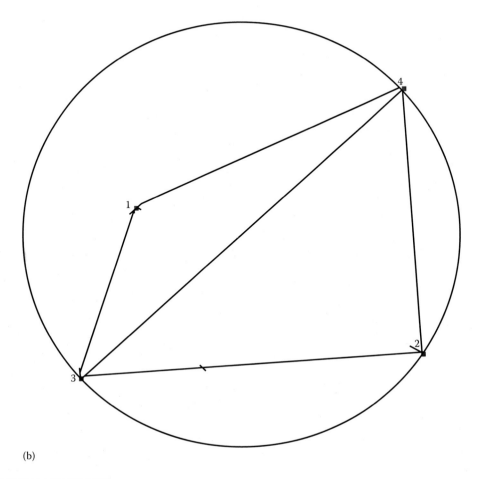

(b)

FIGURE 7.11 (CONTINUED)
(b) Triangular elements that do not satisfy the circumcircle property.

as shown in Figure 7.11b, it can be seen that v_1 lies inside the circumcircle of triangle (v_2, v_3, v_4). The triangles do not satisfy the circumcircle property; therefore, Figure 7.11b does not illustrates the Delaunay triangulation.

In general, the algorithm to construct a Delaunay triangulation can be classified into three classic types, as follows:

1. Gift wrapping
2. Divide-and-conquer
3. Incremental insertion algorithm

The algorithm of gift wrapping constructs a Delaunay triangulation one by one iteratively. New triangles are formed by using the previous computed triangular element as a seed. Gift wrapping is easy to implement; however, it is not that easy to make it fast. The divide-and-conquer algorithm splits the set of points into two halves using a line. The Delaunay triangulation is then constructed in each subset recursively. The two subsets of the Delaunay triangulation are then combined together into one. This algorithm is fast in two-dimensional domains but not in three-dimensional domains. The final one, which is the incremental insertion algorithm for Delaunay triangulation, decides the insertion of nodes one at a time.

7.3.2 Delaunay's Gift Wrapping Algorithm

The gift wrapping algorithm for Delaunay triangulation forms a triangle by adjoining specified edges following a very simple procedure. It start with an oriented edge, let us say $e = v_1v_2$, where v_1 and v_2 are two nodes. During the execution of the algorithm, a circumcircle that contains only e and no other points will be formed. The circumcircle will keep on deforming by expanding in front of e and at the same time shrinking behind e in order to find the node by which to construct a triangle in front of the oriented edge. During the execution of the gift wrapping procedure, the specified oriented edge is classified as unfinished if a triangle in front of the edge is not yet identified. The oriented edge will only be classified as finished once the triangle in front of the edge is constructed or there is no way to construct a triangle since the oriented edge itself is a boundary edge. Figure 7.12 shows the execution of the gift wrapping algorithm to construct a Delaunay triangulation. Note that as the circumcircle keeps on deforming in order to find the node, the center of all the circumcircles lies on the bisector of the oriented edge.

The algorithm of gift wrapping can be summarized as follows:

Given: Set $V = \{v_i\}$ for $i = 1, 2,..n$ of all points in a two-dimensional domain
Problem: To find Delaunay triangulation
Initialization:
Construct $e = v_iv_j//e$ is oriented edge;
Construct circumcircle containing e;
Compute bisector of e;
Process:
While (e is unfinished)
   For $k = 1$ to n
      Deformed circumcircle until it touches any available node, v_k in front of e;
      A triangle $v_iv_jv_k$ will be constructed and the current e will be identified as finished;
Endwhile
Output: A set of Delaunay triangulation, T.

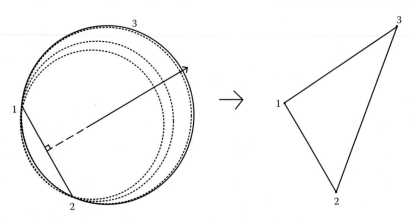

FIGURE 7.12
Execution of the gift wrapping algorithm.

7.3.3 Bowyer-Watson's Algorithm

A very efficient and convenient method by which to implement Delaunay triangulation is the Bowyer-Watson [5] algorithm. For example, if there are existing points forming a Delaunay triangulation, it is necessary to insert further points one by one iteratively. With this method, the iterative procedure is initiated by having a large equilateral triangle that can contain all the points for triangulation. This means that the initial triangulation can be formed by only three nodes as initial points. Once all the points are incorporated and completely triangulated, then the initial triangle containing the initial three points will be deleted.

For example, there is a set of points, $v_1, v_2, v_3...v_n$, which has already formed a Delaunay triangulation. The union of the triangle in the existing Delaunay triangulation is denoted as T_i. Since the triangulation is in the form of a Delaunay triangulation, hence there should be a circumcircle for each triangle. The union of the entire existing circumcircles of the triangles is denoted as B_i. The intention now is to insert a new point, v_{i+1}, into the existing triangulation. According to the Bowyer-Watson algorithm, when inserting the new points, there are three different conditions that need to be considered. The first condition is when $v_{i+1} \in T_i$, the second condition is $v_{i+1} \notin T_i$ but $v_{i+1} \in B_i$, and finally, the third condition is $v_{i+1} \notin B_i$.

Case 1: The New Point Is in the Existing Triangulation

The first thing that needs to be determined is whether the new inserted point lies in the existing triangle. If it lies in the existing triangle, then it is necessary to determine all the circumcircles that contain the new points. The union of triangles that correspond to the circumcircles detected with the new point is classified as set S. The internal edge in S is then removed, which will create a temporary hole. Every vertex of S will be connected to the new point. A new Delaunay triangulation, T_{i+1}, will be produced with this process.

An example of a Delaunay triangulation is shown in Figure 7.13a. There are five existing points, $v_1, v_2...v_5$, and it can be seen that the new inserted point, v_6, is lying in one of the

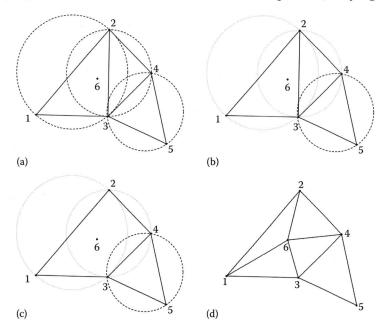

(a) (b)

(c) (d)

FIGURE 7.13

Execution of case 1 in the Bowyer-Watson insertion strategy. (a) v_6 lies in existing triangle, (b) the circumcircles that contain v_6, (c) internal edge in S is deleted, and (d) all vertices of S are connected to v_6.

triangles. It is obvious that the new point is within two circumcircles, as highlighted in (b). The union of the triangles corresponding to the two circumcircles is called S with nodes v_1, v_2, v_3, v_4. The edge that is connected by v_2 and v_3 is the only internal edge in S and it will be deleted, hence creating a temporary hole in the tessellation, as illustrated in (c). The new edges are then created by connecting all the nodes in S to the new inserted point, as shown in (d).

Case 2: The New Point Is Not in the Triangle but Is within at Least One Circumcircle
The second case is if the new inserted point is not in the triangle but is located in at least one circumcircle. Note that a similar definition of set S in the previous case is still used here. The edge in S that is the nearest to (and visible from) v_{i+1} will be removed. All the nodes of S as well as any existing nodes of T_i (which is visible from v_{i+1}) will be connected to v_{i+1} in order to form a new Delaunay triangulation.

An example of Delaunay triangulation with five points, v_1, v_2...v_5, is shown in Figure 7.14a. It can be seen that the new point, v_6, is outside of the triangles but is lying in the circumcircle of triangle v_1, v_2, v_3. Therefore, the set S for this example is the one formed by nodes v_1, v_2, v_3. It can be seen that the only edge in S connected by v_1 and v_3 is the one that is nearest and visible to v_6. Hence, this edge will be deleted, as illustrated in (b). Finally, all the vertices of S will be connected to v_6, and at the same time, the existing point in the triangulation that is visible from v_6, which is v_5, will also be connected to v_6, as illustrated in (c).

Note that with the Bowyer-Watson algorithm, the resulting Delaunay triangulation is maintained every time new points are inserted. In other words, the triangulation still satisfies the circumcircle property.

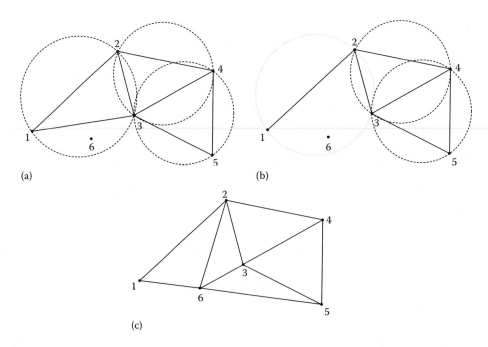

(a)   (b)

(c)

FIGURE 7.14
Execution of case 2 in the Bowyer-Watson insertion strategy. (a) v_6 is outside of triangles but lying in circumcircle of triangle v_1, v_2, v_3, (b) the nearest and visible edge to v6 is deleted, and (c) v_1, v_2, v_3 and v_5 are connected to v_6.

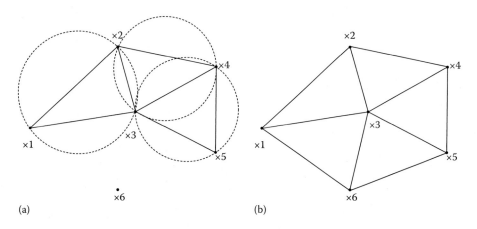

FIGURE 7.15
Execution of case 3 in the Bowyer-Watson insertion strategy. (a) v_6 is lying outside the triangulation and circumcircles and (b) v_1, v_3, and v_5 are connected to v_6.

Case 3: The New Point Is Lying outside Both the Triangulation and Circumcircle

The final case for the Bowyer-Watson algorithm is when the new inserted point, v_{i+1}, is lying outside the triangulation and also outside all the circumcircles. There is no removal of the existing edge in this case; however, it is necessary to locate the external edge of T_i that is visible to the new point. The vertices that are associated with the external edge will be connected to the new point to form a Delaunay triangulation.

An example of Delaunay triangulation with five points, v_1, v_2...v_5, is shown in Figure 7.15a. From the figure, it can be seen that the new point, v_6, is lying outside the Delaunay triangulation as well as outside all the circumcircles. It is also obvious that the external edges of the Delaunay triangulation that are visible to v_6 are the edge connecting v_1 to v_3 and the edge connecting v_3 to v_5. In order to form the Delaunay triangulation, all nodes of these edges, v_1, v_3 and v_5, will be connected to v_6, as illustrated in (b).

The Bowyer-Watson algorithm can be summarized as follows:

> **Given:** Set of nodes $V = \{v_i\}$, and its existing triangulation, T_{i-2} for $i = 1, 2,..n$ in a two-dimensional domain.
>
> **Problem:** To form Delaunay triangulation using Bowyer-Watson.
>
> **Process:**
> Define T_i and B_i;
> Determine the condition of the new node inserted, v_{i+1};
> If $v_{i+1} \in T_i$
>   Define S;
>   Remove all internal edges in S;
>   Connect the nodes of S to v_{i+1};
> If $v_{i+1} \notin T_i$ and $v_{i+1} \in B_i$
>   Define S;
>   Determine the edge in S that is nearest and visible to v_{i+1};
>   Remove the edge;
>   Connect the nodes in S and nodes in T_i that are visible to v_{i+1};
> If $v_{i+1} \notin B_i$
>   Determine the external edges of T_i that are visible to v_{i+1};
>   Connect the nodes of the external edges to v_{i+1};

Update T_i, B_i, and v_i;
Output: Delaunay triangulation, T.

7.3.4 *Code7B*: Implementing Delaunay's Gift Wrapping Algorithm

Code7B is the implementation of the gift wrapping algorithm for computing a Delaunay triangulation problem. The output of *Code7B* is shown in Figure 7.16, which displays 30 randomly scattered nodes inside a box. The coordinate of each node is displayed at the right side of the window. One edit box for collecting the number of nodes serves as input. The Delaunay triangulation is computed once the *Start* button is clicked. The interface also allows a new set of nodes to be generated through the *Reset* button.

Code7B has two source files, *Code7B.h* and *Code7B.cpp*. In *Code7B.h*, *n* represents the number of nodes specified by the user and the value is obtained from the edit box. The nodes and triangular element of the triangulation are represented by the structure *PT* and *TRIANGLE*. There are also two button objects called *Start* and *Reset*. Other important objects and variables are briefly described in Table 7.2.

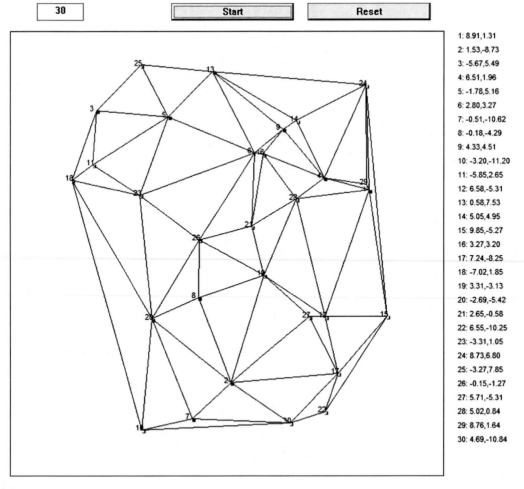

FIGURE 7.16
Output of *Code7B*.

TABLE 7.2

Important Objects/Variables in *Code7B* and Their Descriptions

Variable	Type	Description
N	constant	Maximum number of nodes
n	int	Number of nodes
P[i]	int	Node i
T[i]	int	Triangular element i
Home, End	CPoint	Top-left and bottom-right corners of the drawing area
Start	CButton	Button for generating Delaunay triangulation
Reset	CButton	Button for reset

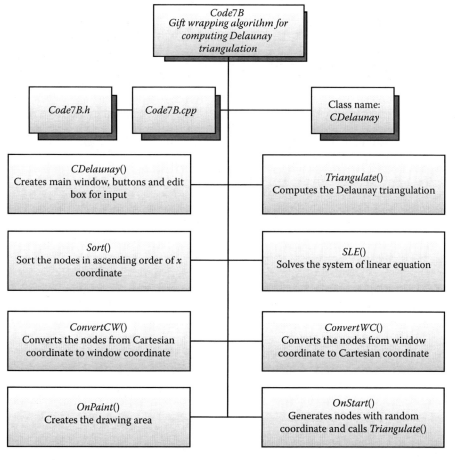

FIGURE 7.17
Organization of *Code7B*.

The organization of *Code7B* is shown in Figure 7.17. The figure shows all the functions involved in the program. The processing starts with the constructor *CDelaunay()*, which draws the main window and the *Start* and *Reset* buttons. The number of nodes, *n*, are read from the edit box *eN*. The edit box is created inside *CDelaunay()* and its values are read inside *OnStart()*, which then calls *Triangulate()* to compute and draw the Delaunay triangulation.

The main engine is *Triangulate()*, which applies the gift wrapping algorithm to compute the Delaunay triangulation. In this function, the combination of the three points for a triangle is first generated in the first nested loop. Note that the triangle with the combination points will initially be termed to be valid since it is only a temporary triangle before the circumcircle property is determined. This is achieved by having the following flag:

$$T[r].flag = 1.$$

The initial combination of the three points for the temporary triangle is then sorted in ascending order of *x*-coordinates in *Sort()*. Once the points have been sorted in ascending order of *x*-coordinates, both the center and the radius of the circumcircle of the temporary triangle are computed using the system of linear equations. The function then calls *SLE()* to solve the system of linear equations. In order to solve the system of linear equations, the square matrix is reduced to its upper triangular form followed by finding the inverse of the matrix. The augmented matrix $A|b$ is reduced to $U|v$ using row operation and finally the backward substitutions. By solving the system of linear equations, the *x*- and *y*-coordinates of the center of the circumcircle for the temporary triangle will be obtained. At the same time, the radius of the circumcircle of the temporary triangle needs to be computed.

Once the center and the radius of the circumcircle of the temporary triangle have been obtained, it is necessary to figure out whether there are any other nodes lying within or at the circumcircle. This is actually the core part of the Delaunay triangulation, since every triangle in the Delaunay mesh must satisfy the circumcircle property, as explained in Section 7.3.1. In order to determine whether each point is lying inside the circumcircle of the temporary triangle, it is necessary to compute the distance between the points and the center of the circumcircle. If the distance is equal to or less than the radius of the circumcircle, then the points are definitely lying within or at the circumcircle. If the distance is equal or less, it means that the circumcircle property is being violated and the temporary triangle is not valid anymore; hence, the value of triangle validity becomes $T[r].flag = 0$. However, if all the points are determined not to be in the circumcircle of the temporary triangle, which means the circumcircle property is satisfied, then the temporary triangle becomes a valid triangular element. Therefore, it is necessary to generate the lines or edges in order to connect the three points as a triangular element.

A new set of points with randomly located positions in windows is produced by *OnStart()*. Note that the codes first generate points based on the Window coordinates; therefore, when it comes to computing the triangles, these Windows coordinates need to be transformed into Cartesian coordinates using the linear relationship $Y = m_1x + c_1$, which is done in *ConvertWC()*. A triangle is computed in Cartesian coordinates; therefore, when it comes to displaying the triangle in the window, each node of the triangle needs to be converted into Windows coordinates by using *ConvertCW()*. Another function, *OnPaint()*, serves to draw the window in its argument.

The codes for *Code7B* are given as follows:

```
//Code7B.h
#include <afxwin.h>
#include <math.h>
#include <time.h>
#define IDC_START 500
#define N 100
typedef struct            //Cartesian points
```

```
{
      double x,y;
} PT;
typedef struct
{
      PT a,b,c;
      PT center;
      double radius;
      bool flag;
} TRIANGLE;
class CCode7B: public CFrameWnd
{
private:
      int idc,n;
      CFont fArial;
      CEdit eN;
      CButton Start,Reset;
      CPoint WHome,WEnd;   //Windows home, end
      PT P[N+1];
      TRIANGLE T[N*N+1];
      PT CHome,CEnd;        //CHome,CEnd=Cartesian home,end
public:
      CCode7B();
      ~CCode7B()                {}
      void Triangulate(),sort(int,int,int,int),SLE(int);
      PT ConvertWC(CPoint);
      CPoint ConvertCW(PT);
      afx_msg void OnStart();
      afx_msg void OnPaint();
      DECLARE_MESSAGE_MAP();
};
class CMyWinApp: public CWinApp
{
public:
      virtual BOOL InitInstance();
};
CMyWinApp MyApplication;

BOOL CMyWinApp::InitInstance()
{
      m_pMainWnd=new CCode7B;
      m_pMainWnd->ShowWindow(m_nCmdShow);
      return TRUE;
}
//Code7B.cpp
#include "Code7B.h"
#define PI 3.142
BEGIN_MESSAGE_MAP(CCode7B,CFrameWnd)
      ON_BN_CLICKED(IDC_START,OnStart)
      ON_WM_PAINT()
END_MESSAGE_MAP()
      CCode7B::CCode7B()
      {
      idc=400;
```

```
        Create(NULL,L"Delaunay Triangulation",WS_OVERLAPPEDWINDOW,CRect
(0,0,1200,800));
        WHome=CPoint(50,50); WEnd=CPoint(600,600);
        CHome.x=-10; CHome.y=10; CEnd.x=10; CEnd.y=-10;
        eN.Create(WS_CHILD | WS_VISIBLE | WS_BORDER | SS_CENTER,
                CRect(CPoint(WHome.x+140,WHome.y-40),CSize(70,25)),
                this,idc++);
        Start.Create(L"Generate",WS_CHILD | WS_VISIBLE | BS_DEFPUSHBUTTON,
                CRect(CPoint(WHome.x+250,WHome.y-40),CSize(150,25)),
                this,IDC_START);
        fArial.CreatePointFont(80,L"Arial");
}
void CCode7B::OnPaint()
{
        CPaintDC dc(this);
        CPen pBlack(PS_SOLID,2,RGB(0,0,0));
        dc.SelectObject(pBlack);
        CRect rct=CRect(WHome.x,WHome.y,WEnd.x,WEnd.y);
        dc.Rectangle(rct);
        dc.SelectObject(fArial);
        dc.TextOutW(WHome.x,WHome.y-35,L"#Nodes (3-20, Default: 15)");
}
void CCode7B::OnStart()
{
        CString s;
        CClientDC dc(this);
        CRect rct;
        int i,rx,ry;
        CPoint Q;
        CBrush bkBrush(RGB(255,255,255));
        rct=CRect(WHome.x+2,WHome.y+2,WEnd.x-2,WEnd.y-2);
        dc.FillRect(rct,&bkBrush);

        eN.GetWindowText(s); n=_ttoi(s);
        n=((n<3 || n>20)?15:n);
        srand(time(0));
        dc.SelectObject(fArial);
        for (i=1;i<=n;i++)
        {
                rx=WEnd.x-WHome.x-30; ry=WEnd.y-WHome.y-30;
                Q.x=WHome.x+15+rand()%rx; Q.y=WHome.y+15+rand()%ry; P[i]
                =ConvertWC(Q);
                rct=CRect(Q.x,Q.y,Q.x+5,Q.y+5);
                dc.FillSolidRect(rct,RGB(100,100,100)); dc.SetBkColor
                        (RGB(255,255,255));
                s.Format(L"%d",i); dc.TextOutW(Q.x-10,Q.y-10,s);
                s.Format(L"%d:%.2lf,%.2lf",i,P[i].x,P[i].y);
                dc.TextOutW(WEnd.x+100,WHome.y+10+20*(i-1),s);
        }
        Triangulate();
}
void CCode7B::Triangulate()
{
        CClientDC dc(this);
```

```
            CPoint Q;
            double rDistance;
            int i,j,k,ii,r=1;
            for (i=1;i<=n;i++)
                    for (j=i+1;j<=n;j++)
                            for (k=j+1;k<=n;k++)
                            {
                                    T[r].flag=1;
                                    sort(r,i,j,k);
                                    SLE(r);
                                    for (ii=1;ii<=n;ii++)
                                            if (ii!=i && (ii!=j && ii!=k))
                                            {
                                                    rDistance=sqrt(pow(T[r].center.x-
P[ii].x,2)+pow(T[r].center.y-P[ii].y,2));
                                                    if (rDistance<=T[r].radius)
                                                    {
                                                            T[r].flag=0; break;
                                                    }
                                            }
                                    if (T[r].flag)
                                    {
                                            Q=ConvertCW(T[r].a); dc.MoveTo(Q);
                                            Q=ConvertCW(T[r].b); dc.LineTo(Q);
                                            Q=ConvertCW(T[r].c); dc.LineTo(Q);
                                            Q=ConvertCW(T[r].a); dc.LineTo(Q);
                                    }
                                    r++;
                            }
}
void CCode7B::sort(int r,int u,int v,int z)
{
        double tmp1,w[4];
        int s[4],tmp2,i,k;
        s[1]=u; w[1]=P[u].x;
        s[2]=v; w[2]=P[v].x;
        s[3]=z; w[3]=P[z].x;
        for (k=1;k<=3;k++)
                for (i=1;i<=2;i++)
                        if (w[i]>=w[i+1])
                        {
                                tmp1=w[i]; w[i]=w[i+1]; w[i+1]=tmp1;
                                tmp2=s[i]; s[i]=s[i+1]; s[i+1]=tmp2;
                        }
        T[r].a=P[s[1]]; T[r].b=P[s[2]]; T[r].c=P[s[3]];
}
void CCode7B::SLE(int r)
{
        int i,j,k,nsle=2;
        double a[3][3],b[3],xsle[3],Sum,p;
        a[1][1]=2*(T[r].a.x-T[r].b.x); a[1][2]=2*(T[r].a.y-T[r].b.y);
        a[2][1]=2*(T[r].a.x-T[r].c.x); a[2][2]=2*(T[r].a.y-T[r].c.y);
        b[1]=pow(T[r].a.x,2)+pow(T[r].a.y,2)-pow(T[r].b.x,2)-pow
```

```
        (T[r].b.y,2);
        b[2]=pow(T[r].a.x,2)+pow(T[r].a.y,2)-pow(T[r].c.x,2)-pow
        (T[r].c.y,2);
        for (k=1;k<=nsle-1;k++)
                for (i=k+1;i<=nsle;i++)
                {
                        p=a[i][k]/a[k][k];
                        for (j=1;j<=nsle;j++)
                                a[i][j]-=p*a[k][j];
                        b[i] -=p*b[k];
                }
for (i=nsle;i>=1;i--)
{
        Sum=0;
                xsle[i]=0;
                for (j=i;j<=nsle;j++)
        Sum +=a[i][j]*xsle[j];
                xsle[i]=(b[i]-Sum)/a[i][i];
}
        T[r].center.x=xsle[1]; T[r].center.y=xsle[2];
        T[r].radius=sqrt(pow(T[r].center.x-T[r].a.x,2)+pow(T[r].
        center.y-T[r].a.y,2));
}
PT CCode7B::ConvertWC(CPoint p)           //converts W to C
{
        double m1,c1,m2,c2;
        PT q;
        m1=(CEnd.x-CHome.x)/((double)(WEnd.x-WHome.x));
        c1=(CHome.x-(double)WHome.x*m1);
        m2=(CEnd.y-CHome.y)/(WEnd.y-WHome.y);
        c2=(CHome.y-(double)WHome.y*m2);
        q.x=(double)p.x*m1+c1; q.y=(double)p.y*m2+c2;
        return q;
}
CPoint CCode7B::ConvertCW(PT p)           //converts C to W
{
        CPoint px;
        double m1,c1,m2,c2;
        m1=(double)(WEnd.x-WHome.x)/(CEnd.x-CHome.x);
        c1=(double)(WHome.x-CHome.x*m1);
        m2=(double)(WEnd.y-WHome.y)/(CEnd.y-CHome.y);
        c2=(double)(WHome.y-CHome.y*m2);
        px.x=p.x*m1+c1; px.y=p.y*m2+c2;
        return px;
}
```

8

Scheduling Application

8.1 Scheduling Problem

Scheduling is one common application of graph theory. The problem is obvious in management where human power and the available resources need to be utilized to the maximum in order to increase productivity in the organization, save costs, and maximize profit. In general, most problems of scheduling are nonlinear in nature and classified as NP-complete or NP-hard [1–4].

Scheduling is the assignment of customers to a set of servers or counters according to the available resources so as to fulfill one or more objectives. A *scheduler* is a program or software that maps the customers to the servers. A *customer* here refers to a piece of job that needs to be serviced at a *server*. A server in this problem can be a service counter like in the bank, a processor in a parallel computing network, or a classroom in a school. The jobs can be independent of each other, which suggest they can be scheduled freely in any order. In many cases, the jobs are related to each other in a sequence according to a partial order that requires scheduling to be performed according to that order.

One example of scheduling is customers who arrive at a bank waiting for their turn to be served at a counter. Another example is the telephone calls (customers) that arrive at a public switched telephone network (PSTN) exchange waiting to be assigned with a channel (counter). In these two examples, the number of counters is limited and they may not be able to cater to all the needs of the arriving customers simultaneously. Hence, only a handful of the customers will be assigned to the counters immediately upon arrival. The rest either have to wait or are blocked from being served, depending on the type of system implemented. If the system allows waiting, then a queue is formed where the unsuccessful customers still have the chance to be served. Otherwise, in a system that doesn't allow waiting, the customers are simply turned away or blocked.

Scheduling is always associated with optimization where the main objective is minimization or maximization of a performance. There are many objectives in scheduling. The most common is to maximize the *throughput* or the total amount of work per unit time. One way to work on this objective is to achieve the *makespan*, or the optimum completion time for all jobs. Another common objective is to distribute the jobs evenly on the servers, referred to as load balancing. There are also the objectives of minimizing the response time and minimizing the latency for jobs. In terms of sales, the most obvious objective is to produce a working schedule that minimizes the operating costs and maximizes the profit.

In any scheduling problem an optimum schedule is not easy to obtain. Many unpredictable factors affect the performance of the system that may cause some delay to its execution. In this case, a *feasible schedule* is good enough for most systems. A feasible schedule is one where the completion time may not be optimum, but at the same time, it does not cause any discomfort to the system. In this case, a slight delay in its completion is still tolerable and the system may function normally.

Figure 8.1 shows a scheduling problem involving eight jobs and three servers. The jobs in the figure are represented by a graph (top left) while the servers are arranged in a vertical column. The output is shown at the bottom of the figure as the assignment of the jobs to the servers according to time t from $t = 0$ to $t = 3$.

A schedule can be prepared in the form of a graph where the nodes represent the jobs to be scheduled while the edges denote the relationship between the nodes that may incur some constraints. Scheduling can be performed in two ways. In *static scheduling*, the problem is modeled as a graph where the *a priori* information about the nodes and edges is determined beforehand. The schedule is produced offline based on the properties of the graph.

In contrast, *dynamic scheduling* involves jobs that arrive on the fly, which require immediate assignment on the servers. Dynamic scheduling is always associated with real-time scheduling where the instant output produced is derived from the processing of current data. In most cases, it is not possible to model the problem as a graph as its shape cannot be determined before scheduling. The information about the job is only known at the time of processing. Dynamic scheduling may involve blocking and no blocking. A job is said to be blocked if it fails to be scheduled upon arrival because all the servers are busy. No blocking means every job that arrives is put in a queue to wait for the first available server. Hence, blocking suggests some jobs may not be scheduled while no blocking guarantees all the jobs will be served.

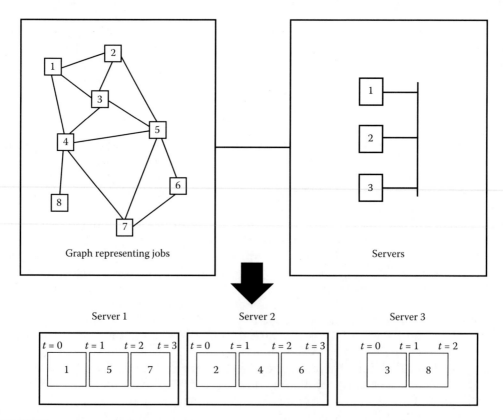

FIGURE 8.1
Scheduling eight jobs on a set of three servers.

Scheduling has deadlines. A schedule with a *hard deadline* requires all jobs to meet some strict and inflexible dates in their execution, or else the whole system will collapse. Scheduling with a *soft deadline* has important dates for the jobs to meet but failure to meet some or all the dates will not cause the system to crash. The system may still function with its deadline missed but it may cause some inconveniences to the system. In many cases, proper scheduling requires the imposition of both hard and soft deadlines. The soft deadline serves as a first reminder about the jobs that are not completed. Top attention and priority are given to these jobs so that they can be completed before the hard deadline.

Figure 8.2 illustrates scheduling with deadlines. A job starts at the earliest possible time at time $t = t_0$. With this starting time, the job may be completed optimally at t_{opt}. An earlier completion time than t_{opt} may result in a premature ending, which is not good for the system because the job may have not been executed properly. Some slight delay to the starting time may force the job to be completed at a later time. Completion before the soft deadline t_s indicates a feasible schedule, which is tolerable to the system. A hard deadline is indicated by t_h. Completion between t_s and t_h is considered critical to the system although the system may still function normally. The completion time above t_h is no longer tolerable as it may bring a disastrous effect on the system.

Some common scheduling problems are discussed here. *Timetabling* in a school or college is the scheduling of classes for students in order to fill all the teaching timeslots and utilize all the available classrooms on the campus. In addition, the timetable has to fulfill other requirements such as reasonable workload for teachers or lecturers and balanced contact hours for students. A lot of human factors are involved in the preparation of a timetable,

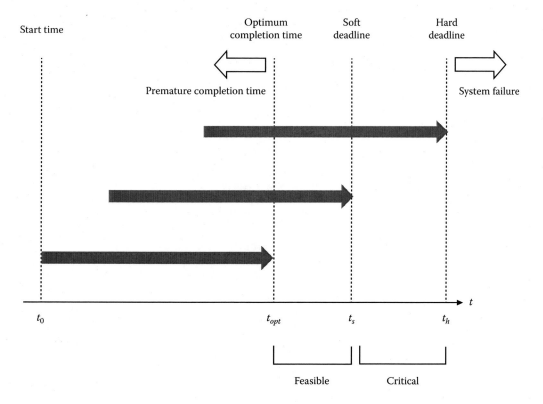

FIGURE 8.2
Scheduling with hard and soft deadlines.

which makes the automation process difficult. Referring to Figure 8.2, the classes in this problem are the nodes of the graph while the classrooms are the servers. The eight classes are assigned to the rooms with time slots from $t = 0$ to $t = 3$.

Machine job scheduling is the mapping of jobs to a set of machines where each machine can process one job at any given time. This type is scheduling is common in manufacturing and assembling factories that have up to several hundred machines on their premises. The machines may be arranged in an orderly sequence so that jobs completed in one machine can continue to the next machine easily. A delay in processing a job in one machine will definitely affect the whole production. A good discussion of the problem and its solution can be found in [1].

Personnel scheduling is the scheduling of workers in an organization so that the workplace has sufficient manpower to function. In a fast food restaurant that operates for 18 hours a day, for example, each employee works an 8-hour shift per day. Hence, the timetable needs to be drawn in such a way that each department has the required minimum number of workers to serve customers and each employee is assigned 8 hours of duty.

Airline scheduling is a broad system in the management of aircraft, passengers, and resources belonging to an airline. An aircraft is expensive to maintain; therefore, it is necessary for it to be in the air transporting passengers for as much time as possible with very little idle time. An aircraft arriving at an airport has only 1 or 2 hours of rest before continuing to the next destination. This is necessary because its stay on the ground incurs high rental costs that are proportional to the duration. Airline scheduling also includes the timetabling of flights for passengers with connections from the source cities to their destinations. Refer to [2] for some ideas on this challenge.

Task scheduling is the challenge of mapping a set of tasks or jobs onto a set of processors in a parallel computing system in order to meet some performance objectives. The most common objective is to assign the tasks to processors to minimize completion time. Another common objective is *load balancing*, which is to distribute the tasks evenly among the processors so that no processors will overwork while some others are idle. The tasks may be independent or dependent of each other and can be modeled as a directed or undirected graph.

Project management is a form of scheduling popularly implemented to manage a project with jobs arranged in a sequence involving several important deadlines. A good example is the construction of a high-rise condominium in a city. The jobs are outlined as individual steps whose completion according to the given deadlines guarantees the success of the whole project. Meeting the deadline in each job is crucial as a delay in one or more jobs will cause delays in the jobs that follow which, in turn, may delay the whole project.

Channel assignment is another common form of scheduling that applies to wireless cellular telephone and wireless mesh networks. The challenge is about assigning the minimum number of radio channels or frequencies to the mobile users in the network. Two users in the network can communicate with each other if each of them has been assigned to a radio channel. The channels for assignment cannot be too close to each other in order to avoid electromagnetic interferences. Hence, channel assignment is a form of a constrained optimization problem. Further discussion on this problem and its solution can be found in [3].

8.1.1 Gantt Chart

A Gantt chart displays jobs as horizontal or vertical bars whose durations are the lengths of the bars. For a job v_i the horizontal bar in a Gantt chart has its left side representing its

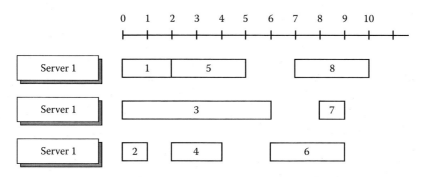

FIGURE 8.3
Gantt chart with a makespan of 10 units.

start time $v_i.start$ while the right side is its end $v_i.end$. The length of the job $v_i.len$ corresponds to the difference between its start time $v_i.start$ and the ending time $v_i.end$, or

$$v_i.len = v_i.end - v_i.start.$$

The jobs for scheduling normally have different start and ending times. The bar representing the length of the job is drawn in such a way that it will not cross another bar. If one bar clashes with another bar then the two bars are separated by placing them on different servers, channels, or processors.

Figure 8.3 is a typical Gantt chart of the output from scheduling eight jobs on a set of three servers. The Gantt chart has a makespan of 10 units, which is the length from the starting time to the ending time. The jobs are well distributed on the servers with each server getting two or three jobs. They have varying lengths that show their differences in terms of execution time. The chart clearly displays the starting time and ending time for each job. As well, there are empty gaps between some of the jobs that represent idle time for the servers. The gaps are unavoidable because some of the jobs may have to wait for some specific jobs to complete before they can start.

8.2 Dynamic Job Scheduling

Job shop scheduling refers to the assignment of a set of n jobs J_i for $i = 1, 2,\ldots,n$ on M machines m_k for $k = 1, 2,\ldots,M$. The jobs are of different lengths whose values correspond to their execution times. The jobs may be independent of each other, which suggest they can be assigned to the machines in any order. If the jobs are dependent on each other then their assignments to the machines will follow a partial order according to the priorities given to them.

The machines can be placed according to a sequence or free arrangement. Machines placed in a sequence represent an orderly flow of the jobs they are executing. They need to be positioned close to each other according to the sequence in order to minimize the transportation costs. In the free arrangement, machines can be placed anywhere independent of the order of the jobs that they execute.

In this section we discuss a simple system with blocking. In this system, jobs arrive at random at time t to make up a discrete event and they are assigned to a limited pool of servers on a first-come, first-served basis. We assume a system with no waiting where a job that fails to find a server at time t will be blocked. Therefore, this simple system has no queue. There is no tolerance in the timeslot where an arriving job that fails to find a server is immediately blocked. An example of this type of system is the process of uploading boxes of perishable fruits into a number of trucks for shipment. Since the fruits cannot wait for the next few trucks to arrive, they are immediately discarded if the waiting trucks do not provide enough spaces for them for shipment.

8.2.1 Scheduling Model

The objective of the scheduling model application is to design a *dynamic scheduler* for mapping the randomly arrived jobs to the servers. A scheduler is a component of the system that maps the jobs to the available servers. In doing this, the scheduler first checks the state of each server and determines whether it is busy or available. A job is assigned to a server if the server is available. Otherwise, the job is blocked since no waiting is allowed. The progress of the services at the server is shown in the form of Gantt charts that clearly show the start and ending times of the execution of each job.

We refer to Figure 8.4 to illustrate the scheduling model. The figure depicts an instance of time $t = t_j$ where five jobs arrive for mapping on six servers. Job i is referred as v_i and server k

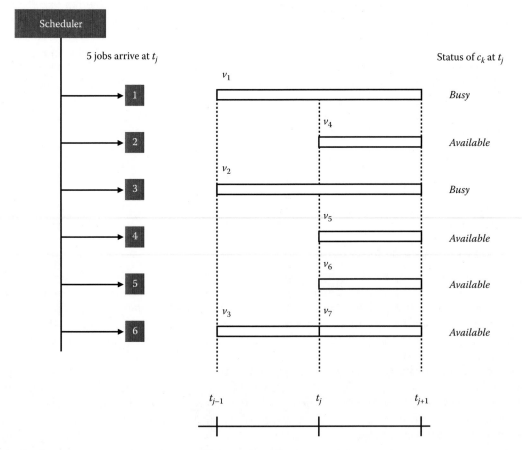

FIGURE 8.4
Five arrivals with four assigned and one blocked at time t_j.

as c_k. The scheduler assigns the arriving jobs on a first-come, first-served basis where the earliest arriving job goes to the first available server. Prior to $t = t_j$, three jobs marked as v_1, v_2, and v_3 have been preassigned to c_1, c_3, and c_6, respectively. The job v_1 has been assigned earlier than $t = t_{j-1}$ and is scheduled to complete its execution at $t = t_{j+1}$, while v_3 is assigned at $t = t_{j-1}$ and completes at $t = t_j$. Hence, at $t = t_j$, only servers v_2, v_4, v_5, and v_6 are available. They are immediately assigned with v_4, v_5, v_6, and v_7, respectively, leaving v_8 unassigned. Since the model does not allow a queue, this unassigned job is blocked and discarded.

The scheduler in this model is a C++ program that manages the assignment of jobs to the servers according to the requirement of the system. The scheduler responds to the arrival of several jobs generated randomly at time t. The full duties and responsibility of the scheduler at each timeslot are outlined as follows:

a. Determines the state of the servers

b. Maps the arriving jobs to the servers

c. Blocks the jobs that fail to be assigned to the servers

d. Produces Gantt charts for the assigned jobs

e. Records and displays the blocked jobs

8.2.2 *Code8A*: Dynamic Machine Scheduling

Code8A depicts dynamic scheduling with blocking where not all of the arriving jobs can be assigned to machines. The program illustrates dynamic scheduling for jobs of varying lengths for mapping onto machines arranged in a rectangle. The jobs are independent of each other and they can be assigned in any order. Jobs that arrive at a given timeslot are not guaranteed to be assigned to machines depending on the availability of the machines.

A machine can be in one of two states: available or busy. The machine is said to be available if it is not executing a job; otherwise, it is busy. If the number of arriving jobs is lower or the same as the number of available machines then all the jobs will be assigned to the machines; otherwise, some jobs cannot find available machines and will be blocked. At time t, if three jobs arrive and there are five machines available then all three jobs will be assigned to three of the five machines. If three jobs arrive and there is only one machine available, then only one job will be assigned while the other two will be blocked.

The simulation in *Code8A* is about the discrete-time assignment of jobs onto 16 machines with a press on the spacebar key representing a timeslot. Figure 8.5 shows an output from *Code8A*. The output consists of three areas: a Gantt chart at the top, the state of the machines in the middle, and a table for displaying the status of the incoming jobs. The Gantt chart describes the assigned jobs to their machines with each bar shown with its start and ending time. The difference between the start and ending time gives the length of the job. The state of the machines is shown on Windows as black, blue, and red rectangles (they appear as shades of gray in the book). Black means the machine is available, blue denotes the machine has just been assigned with a task, while red indicates the machine is busy executing a job that arrived from an earlier timeslot.

A total number of 300 independent jobs arrive at different timeslots from $t = 0$ to $t = 65$. There are 16 machines and each can only service one job at any given time. The length of each job is set at between one to eight time units, also determined randomly. The time countdown begins at $t = 0$, and this value increases by one unit with every press on the spacebar key. At every timeslot between zero to 10 jobs arrive, whose numbers are also determined randomly.

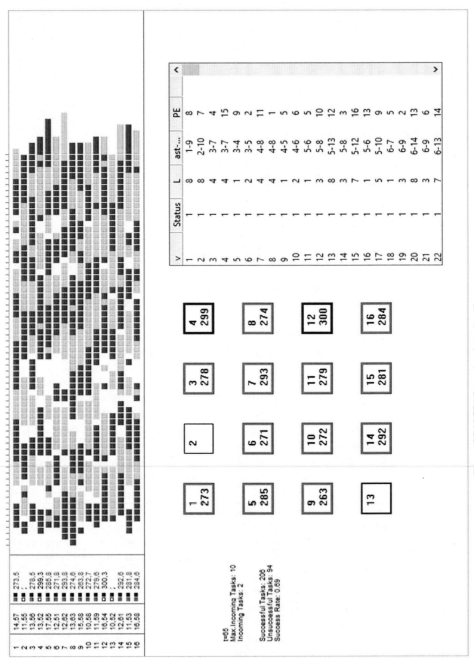

FIGURE 8.5
Output from *Code8A* showing discrete-time dynamic scheduling.

Code8A consists of the files *Code8A.h* and *Code8A.cpp*. The application class is CCode8A. The jobs in this application are the array *v* whose properties are represented by the structure called *JOB*. The machines are represented by two structures, *MACHINE* and *MESH*.

Dynamic scheduling involves the movement of jobs and their mapping on machines according to timeslots. Table 8.1 summarizes some of the most important global variables and objects in the dynamic scheduling model. The time *t* starts at *t* = 0 and it makes one unit per timeslot. A random number of jobs arrive at time *t* whose maximum is *nIncJobs-Max*, where not all of them are assigned to the machines. The cumulative number of jobs at time *t* is *n* where *nSuccessfulJobs* is the number that are successfully assigned.

In this model a job v_i that arrives at each timeslot is represented by a structure called *JOB*. Each job has its actual arrival time *ast*, actual completion time *act*, length *len*, and assigned machine *aPE* recorded in the structure. A machine is initially represented by *MACHINE*. The structure defines the jobs assigned to it as the array *v*. The total number of jobs assigned to each machine is represented by *totJobs* while *lastJob* is the latest job it is executing. At time *t*, a machine may be available or busy, and this state is recorded by *status*. Another variable *totExecTime* is the total execution time for the machine. In terms of management, the machines in this application are arranged in a rectangular mesh and represented as a structure called *MESH*.

Figure 8.6 is the program structure showing the functions in *Code8A*. The program starts with the constructor CCode8A() that creates the main window. Next, the function initializes the machine variables including the total number of jobs and total execution time, each to zero. Windows coordinates are given for the location of the machines in the rectangular mesh.

The initial display of the rectangular mesh and the Gantt chart area is handled by *OnPaint()*. The machines are drawn according to the Windows coordinates assigned earlier in the constructor.

The main event in *Code8A* is the spacebar key press that increases time value *t* by one unit. The event is handled by *OnKeyDown()*. The key press causes a random number of jobs to arrive. For each arriving job *vi* a check is performed on the machines. If a machine is available the job is immediately assigned to it. With the assignment, the information about its arrival time, its completion time, and its length are updated. At the same time the information about the assigned task is recorded in the machine information. The successful jobs are shown as bars in the Gantt chart area. The whole operations stops at *t* = 65.

DrawMesh() draws the initial display of the machines in the rectangular mesh, while their available are busy status at time *t* is shown in different colors through *UpdateMesh()*. By displaying the machines in a rectangular mesh, the latter function performs the mapping

TABLE 8.1

Important Global Variables and Objects in *Code8A*

Variable	Type	Description
home, end	CPoint	The top-left and bottom-right locations of the Gantt chart area
JobLengthMax	constant	Maximum length of each job
M	constant	Size of rectangular mesh given by $M \times M$
n	int	Current number of jobs where *N* is its maximum
nSuccessfulJobs	int	Number of jobs successfully assigned to machines
nIncJobsMax	constant	Maximum number of incoming job at time *t*
t	int	Time that is activated by pressing the spacebar key

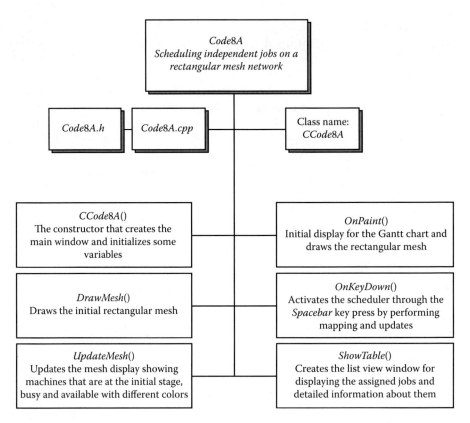

FIGURE 8.6
Organization of functions in *Code8A*.

of the machine number (1 to 16) to the rectangular format of row and column numbers. Finally, *ShowTable*() displays detail information about the assigned jobs on the machines.

The full listings of the codes are given below:

```cpp
// Code8A.h
#include <afxwin.h>
#include <afxcmn.h>
#include <fstream>
#define IDC_TABLE 401
#define M 4                      // M x M processors
#define N 500                    // max #Jobs
#define nIncJobsMax 10           // max #incoming Jobs per unit time
#define JobLengthMax 8
using namespace std;

class CCode8A : public CFrameWnd
{
private:
        CPoint home,end;
        int t,n;               // t=time, n=#tasks
        int nSuccessfulJobs,hBar,sBar,BarXSta;
                // hBar=bar height, sBar=bar spacing
        CFont fArial;
```

```
        CListCtrl table;

        typedef struct
        {
                CRect rct;
                CPoint home;
        } MESH;
        MESH mPE[M+1][M+1];

        typedef struct
        {
                bool status;
                int v[N+1],lastJob,totJobs,totExecTime;
        } MACHINE;
        MACHINE PE[M*M+1];

        typedef struct
        {
                bool status;
                int length,ast,act,aPE;    // aPE=assigned PE
        } JOB;
        JOB v[N+1];
public:
        CCode8A() ;
        ~CCode8A()                              {}
        afx_msg void OnPaint();
        afx_msg void OnKeyDown (UINT nChar,UINT nRep,UINT nFlags);
        DECLARE_MESSAGE_MAP();
        void ShowTable();
        void PEInfoColumn();
        void DrawMesh(),UpdateMesh(int,int,int [M*M+1]);
};

class CMyWinApp : public CWinApp
{
public:
        virtual BOOL InitInstance();
};
CMyWinApp MyApplication;

BOOL CMyWinApp::InitInstance()
{
        m_pMainWnd=new CCode8A;
        m_pMainWnd->ShowWindow(m_nCmdShow);
        return TRUE;
}

// Code8A.cpp: Jobs on rectangular mesh network
#include "Code8A.h"

BEGIN_MESSAGE_MAP(CCode8A, CFrameWnd)
     ON_WM_PAINT()
     ON_WM_KEYDOWN()
END_MESSAGE_MAP()
```

```
CCode8A::CCode8A()
{
        hBar=8; sBar=3;
        home=CPoint(150,0); end=CPoint(900,200);
        Create(NULL,L"Dynamic Machine Scheduling Model",
                WS_OVERLAPPEDWINDOW,CRect(0,0,900,700));
        t=0; n=0; nSuccessfulJobs=0;
        for (int k=1;k<=M*M;k++)
        {
                PE[k].status=0;
                PE[k].totJobs=0;
                PE[k].totExecTime=0;
        }
        mPE[1][1].home=CPoint(home.x+50,end.y+50);
        mPE[1][1].rct=CRect(CPoint(mPE[1][1].home),CSize(50,50));
        for (int j=1;j<=M;j++)
                for (int i=1;i<=M;i++)
                {
                        mPE[i][j].home=mPE[1][1].home+CPoint(80*(i-1),
                        80*(j-1));
                        mPE[i][j].rct=CRect(mPE[i][j].home,mPE[i][j].
                        home+CPoint(40,40));
                }
        fArial.CreatePointFont(70,L"Arial");
}

void CCode8A::OnPaint()
{
        CPaintDC dc(this);
        CString s;
        dc.TextOut(home.x+100,640,
                L"Press SPACEBAR to simulate the discrete event");
        CPen pBlue(PS_SOLID,1,RGB(100,100,100));
        dc.SelectObject(&pBlue);
        dc.Rectangle(CRect(home.x-140,home.y+10,end.x-30,635));
        dc.MoveTo(home.x-140,home.y+195); dc.LineTo(end.x-31,home.y+195);
        dc.MoveTo(home.x-110,home.y+10); dc.LineTo(home.x-110,home.y+195);
        dc.MoveTo(home.x-70,home.y+10); dc.LineTo(home.x-70,home.y+195);
        dc.MoveTo(home.x-5,home.y+10); dc.LineTo(home.x-5,home.y+195);
        dc.SelectObject(&fArial);
        for (int i=1;i<=M*M;i++)
        {
                s.Format(L"%d",i);
                dc.TextOut(20,3+(hBar+sBar)*i,s);
        }
        DrawMesh();
}

void CCode8A::OnKeyDown(UINT nChar, UINT nRep, UINT nFlags)
{
        CClientDC dc(this);
        CString s;
        CRect rct;
        CBrush bWhite(RGB(255,255,255));
```

```
CPen pGray(PS_SOLID,1,RGB(255,255,255));
CPen pRed(PS_SOLID,1,RGB(255,0,0));
CPen pBlack(PS_SOLID,1,RGB(50,50,50));
CPen pBlue(PS_SOLID,1,RGB(0,0,250));
int acc[M*M+1],BusyPE[M*M+1];
int aPE,nIncJobs;    // aPE=assigned PE
int i,k,q,r,w,nAvailPE,nBusyPE,nNewBusyPE;
dc.SelectObject(&fArial);
srand(time(0));
if (nChar==VK_SPACE && (n<=N && t<=65))
{
        dc.SelectObject(&pRed);
        dc.MoveTo(home.x+(t+1)*8,home.y+6);
        dc.LineTo(home.x+(t+1)*8,home.y+10);
        dc.SelectObject(&pGray);
        dc.MoveTo(home.x+(t+1)*8,home.y+12);
        dc.LineTo(home.x+(t+1)*8,home.y+15+(hBar+sBar)*M*M);
        dc.FillRect(CRect(home.x-48,home.y+12,home.x-7,home.
        y+190),&bWhite);
        s.Format(L"t=%d",t++);
        dc.TextOut(30,300,s);
        nIncJobs=rand()%nIncJobsMax;
        s.Format(L"Max.Incoming Jobs :%d ",nIncJobsMax);
        dc.TextOut(30,310,s);
        s.Format(L"Incoming Jobs :%d ",nIncJobs);
        dc.TextOut(30,320,s);
        w=0;
        for (k=1;k<=M*M;k++)
        {
                if (PE[k].totJobs>=1)
                {
                        r=PE[k].lastJob;
                        if (v[r].act<=t)
                                PE[k].status=FALSE;
                }
                if (!PE[k].status)
                {
                        rct=CRect(CPoint(home.x-66,5+(hBar+sBar)*k),
                        CSize(5,5));
                        dc.SelectObject(&pBlack);
                        dc.Rectangle(&rct);
                        acc[++w]=k;
                }
                else
                {
                        rct=CRect(CPoint(home.x-66,5+(hBar+sBar)*k),
                        CSize(5,5));
                        dc.FillSolidRect(&rct,RGB(255,0,0));
                        dc.SetTextColor(RGB(255,0,0));
                        dc.SetBkColor(RGB(255,255,255));
                        r=PE[k].lastJob;
                        s.Format(L"%d,%d",r,v[r].length);
                        dc.TextOut(home.x-50,3+(hBar+sBar)*k,s);
                }
```

```
        }
        nBusyPE=0;
        for (k=1;k<=M*M;k++)
                if (PE[k].status)
                        BusyPE[++nBusyPE]=k;
        nAvailPE=w;
        dc.SetBkColor(RGB(255,255,255));
        nNewBusyPE=0;
        for (i=1;i<=nIncJobs;i++)
        {
                if (++n>=N)
                        break;
                v[n].status=FALSE;
                v[n].length=1+rand()%JobLengthMax;
                v[n].ast=t;
                v[n].act=v[n].ast+v[n].length;
                for (k=1;k<=M*M;k++)
                        if (nAvailPE!=0)
                        {
                                q=1+rand()%nAvailPE;
                                if (!PE[acc[q]].status)
                                {
                                        aPE=acc[q];
                                        ++nNewBusyPE;
                                        BusyPE[++nBusyPE]=aPE;
                                        nSuccessfulJobs++;
                                        v[n].aPE=aPE;
                                        v[n].status=TRUE;
                                        PE[aPE].status=TRUE;
                                        PE[aPE].lastJob=n;
                                        r=++PE[aPE].totJobs;
                                        PE[aPE].v[r]=n;
                                        PE[aPE].totExecTime+=v[n].
                                        length;
                                        dc.SetTextColor(RGB(0,0,250));
                                        s.Format(L"%d,%d",r,PE[aPE].
                                        totExecTime);
                                        dc.TextOut(45,3+(hBar+sBar)*aP
                                        E,s);
                                        s.Format(L"%d,%d",n,v[n].length);
                                        dc.TextOut(home.x-50,3+
                                        (hBar+sBar)*aPE,s);
                                        rct=CRect(CPoint(home.
                                        x+t*8,5+(hBar+sBar)*aPE),
                                                CSize(8*v[n].
                                        length,hBar));
                                        if (PE[aPE].totJobs%2==0)
                                                dc.FillSolidRect(&rct,
                                                RGB(0,235,235));
                                        else
                                                dc.FillSolidRect(&rct,
                                                RGB(0,100,200));
                                        rct=CRect(CPoint(home.x-60,5+
                                        (hBar+sBar)*aPE),
```

```
                                                CSize(5,5));
                                        dc.FillSolidRect(&rct,
                                        RGB(0,0,180));
                                        break;
                                }
                        }
                        if (!v[n].status)
                                v[n].aPE=0;
                        dc.SetBkColor(RGB(255,255,255));
                }
                dc.SetTextColor(RGB(0,0,0));
                s.Format(L"Successful Jobs:%d",nSuccessfulJobs);
                dc.TextOut(30,350,s);
                s.Format(L"Unsuccessful Jobs:%d",n-nSuccessfulJobs);
                dc.TextOut(30,360,s);
                double q=(double)nSuccessfulJobs/(double)n;
                s.Format(L"Success Rate:%.2lf",q);
                dc.TextOut(30,370,s);
                if (t==65)
                        ShowTable();
                UpdateMesh(nBusyPE,nNewBusyPE,BusyPE);
        }
}

void CCode8A::ShowTable()
{
        CString s;
        table.DestroyWindow();
        table.Create(WS_VISIBLE | WS_CHILD | WS_BORDER | LVS_REPORT
                        | LVS_NOSORTHEADER,CRect(home.x+380,home.y+230,
                        home.x+650,home.y+600),this,IDC_TABLE);
        table.InsertColumn(0,L"v",LVCFMT_CENTER,45);
        table.InsertColumn(1,L"Status",LVCFMT_CENTER,45);
        table.InsertColumn(2,L"L",LVCFMT_CENTER,45);
        table.InsertColumn(3,L"ast-act",LVCFMT_CENTER,45);
        table.InsertColumn(4,L"PE",LVCFMT_CENTER,45);
        for (int i=1;i<=n;i++)
        {
                s.Format(L"%d",i); table.InsertItem(i-1,s,0);
                s.Format(L"%d",v[i].status); table.SetItemText(i-1,1,s);
                s.Format(L"%d",v[i].length); table.SetItemText(i-1,2,s);
                s.Format(L"%d-%d",v[i].ast,v[i].act); table.
                        SetItemText(i-1,3,s);
                s.Format(L"%d",v[i].aPE); table.SetItemText(i-1,4,s);
        }
}

void CCode8A::DrawMesh()
{
        CClientDC dc(this);
        CRect mRegion,cPort;
        CPoint a,b;
        CString s;
        int i,j,k=1;
```

```
        // draw the PEs
        CBrush bWhite(RGB(255,255,255));
        mRegion=CRect(home.x+110,home.y+200,home.x+360,home.y+450);
        dc.FillRect(mRegion,&bWhite);
        CPen pBlack(PS_SOLID,1,RGB(0,0,0));
        dc.SelectObject(&pBlack);
        for (j=1;j<=M;j++)
                for (i=1;i<=M;i++)
                {
                        dc.Rectangle(&mPE[i][j].rct);
                        dc.SetTextColor(RGB(0,0,0));
                        s.Format(L"%d",k++);
                        dc.TextOut(mPE[i][j].home.x+10,mPE[i][j].home.y+4,s);
                }
}


void CCode8A::UpdateMesh(int nBusyPE,int nNewBusyPE,int BusyPE[M*M+1])
{
        CClientDC dc(this);
        CString s;
        int i,j,k,p,q;
        CPen pRed(PS_SOLID,3,RGB(255,0,0));
        CPen pBlue(PS_SOLID,3,RGB(0,0,180));
        CRect rct;
        DrawMesh();
        for (k=1;k<=nBusyPE;k++)
        {
                p=BusyPE[k];
                switch(p)
                {
                        case 1:
                                i=1; j=1; break;
                        case 2:
                                i=2; j=1; break;
                        case 3:
                                i=3; j=1; break;
                        case 4:
                                i=4; j=1; break;
                        case 5:
                                i=1; j=2; break;
                        case 6:
                                i=2; j=2; break;
                        case 7:
                                i=3; j=2; break;
                        case 8:
                                i=4; j=2; break;
                        case 9:
                                i=1; j=3; break;
                        case 10:
                                i=2; j=3; break;
                        case 11:
                                i=3; j=3; break;
                        case 12:
                                i=4; j=3; break;
```

```
                    case 13:
                            i=1; j=4; break;
                    case 14:
                            i=2; j=4; break;
                    case 15:
                            i=3; j=4; break;
                    case 16:
                            i=4; j=4; break;
            }
            if (k<=nBusyPE-nNewBusyPE)
                    dc.SelectObject(&pRed);
            else
                    dc.SelectObject(&pBlue);
            dc.Rectangle(&mPE[i][j].rct);
            dc.SetTextColor(RGB(0,0,0));
            s.Format(L"%d",BusyPE[k]);
            dc.TextOut(mPE[i][j].home.x+10,mPE[i][j].home.y+4,s);
            q=PE[p].lastJob;
            dc.SetTextColor(RGB(0,0,220));
            s.Format(L"%d",q);
            dc.TextOut(mPE[i][j].home.x+10,mPE[i][j].home.y+20,s);
    }
}
```

8.3 Task Scheduling on Multiprocessor Systems

A complete computer program has dozens of codes that represent instructions to the computer in solving a problem. The codes may be written in primary languages like C++, C#, and Java, or secondary languages such as MATLAB®, Mathematica, and Maple. A big program may have thousands of lines of codes that require a fast computer for processing. The execution of codes in the program may become slow especially if they involve many mathematical routines for handling large arrays, images, and graphics. A sequential computer, such as a desktop or a laptop, may not be suitable for this task because the computer only has a single processor to handle all the complex operations. The processing may be too slow and not good enough to meet the program requirement. A computer with many processors can do the job better as the job can be distributed to many processors instead of one for a faster and reliable result.

The codes in a complete computer program can be partitioned into several tasks or routines. A *task*, in this case, may represent a single job that is independent of other tasks. The task is a set of codes created specifically to tackle a small part of the problem. For example, the computer routine for finding the inverse of a matrix is a task while a routine for determining a square array is diagonal is another task. The magnitude of the task is called its *length* and is measured in unit of time of execution. Obviously, the tasks in a program have different lengths depending on the complexity of the job they are doing. A task for computing the inverse of a matrix of size 100×100 will definitely take longer to complete than another task that adds two matrices. Therefore, the former has a larger length than the latter. A single big program may have hundreds of small tasks where each is specialized to do one piece of a job.

A task may call another task in completing the given solution. Furthermore, a task may depend on another task for data before it can function properly. Hence, a dependency relationship is created between a pair of tasks. This dependency is represented as an edge of a graph where the tasks are the nodes of the graph. The graph with complete tasks and edges is called a *task graph*.

8.3.1 Multiprocessor Systems

A multiprocessor or parallel computing network has a set of processors or processing elements that work independently in processing a set of tasks. Each processor in the network has the capability to read the assigned program codes and execute them to produce results that can then be passed to another processor. By working in a group cooperatively, a multiprocessor system executes and delivers the results faster than a single processor system.

The processors in a parallel computing system are arranged in many different ways. Some common topologies for the processor arrangement are the bus, star, ring, and hypercube. Each processor in the system has limited or large built-in memories that are needed at the start-up and for processing some critical jobs. The cable links between the processors enable them to communicate with each other and perform data transfer as well.

The working memory in a parallel computing system is provided in the form of parallel random-access memory (PRAM). There are two types of parallel computing systems: shared-memory and distributed-memory systems. A *shared-memory parallel system* has a single or several memory modules that are shared by all processors in the system. The network is organized in such a way that the processors have fast access to these memory modules. Figure 8.7 (left) shows a shared-memory parallel computing system that has eight processors P_k for $k = 1, 2,...,8$ sharing a single memory module. The processors are arranged in a bus topology that allows them to refer to the memory module independently of other processors.

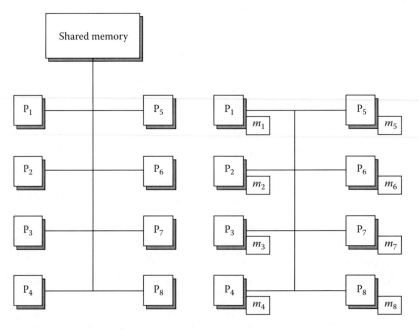

FIGURE 8.7
Shared-memory (left) and distributed-memory (right) parallel systems.

A *distributed-memory parallel system* has processors that are equipped with their own memory modules. Figure 8.7 (right) shows a distributed system of eight processors P_k with their own memory modules m_k for $k = 1, 2,...,8$. The distributed-memory system has the advantage over the shared-memory system as jobs can be executed faster at the individual processors. However, a distributed system always encounters heavier communication costs in transferring data from one processor to another compared to the shared-memory system.

8.3.2 Task Graph

Task scheduling is an NP-complete combinatorial optimization problem that is about mapping a set of tasks onto a set of processors on the given timeslots. In static scheduling, the tasks are normally represented as a task graph where the nodes are the tasks and the edges between the nodes represent their relationship. All the information about the tasks is available prior to scheduling. Dynamic scheduling deals with arbitrary tasks that are created at the time of processing. It is not possible to form a task graph in this case as the information about the tasks is not available at the time of processing.

Task scheduling has been an active research area. A good discussion on different algorithms proposed for solving the task scheduling problem can be found in [4–6]. The problem is highly nonlinear in nature, so most methods proposed to solve the problem involve metaheuristics. Among the methods are greedy algorithms [7], genetic algorithm [8], and meanfield annealing that combines simulated annealing with Hopfield neural networks [9].

In static scheduling, the assignment involves all the tasks to the processors. There is no blocking as the assignment is made based on the fact that all the information about the tasks is known beforehand. In the case of dynamic scheduling, very little information about the tasks is known at the time of scheduling. Dynamic scheduling can be performed with blocking and no blocking. Blocking occurs if the arriving tasks are not allowed to wait in a queue. Therefore, the arriving tasks that cannot find available processors are blocked and discarded. If the arriving tasks are allowed to wait in a queue then they get the opportunity to be assigned to the available processors. In this case, the multiserver Markovian queuing model can be applied to model the task assignment.

The edges in a task graph may be undirected or directed. In an *undirected task graph* $G(V, E)$, the edges represent the flow of instructions that can go in either way. There is no precedence relationship between the two paired nodes, which suggest the two tasks (nodes) can be scheduled independently of each other.

A *directed task graph* $G(V, \vec{E})$ has a set of n nodes $V = \{v_i\}$ and directed edges $\vec{E} = \{\vec{e}_{ij}\}$ for $i, j = 1, 2,..., n$. Each directed edge $\vec{e}_{ij}$ is the task precedence relationship that describes the partial order or flow from v_i to v_j, which is written as

$$\vec{e}_{ij} = v_i \rightarrow v_j$$

The partial order means v_i must complete its execution first before v_j can start. For $\vec{e}_{ij} = v_i \rightarrow v_j$, the tail node v_i is called the *predecessor* of the head node, v_j. Also, v_j is called the *successor* of v_i. A node v_j can have several predecessors v_i for $i = 1, 2,..., m$. With the presence of m predecessor tasks v_j can only start its execution when all its m predecessor tasks have completed their execution.

Figure 8.8 shows two directed task graphs of size $n = |V| = 4$. A task in the graph is marked as v: c where v is the task number and c is its length. The graph on the left has

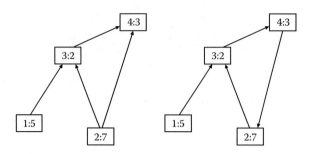

FIGURE 8.8
Directed acyclic (left) and cyclic (right) graphs.

$\vec{E} = \{\vec{e}_{13}, \vec{e}_{23}, \vec{e}_{24}, \vec{e}_{34}\}$ where v_1 and v_2 are the predecessors of v_3, while v_2 and v_3 are the predecessors of v_4. The process in this graph starts at v_1 and v_2, continues at v_3, and ends at v_4. Any delay at v_1 and v_2 will cause a delay in v_3, which then delays v_4. Also, a failure in v_1 or v_2 will cause a disruption of the whole graph because v_3 and v_4 cannot continue.

The task graph (left) in Figure 8.8 is a *directed acyclic graph*. A directed acyclic graph does not have a cycle, which suggests every edge is visited at most once. The graph becomes a tree and possesses all the properties of a tree if a node in it has at most one successor, although it can also have many predecessors. In contrast, a *directed cyclic graph* is a graph having at least one cycle that represents the repeat of a process. An edge in the graph may be visited two or more times. The cycle may repeat several times before the process continues with the next task. In many cases, the cycle has a period that is part of the whole process according to the program requirement.

Figure 8.8 (right) shows a directed cyclic graph. In the cyclic graph, a cycle exists in $\{\vec{e}_{23}, \vec{e}_{34}, \vec{e}_{42}\}$ that suggests one or more repeats before the process terminates in one of the three nodes. The graph starts at v_1 and terminates at either v_2, v_3, or v_4.

In a partial order $\vec{e}_{ij} = v_i \rightarrow v_j$, the earliest time v_j can start executing is at the completion time of v_i. If v_j is scheduled in the same processor as v_i then the completion time of v_i is the starting time of v_j, provided the timeslot is available. Figure 8.9 shows the result when scheduling is performed on two processors according to this rule on the task graph in Figure 8.8 (left).

The directed edges of the task graph discussed so far do not have weights. The directed edges of a task graph can also have weights. The weights in this case represent the communication costs, for example, for transferring data from the tail node to the head node. This graph is called a directed graph with communication costs. The communication costs may be expressed in quantities such as time, monetary value, and distance. In a task graph

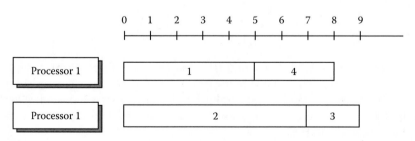

FIGURE 8.9
Gantt chart of the graph from Figure 8.8 (left).

with communication cost, the presence of costs requires synchronization in the execution flow. A node having two or more predecessors can only start when its predecessor nodes have completed their executions plus the communication costs. If the costs from their predecessor nodes are different then synchronization is performed by considering the edge with the maximum cost. The node can start when its predecessor node with the maximum weight completes its execution.

Figure 8.10 is a weighted directed acyclic graph with communication costs shown on their directed edges. The edges of the graph are described as follows:

$$\vec{e}_{13} = 7 : v_1 \rightarrow v_3$$

$$\vec{e}_{23} = 4 : v_2 \rightarrow v_5$$

$$\vec{e}_{24} = 8 : v_2 \rightarrow v_4$$

$$\vec{e}_{34} = 2 : v_3 \rightarrow v_4$$

From the figure, v_3 has v_1 and v_2 as its predecessors with the communication costs of 7 and 5 units, respectively. The node can only start its execution when both its predecessors have completed their executions as well as satisfying the communication costs on their edges. The other node v_4 has to wait for the completion of v_3 and its communication cost of 2 units as well as v_2 and its communication cost of 8 units.

The rules for scheduling a task v_j, whose predecessor is v_i in a directed acyclic graph where $\vec{e}_{ij} = c : v_i \rightarrow v_j$ with communication cost c onto processor P_k, are given as follows:

1. If v_j is assigned to the same processor as v_i, then $v_j.start = v_i.end$ as the communication cost within the same processor is negligible, or $\vec{e}_{ij} = 0$.
2. Otherwise, if v_j is scheduled in a different processor, then $v_j.start = v_i.end + \vec{e}_{ij}$.
3. If more than one predecessor exists, then synchronization is performed. The earliest time the task can start is the maximum of the sum of the execution time of the predecessor tasks and their communication costs. For example, if v_j has two predecessors v_i and v_r that are assigned to processors P_1 and P_2, respectively, then

$$v_j.start = \max(v_i.end + \vec{e}_{ij}, v_r.end + \vec{e}_{rj}).$$

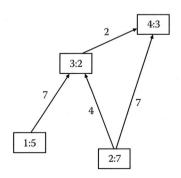

FIGURE 8.10
Directed acyclic graph with communication costs.

From Rule 1, the assignment of v_j to P_1 causes $\vec{e}_{ij} = 0$. Then

$$v_j.start = \max(v_i.end, v_r.end + \vec{e}_{rj}).$$

Similarly, the assignment of v_j to P_2 results in $\vec{e}_{rj} = 0$, to produce

$$v_j.start = \max(v_i.end + \vec{e}_{ij}, v_r.end).$$

It follows that the choice of processor for v_j is one that provides the minimum of the two values.

4. If a timeslot is occupied, then the scheduled task on this slot will have to wait until the currently executing task completes its execution.

Figure 8.11 shows the Gantt chart for the task graph from Figure 8.9. Compared to the task graph from Figure 8.8, the presence of communication costs in the current task graph greatly increases the makespan. In general, communication incurs heavy costs to the system. The mapping of tasks start with v_1 and v_2 at P_1 and P_2, respectively. The next task v_3 is dependent on v_1 and v_2. The start time for v_3 is calculated as follows:

On P_1: $v_3.start = \max(v_1.end, v_2.end + \vec{e}_{23}) = \max(5, 7 + 4) = 11$
On P_2: $v_3.start = \max(v_1.end + \vec{e}_{13}, v_2.end) = \max(5 + 7, 7) = 12$

Obviously, v_3 is assigned to the minimum of the two values; that is, $v_3.start = 11$ on P_1. The last task v_4 has v_2 and v_3 as its predecessors. Its start time is calculated in a similar manner:

On P_1: $v_4.start = \max(v_2.end + \vec{e}_{24}, v_3.end) = \max(7 + 7, 13) = 14$
On P_2: $v_4.start = \max(v_2.end, v_3.end + \vec{e}_{34}) = \max(7, 13 + 2) = 15$

The final result is obtained with v_4 assigned to P_1 where $v_4.start = 14$ by taking the minimum of the two values.

We present a greedy algorithm approach to static scheduling. The algorithm was originally proposed in our earlier work in [10]. The algorithm is based on the rules outlined above with regard to scheduling tasks in a task graph with communication costs. The algorithm is greedy in nature as scheduling is performed from the bottom up where a task is assigned to a processor immediately without considering its successor tasks once it obeys all the said scheduling rules.

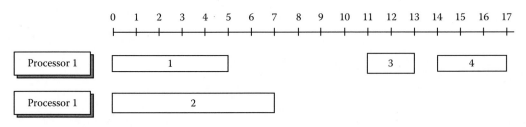

FIGURE 8.11
Gantt chart of the task graph from Figure 8.9.

For *n* tasks in the graph, they are first sorted based on their path magnitude values. The *path magnitude* of a task is defined as the sum of all the lengths of the tasks and the associated communication costs along the path from the task to the root node. Let Q be a path from the task vj to a root node. The path magnitude of v_j is given by

$$\sum_{i=1, v_i \in Q}^{n} (v_i.len + \vec{e}_i).$$

In the above equation, v_i is the node along Q, $v_i.len$ is its length, and $\vec{e}_i$ is the communication cost with its successor task in the path. There can be more than one path from the node to its root. There can also be more than one root for the task graph. The one we are interested in is the maximum of the path magnitude of the task, given by

$$v_j.pmm = \max \sum_{i=1, v_i \in Q}^{n} (v_i.len + \vec{e}_i).$$

Tasks are sorted in a list according to their maximum path magnitude values in ascending order. A task with the lowest maximum path magnitudes is assigned to the first available processor. Once assigned, the task is immediately removed from the list and the next task with the lowest value is chosen for the next assignment.

```
//Greedy scheduling algorithm
Given:
    Task vi information for i = 1, 2,...,n and processors Pk for k = 1, 2,...,M.
Initialization:
    Set time t = 0.
    Assign v1 to P1.
Process:
    For j = 1 to n
        For i = 1 to n
            If vi in the path to the root
            Identify the predecessors vi.pred[j];
            Compute vj.start = max(vi.end + eij, vk.end + ekj);
        Endfor
        Compute the path maximum magnitude vi.pmm;
        Choose vj
    Endfor
Output: Gantt chart and detail scheduling information.
```

8.3.3. *Code8B*: Task Scheduling on a Four-Processor System

The task scheduling problem and its solution using a greedy algorithm is illustrated in *Code8B*. The program is an interactive one that allows the user to draw the task graph using a mouse. Interaction is one of the most important features in a program design as it provides the user an opportunity to test the simulation data through a set of input values.

Different results are produced from different input values, and the results can be used to test the effectiveness of the simulation model.

Code8B illustrates static scheduling using four processors that are assumed to be interconnected with each other. The task graph has communication costs whose values from one to four units are determined randomly. The nodes are created by the user through the left clicks of the mouse while the edges are drawn through right clicks. The arrow in an edge denotes the partial order for the precedence relationship between the two tasks, the tail, and its head.

Figure 8.12 shows the output from *Code8B*. It consists of the drawing area, a Gantt chart for displaying the task bars, and a list view window for disclosing detailed information about the assigned tasks. There are two buttons called *Task Scheduler* and *New Graph*. *Task Scheduler* activates the processing of tasks by reading the information from the task graph and applies the rules of scheduling to produce the output. *New Graph* refreshes the graph by clearing the drawing area for a new graph.

The input from *Code8B* is a complete task graph drawn in a drawing area. The first input is produced when the user creates a node through the left click of the mouse. A task is drawn as a small rectangle with the task number and its length. The length of the task is a value from one to nine units, determined randomly. The second input is the directed edge that is produced when the user right-clicks on two tasks consecutively using the mouse. The task start time for execution is at the tail node and ends at the head node, but their actual execution times are determined in the overall schedule

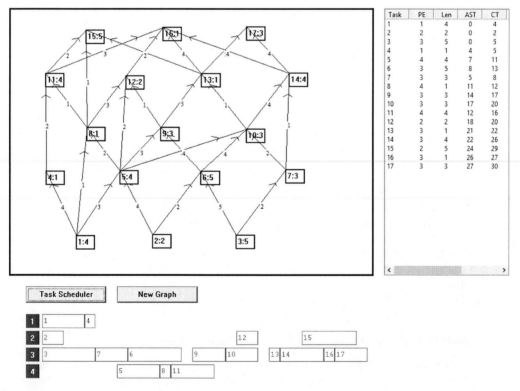

FIGURE 8.12
Output from *Code8B* showing the mapping of 17 tasks on four processors.

when the *Task Scheduler* button is pressed. The communication cost from the precedence relationship is also determined randomly whose value is from one to five time
units.

The input in the drawing area displays the task graph with complete information on
the length of each node, the precedence relationship between the tasks, and the communication costs incurred. This visual display represents a structured way of understanding the features of a task graph. In adding values to this facility, the user can modify
the program by providing other features such as allowing direct input to the length of
the tasks and communication costs between the tasks instead of the generated random
numbers.

The output from scheduling is shown as a Gantt chart and a table that displays detailed
information about the assigned tasks. In Figure 8.12, there are 17 tasks and 31 directed
edges in the task graph. The graph has its roots at v_1, v_2, and v_3. The Gantt chart and
the information on the assigned tasks are computed using the greedy algorithm once the
Task Scheduler button is pressed. The Gantt chart provides the graphical display of each
assigned task by displaying their actual start and end execution times. The schedule has
the makespan value of 30 units with P_3 having the most number of tasks at nine. The
results can be improved further by modifying the codes to achieve another objective such
as even distribution of the workload between the processors. We leave it to the reader to
continue from this program.

The list view window displays detailed information about the assigned tasks. The information includes the task lengths, the actual start and ending execution times, the predecessor tasks and their communication costs, the priority order from sorting, and the
assigned processor number.

Code8B has a single class called *CCode8B*. The source files are *Code8B.h* and *Code8B.cpp*.
The tasks and processors in this project are represented by structures called *NODE* and
MACHINE, respectively. These two structures define the properties of their respective
members as variables and objects.

In *NODE*, tasks are represented by the array *v*. A task is created through the left click of
the mouse. Its Windows position in the drawing area is recorded as *WHome* while its corresponding Cartesian coordinates is *CHome*. When assigned to a processor, the task has
the actual starting time *ast* and completion time *act* that gives the difference as its length
len. In waiting for assignment to the processor, a task has low ready time *lrt* and high ready
time *hrt*. The maximum path magnitude for the task is *pmm*, which is computed in order
to determine its position from the root of the graph. The sorted value according to the
colevel is *sort*. The predecessors of a task is marked as *pre* while the communication costs
involved for data transfer from the predecessors is *preCom*. The assigned processor for the
task is *aPE*.

MACHINE defines the structure for the processors. A processor *PE* is displayed in the
Gantt chart with coordinates given by *WHome*. The tasks assigned to each processor are
marked as the array *v*. The processor marks the time it is going to be available for the next
task as *prt*. The length of the task under execution is *pel*.

Table 8.2 lists some of the most important global variables, objects, and constants in
Code8B. A task is created through the left click of the mouse while two consecutive right
clicks produce a directed edge. The number of tasks is *n* with *N* as its maximum. The
drawing area has *WHome* and *WEnd* as its top-left and bottom-right locations, respectively.
The points have *CHome* and *CEnd* as their corresponding Cartesian coordinates.

A task is drawn as a small rectangle in the drawing area through the left click of the
mouse. A directed edge is drawn from two consecutive right clicks marked as *pt*1 and

TABLE 8.2

Important Objects/Variables in *Code4A* and Their Descriptions

Variable	Type	Description
bNGraph	*CButton*	The new graph button
CHome, CEnd	*PT*	Cartesian top left and bottom right of the drawing area
n	*int*	Number of nodes in the graph with maximum *N*
pt1, pt2	*CPoint*	First and second node from right clicks of the mouse
RButtonFlag	*int*	Flag value from right click of the mouse
table	*CListCtrl*	Table for displaying the assigned task information
TSbutton	*CButton*	The scheduler button
WHome, WEnd	*CPoint*	Windows top left and bottom right of the drawing area

pt2. Before drawing the directed edge, a control flag called *RButtonFlag* changes its value, with *RButtonFlag* = 0 denoting no activity, *RButtonFlag* = 1 indicating the first node, and *RButtonFlag* = 2 as the second node that then draws the line.

TSbutton starts the scheduling process by calling all the related functions in the program. A press of the button computes the schedule by reading the information of the tasks in the task graph, mapping them to the processors, and then displaying the results. The other object called *table* creates a list view table for displaying detailed information about the assigned tasks.

The functions in *Code8B* are summarized in Figure 8.13. The constructor creates the main window and the buttons as well as initializing some variable and object values. This is followed by control of the whole program through five events: initial display through *OnPaint()*, left click of the mouse through *OnLButtonDown()*, right click of the mouse through *OnRButtonDown()*, and two buttons.

The initial display consists of two buttons, a graph drawing area, and the Gantt chart area. The drawing area serves as the input for the program. A task graph is produced from the nodes and directed edges through the left and right clicks of the mouse, respectively. The left click also sets the initial values of the task: the start time, completion time, its predecessor tasks, and status. Two consecutive right clicks on two nodes create the precedence relationship between the two tasks. The tail node becomes the predecessor task of the head node. Communication cost is then added to the edge, and this value between one to five units is generated randomly.

A graph is considered completely drawn and ready for execution when the *Task Sheduler* button is pressed. The button serves as another event that is handled by *OnCompute()*. This function is the main engine in this program as it manages the whole task scheduling process by calling four functions, as follows:

- *PMM()* to read the graph values and compute the maximum path magnitude of each task
- *PreScheduler()*, which sorts the tasks according to the maximum path magnitude values
- *Scheduler()*, which assigns tasks to processors according to the sorted maximum path magnitude values and displays the results as a Gantt chart
- *ShowTable()*, which displays detailed information of the assigned tasks

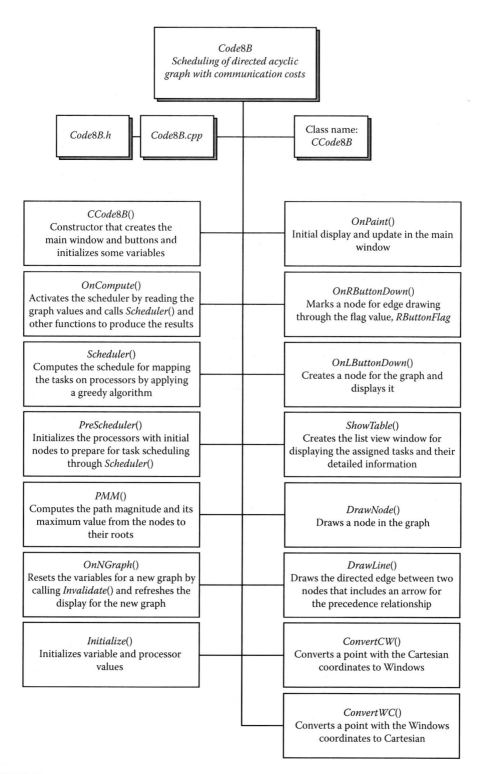

FIGURE 8.13
Program design and organization of functions in *Code8B*.

The full listing of the codes in *Code8B* are given as follows:

```
// Code8B.h
#include <afxwin.h>
#include <afxcmn.h>
#include <fstream>
#define N 20                            // Max #nodes
#define nPE 4                  //#PE
#define IDC_SCHEDULE 500
#define IDC_table 501
#define IDC_NGRAPH 502
#define nFIELDS 7
#define L (double)0.2
using namespace std;

typedef struct                 // Cartesian points
{
       double x,y;
} PT;

class CCode8B : public CFrameWnd
{
private:
       CPoint WHome,WEnd;
       PT CHome,CEnd;
       CFont fCourier,fTimes;
       CButton TSbutton,bNGraph;
       CListCtrl table;
       int n,RButtonFlag,pt1,pt2;
       int TextColor,BgColor;

       typedef struct
       {
              int len;                // length
              int aPE;                // assigned Processor
              int pmm;                // path max. magnitude
              int pre[5];             // pred task, pre[0]=#pred tasks
              int preCom[5];          // comm cost of pred tasks
              int ast,act;            // ast=actual start time, act=
                                      completion time
              int hrt,lrt;            // high ready time, low ready time
              int sort;               // sorted nodes according to colevels
              bool sta;               // status
              CPoint WHome,GHome; // node coordinates in text & graphic
              areas
              PT CHome;
              CRect rct;              // node representation as a box
       } NODE;
       NODE v[N+1];

       typedef struct
       {
              int v[N+1];             // task# (aTs[0]=#tasks in Pr)
              int prt,pel;            // proc ready time, execution length
```

```
                CPoint WHome;
        } MACHINE;
        MACHINE PE[nPE+1];
public:
        CCode8B();
        ~CCode8B()              {}
        void PreScheduler(),Scheduler(),Initialize();
        void ShowTable(),PMM(),DrawNode(int),DrawLine();
        CPoint ConvertCW(PT);
        PT ConvertWC(CPoint);
        afx_msg void OnCompute();
        afx_msg void OnPaint();
        afx_msg void OnLButtonDown (UINT,CPoint);
        afx_msg void OnRButtonDown (UINT,CPoint);
        afx_msg void OnNGraph();
        DECLARE_MESSAGE_MAP();
};

class CMyWinApp : public CWinApp
{
public:
virtual BOOL InitInstance();
};
CMyWinApp MyApplication;

BOOL CMyWinApp::InitInstance(void)
{
        m_pMainWnd=new CCode8B;
        m_pMainWnd->ShowWindow(m_nCmdShow);
        return TRUE;
}

// code8B.cpp: task scheduling
#include "Code8B.h"

BEGIN_MESSAGE_MAP(CCode8B,CFrameWnd)
     ON_WM_PAINT()
     ON_WM_LBUTTONDOWN()
     ON_WM_RBUTTONDOWN()
ON_BN_CLICKED(IDC_SCHEDULE,OnCompute)
        ON_BN_CLICKED(IDC_NGRAPH,OnNGraph)
END_MESSAGE_MAP()

CCode8B::CCode8B()
{
        WHome=CPoint(30,20); WEnd=CPoint(700,500);
        CHome.x=-8; CHome.y=10; CEnd.x=8; CEnd.y=-10;
        Create(NULL,L"Multiprocessor Task Scheduling",
            WS_OVERLAPPEDWINDOW,CRect(0,0,1000,800));
        TSbutton.Create(L"Task Scheduler",WS_CHILD | WS_VISIBLE |
        BS_DEFPUSHBUTTON,
                CRect(CPoint(WHome.x+30,WEnd.y+20),CSize(150,30)),
                this,IDC_SCHEDULE);
```

```
bNGraph.Create(L"New Graph",WS_CHILD | WS_VISIBLE |
BS_DEFPUSHBUTTON,
        CRect(CPoint(WHome.x+200,WEnd.y+20),CSize(150,30)),
        this,IDC_NGRAPH);
fCourier.CreatePointFont(60,L"Courier");
fTimes.CreatePointFont(80,L"Times");
BgColor=RGB(255,255,255);
TextColor=RGB(100,100,100);
Initialize();
}

void CCode8B::OnNGraph()
{
    Initialize(); Invalidate();
}

void CCode8B::Initialize()
{
    n=0; RButtonFlag=0;
    for (int k=1;k<=nPE;k++)
{
        PE[k].v[0]=0;
        PE[k].prt=0;
        PE[k].pel=0;
        PE[k].WHome=CPoint(WHome.x+30,WEnd.y+70+(k-1)*30);
}
    table.DestroyWindow();
}

void CCode8B::OnPaint()
{
    CPaintDC dc(this);
    CRect rct;
    CString s;
    CPen pDark(PS_SOLID,3,RGB(0,0,0));

    rct=CRect(WHome,WEnd);
    dc.SelectObject(&pDark);
    dc.Rectangle(rct);
    dc.SetTextColor(RGB(255,255,255));
    for (int k=1;k<=nPE;k++)
    {
        rct=CRect(PE[k].WHome,CSize(25,25));
        dc.FillSolidRect(&rct,RGB(100,100,100));
        s.Format(L"%d",k); dc.TextOut(PE[k].WHome.x+8,PE[k].WHome.
        y+5,s);
    }
    dc.SetBkColor(BgColor);
    dc.TextOutW(WHome.x+40,WEnd.y+20,
        L"left click to draw a node, right click on nodes to draw an
        edge");
}

void CCode8B::OnLButtonDown(UINT nFlags,CPoint pt)
```

```
{
        if (CRect(WHome,WEnd).PtInRect(pt))
                if (n<=N)
                {
                        n++;
                        v[n].WHome=pt;
                        v[n].pre[0]=0;
                        v[n].ast=0;
                        v[n].act=0;
                        v[n].sta=0;              //set Node status to inactive
                        v[n].rct=CRect(CPoint(v[n].WHome),CSize(35,25));
                        DrawNode(n);
                }
}

void CCode8B::OnRButtonDown(UINT nFlags,CPoint pt)
{
        int i,r;
        srand(time(0));
        for (i=1;i<=n;i++)
                if (v[i].rct.PtInRect(pt))
                {
                        RButtonFlag++;
                        if (RButtonFlag==1)
                                pt1=i;
                        if (RButtonFlag==2)
                        {
                                pt2=i;
                                RButtonFlag=0;
                                r=++v[pt2].pre[0];
                                v[pt2].pre[r]=pt1;
                                v[pt2].preCom[r]=1+rand()%5;
                                DrawLine();
                        }
                }
}

void CCode8B::OnCompute()
{
        PMM();                   // sort the nodes according to their colevel
                                 values
        PreScheduler();          // initialize each PE with 1st Node
Scheduler();                     // the scheduler
        ShowTable();             // display task information
}

void CCode8B:: ShowTable()
{
        int i,j;
        table.DestroyWindow();
        table.Create(WS_VISIBLE | WS_CHILD | WS_BORDER | LVS_REPORT
| LVS_NOSORTHEADER,CRect(WEnd.x+20,WHome.y,WEnd.x+250,WEnd.y),
this,IDC_table);
```

```
        CString s,S[nFIELDS+1]={"Task","PE","Len","AST","CT","Pre","pmm",
        "Order"};
        table.DeleteAllItems();
        for (j=0;j<=nFIELDS;j++)
            table.InsertColumn(j,S[j],LVCFMT_CENTER,45);
        for (i=0;i<=n;i++)
            if (i+1<=n && i+1<=N)
            {
                s.Format(L"%d",i+1);
                    table.InsertItem(i,s,0);
                    s.Format(L"%d",v[i+1].aPE); table.SetItemText(i,1,s);
                    s.Format(L"%d",v[i+1].len); table.SetItemText(i,2,s);
                    s.Format(L"%d",v[i+1].ast); table.SetItemText(i,3,s);
                    s.Format(L"%d",v[i+1].act); table.SetItemText(i,4,s);
                    s.Format(L"%d",v[i+1].pre[0]); table.
                    SetItemText(i,5,s);
                    s.Format(L"%d",v[i+1].pmm); table.SetItemText(i,6,s);
                    s.Format(L"%d",v[i+1].sort); table.SetItemText(i,7,s);
            }
}


void CCode8B::PMM() // Sort the tasks according to levels
{
    int i,j,k,r,tmp;
    for (i=1;i<=n;i++)
    {
        v[i].sort=i;
        if (v[i].pre[0]==0)
        v[i].pmm=0;
        else
        {
            r=v[i].pre[1];
            tmp=v[r].len+v[i].preCom[1]+v[r].pmm;
            for (j=1;j<=v[i].pre[0];j++)
            {
                r=v[i].pre[j];
                tmp=(tmp<v[r].len+v[i].preCom[j]+v[r].pmm)?
                        v[r].len+v[i].preCom[j]+v[r].pmm:tmp;
            }
            v[i].pmm=tmp;
        }
        for (k=1;k<=i-1;k++)
        {
            r=v[k].sort;
            if (v[i].pmm<v[r].pmm)
            {
                for (j=i;j>=k+1;j-)
                        v[j].sort=v[j-1].sort;
                v[k].sort=i;
                break;
            }
        }
    }
}
```

```
void CCode8B::PreScheduler()
{
    int j,i,u,k,AsPE,r,tmp;
    for (i=1;i<=n;i++)
    {
        u=v[i].sort;
        if (v[u].pre[0]==0)
            {
            if (i<=nPE)
            {
                AsPE=i;
                v[u].aPE=AsPE;
                v[u].ast=0;
                v[u].act=v[u].ast+v[u].len;
                v[u].sta=1;
                PE[AsPE].v[0]++;
                PE[AsPE].v[1]=u;
                PE[AsPE].prt=v[u].act;
                PE[AsPE].pel+=v[u].len;
            }
            else
            {
                tmp=PE[1].prt;
                AsPE=1;
                for (k=1;k<=nPE;k++)
                    if (tmp>PE[k].prt)
                    {
                        tmp=PE[k].prt;
                        AsPE=k;
                    }
                v[u].aPE=AsPE;
                v[u].ast=tmp;
                v[u].act=v[u].ast+v[u].len;
                v[u].sta=1;
                r=++PE[AsPE].v[0];
                PE[AsPE].v[r]=u;
                PE[AsPE].prt=v[u].act;
                PE[AsPE].pel+=v[u].len;
            }
        }
        else
            if (i<=nPE)
            {
            AsPE=i;
            v[u].aPE=AsPE;
            r=v[u].pre[1];
            tmp=v[r].act+v[u].preCom[1];
            for (j=1;j<=v[u].pre[0];j++)
            {
                r=v[u].pre[j];
                if (tmp<v[r].act+v[u].preCom[j])
                    tmp=v[r].act+v[u].preCom[j];
            }
            v[u].ast=tmp;
```

```
                    v[u].act=v[u].ast+v[u].len;
                    v[u].sta=1;
                    PE[AsPE].v[0]++;
                    PE[AsPE].v[1]=u;
                    PE[AsPE].prt=v[u].act;
                    PE[AsPE].pel+=v[u].len;
                }
        }
}

void CCode8B::Scheduler()
{
    CClientDC dc(this);
    CString s;
    CRect rct;
    int u,i,j,k,r,m,AsPE;
    int HiDom,HiDomIndex,HiDomPE,LoDom,LoDomIndex,tmp;
    CPen penGray(PS_SOLID,1,TextColor);

    dc.SelectObject(penGray);
    for (j=1;j<=n;j++)
    {
        u=v[j].sort;
        if (!v[u].sta)
        {
            //calculate hrt
            HiDomIndex=1;
            HiDom=v[u].pre[HiDomIndex];
            v[u].hrt=v[HiDom].act+v[u].preCom[HiDomIndex];
            for (i=1;i<=v[u].pre[0];i++)
            {
                r=v[u].pre[i];
                if (v[u].hrt<v[r].act+v[u].preCom[i])
                {
                    v[u].hrt=v[r].act+v[u].preCom[i];
                    HiDom=r;HiDomIndex=i;
                }
            }
        }
        // calculate lrt
        if (v[u].pre[0]==1)
            v[u].lrt=v[u].hrt;
        else
            {
            LoDomIndex=((HiDomIndex==1)?2:1);
            LoDom=v[u].pre[LoDomIndex];
            v[u].lrt=v[LoDom].act+v[u].preCom[LoDomIndex];
            for (i=1;i<=v[u].pre[0];i++)
                if (i!=HiDomIndex)
                {
                r=v[u].pre[i];
                if (v[u].lrt<v[r].act+v[u].preCom[i])
                    v[u].lrt=v[r].act+v[u].preCom[i];
                }
            }
```

```
            HiDomPE=v[HiDom].aPE;
            if (PE[HiDomPE].prt<=v[u].hrt)
            {
                AsPE=HiDomPE;
                v[u].aPE=AsPE;
                if (v[u].pre[0]==1)
                    v[u].ast=PE[HiDomPE].prt;
                else
                    v[u].ast=((PE[HiDomPE].prt>=v[u].lrt)?PE[HiDomPE].
                    prt:v[u].lrt);
                v[u].act=v[u].ast+v[u].len;
                v[u].sta=1;
                PE[AsPE].prt=v[u].act;
                r=++PE[AsPE].v[0];
                PE[AsPE].v[r]=u;
                PE[AsPE].pel+=v[u].len;
            }
            else
            {
                AsPE=((HiDomPE==1)?2:1);
                tmp=PE[AsPE].pel;
                for (k=1;k<=nPE;k++)
                        if (k!=HiDomPE)
                            if (tmp>PE[k].pel)
                            {
                                AsPE=k;
                                tmp=PE[k].pel;
                            }
                                v[u].aPE=AsPE;
                                v[u].ast=((PE[AsPE].prt>=v[u].
                                hrt)?PE[AsPE].prt:v[u].hrt);
                                v[u].act=v[u].ast+v[u].len;
                                v[u].sta=1;
                                PE[AsPE].prt=v[u].act;
                                r=++PE[AsPE].v[0];
                                PE[AsPE].v[r]=u;
                                PE[AsPE].pel+=v[u].len;
            }
        }
    }

    dc.SelectObject(fCourier); dc.SetTextColor(TextColor);
    for (i=1;i<=n;i++)
    {
        m=v[i].aPE;
        v[i].GHome=CPoint(30+PE[m].WHome.x+20*v[i].ast,PE[m].WHome.y);
        rct=CRect(v[i].GHome.x,v[i].GHome.y,v[i].GHome.x+20*v[i].
        len,v[i].GHome.y+25);
        dc.Rectangle(rct);
        s.Format(L"%d",i); dc.TextOut(2+v[i].GHome.x,5+v[i].
        GHome.y,s);
    }
}


void CCode8B::DrawNode(int u)
```

```
{
      CClientDC dc(this);
      CString s;
      CPen pBlack(PS_SOLID,2,RGB(0,0,0));
      dc.SelectObject(pBlack);
      dc.Rectangle(v[n].rct);
      v[n].len=1+rand()%5;
      s.Format(L"%d:%d",n,v[n].len);
      dc.TextOut(v[n].WHome.x+3,v[n].WHome.y+3,s);
}

void CCode8B::DrawLine()
{
      CClientDC dc(this);
      double theta;
      CString s;
      PT a,b,c,d;
      CPoint A,B,C,D,mPoint;
      dc.MoveTo(CPoint(v[pt1].WHome));
      dc.LineTo(CPoint(v[pt2].WHome));
      v[pt1].CHome=ConvertWC(v[pt1].WHome);
      v[pt2].CHome=ConvertWC(v[pt2].WHome);

      // Draw arrow
      A=CPoint((v[pt1].WHome.x+3*v[pt2].WHome.x)/4,(v[pt1].WHome.
      y+3*v[pt2].WHome.y)/4);
      a=ConvertWC(A);
      if (v[pt1].CHome.x<=v[pt2].CHome.x && v[pt1].CHome.y<=v[pt2].
      CHome.y)
        {
          theta=atan((v[pt2].CHome.y-v[pt1].CHome.y)/(v[pt2].CHome.x-
          v[pt1].CHome.x));
          b.x=a.x-L*sin(theta); b.y=a.y+L*cos(theta); B=ConvertCW(b);
          c.x=a.x+L*cos(theta); c.y=a.y+L*sin(theta); C=ConvertCW(c);
          d.x=a.x+L*sin(theta); d.y=a.y-L*cos(theta); D=ConvertCW(d);
        }
      if (v[pt1].CHome.x>=v[pt2].CHome.x && v[pt1].CHome.y>=v[pt2].
      CHome.y)
        {
          theta=atan((v[pt1].CHome.y-v[pt2].CHome.y)/(v[pt1].CHome.x-
          v[pt2].CHome.x));
          b.x=a.x-L*sin(theta); b.y=a.y+L*cos(theta); B=ConvertCW(b);
          c.x=a.x-L*cos(theta); c.y=a.y-L*sin(theta); C=ConvertCW(c);
          d.x=a.x+L*sin(theta); d.y=a.y-L*cos(theta); D=ConvertCW(d);
        }
      if (v[pt1].CHome.x>=v[pt2].CHome.x && v[pt1].CHome.y<=v[pt2].
      CHome.y)
        {
          theta=atan((v[pt2].CHome.y-v[pt1].CHome.y)/(v[pt1].CHome.x-
          v[pt2].CHome.x));


          b.x=a.x+L*sin(theta); b.y=a.y+L*cos(theta); B=ConvertCW(b);
          c.x=a.x-L*cos(theta); c.y=a.y+L*sin(theta); C=ConvertCW(c);
```

```
      d.x=a.x-L*sin(theta); d.y=a.y-L*cos(theta); D=ConvertCW(d);
    }
    if (v[pt1].CHome.x<=v[pt2].CHome.x && v[pt1].CHome.y>=v[pt2].
    CHome.y)
    {
      theta=atan((v[pt1].CHome.y-v[pt2].CHome.y)/(v[pt2].CHome.x-
      v[pt1].CHome.x));
      b.x=a.x+L*sin(theta); b.y=a.y+L*cos(theta); B=ConvertCW(b);
      c.x=a.x+L*cos(theta); c.y=a.y-L*sin(theta); C=ConvertCW(c);
      d.x=a.x-L*sin(theta); d.y=a.y-L*cos(theta); D=ConvertCW(d);
    }
    dc.MoveTo(B); dc.LineTo(C); dc.LineTo(D);
    int u,w,r;
    r=v[pt2].pre[0];
    u=(v[pt1].WHome.x+v[pt2].WHome.x)/2;
    w=(v[pt1].WHome.y+v[pt2].WHome.y)/2;
    dc.SelectObject(fTimes);
    s.Format(L"%d",v[pt2].preCom[r]); dc.TextOut(u,w,s);
}


CPoint CCode8B::ConvertCW(PT p)          // converts C to W
{
    CPoint px;
    double m1,c1,m2,c2;
    m1=(double)(WEnd.x-WHome.x)/(CEnd.x-CHome.x);
    c1=(double)(WHome.x-CHome.x*m1);
    m2=(double)(WEnd.y-WHome.y)/(CEnd.y-CHome.y);
    c2=(double)(WHome.y-CHome.y*m2);
    px.x=p.x*m1+c1; px.y=p.y*m2+c2;
    return px;
}


PT CCode8B::ConvertWC(CPoint p)          // converts W to C
{
    double m1,c1,m2,c2;
    PT q;
    m1=(CEnd.x-CHome.x)/((double)(WEnd.x-WHome.x));
    c1=(CHome.x-(double)WHome.x*m1);
    m2=(CEnd.y-CHome.y)/(WEnd.y-WHome.y);
    c2=(CHome.y-(double)WHome.y*m2);
    q.x=(double)p.x*m1+c1; q.y=(double)p.y*m2+c2;
    return q;
}
```

9

Target Detection Application

9.1 Target Detection Problem

One active use of graph theory today is in detecting targets over some critical geographical area. The targets can be intruders and unauthorized vehicles to a security area, visible and invisible rays emitted from electronic devices, or dangerous chemicals produced from a leakage. Early detection of the targets may help in preventing a major catastrophe from happening that can cause heavy damage to properties and people.

In general, the target detection problem can be reduced to a graph model by considering the entities in the problem as a network. In its simplest case, the targets are modeled as the nodes of a graph. The edges are represented from the way the targets interact with each other. For example, an edge is formed when a target communicates with another target through a radio frequency. Hence, the graph is the network of the targets in the region. It follows that the reduction of the target network into a graph means several properties of the graph can be applied for solving the target detection problem.

Target detection is about the use of sensors, computers, cameras, and other technological resources to detect certain targets for security, surveillance, and intelligence purposes. The targets may be hidden deeply in a region that cannot be traced easily through standard means. For example, a high-resolution camera can be used to recognize a target in its captured image within its shooting range. Images taken from the camera can be analyzed for real-time or offline processing. The camera is limited in its capability if the target is not located within its range. As well, some targets cannot be detected by standard cameras. Specialized sensors mounted on cameras for detecting infrared, ultraviolet, and other invisible rays may be used instead.

Despite these limitations, the camera is still an efficient tool to detect a target. For detecting a target outside of its range, the camera can be combined with sensors as part of the target detection system. For example, a series of aerial images of the ground can be captured by mounting the camera on a helicopter. These images are then analyzed to describe the region for purposes such as detecting some unusual objects or patterns. The images can also be organized to produce an aerial animation of flight over the region.

Recently, several high-technology devices have been deployed for target detection. Cameras and sensors that are capable of detecting ultrasound, infrared, and multispectral rays are some of the common devices used. Cameras with sensors have been widely used in unmanned aerial vehicles (UAVs), more commonly called drones, for capturing ground images. An UAV is an unmanned remote control vehicle that is capable of flying several kilometers from its base. The vehicle is controlled from the ground, and its size ranges from as small as a book to as large as a normal aircraft. In many cases, the device has been pre-programmed to fly according to a flight path. Due to their generally smaller size, UAVs are cheaper than conventional aircraft to operate and have been proven to be practical for deployment especially for critical missions such as surveillance and search-and-rescue operations.

Target detection is an active area of research. There are many research works that discuss several models of detecting moving or static targets. In [1], a radar sensor network is deployed to detect target in a foliage environment. Data in the form of Chernoff information is used as the selection criteria in a Bayesian environment. In [2], wireless sensor networks are deployed to detect targets in a *target coverage problem*. The sensors have antennas that can detect targets through directional or angular transmissions. In a target coverage problem, the targets are grouped into cover sets that are formed from the overlapped coverage. The target coverage problem is discussed in [3] where directional sensor networks are deployed to track targets in a region by grouping the sensors into sets called *cover sets*. The solution is obtained by applying a metaheuristic method called *learning automata*.

9.2 Target Detection Using Radar and Antennas

Radar, which stands for radio detection and ranging, is a device for detecting the presence of certain objects in the environment using radio waves. The device was originally developed to detect aircraft during World War II. Today, radar is widely used for monitoring air traffic in airports, military surveillance and antimissile defense systems, ship navigation, weather forecasting, and other useful purposes. Target detection using radar is managed as a centralized system that involves one or more radars, antennas, computers, and other related equipment.

In detecting an object, the transmitter in the radar transmits pulses of electromagnetic waves in circular directions within its transmission range. Electromagnetic waves are radio waves that travel through the medium at a speed close to the speed of light. The advantage of electromagnetic waves can be seen from the fact that they can penetrate media such as air, water, buildings, mountains, and trees. The waves are reflected by objects along their path, where some of these waves return to the transmitter and some others are absorbed by the target surface or lost. The returned waves are picked up by an antenna that is located in the same location as the transmitter. The reception of the bounced waves at the antenna indicates the rough location of the target.

Radar is used to track information about a target. A target can be static or moving. The information received over some interval of time is analyzed to determine the distance, speed, direction, altitude, and range of the target. For example, the distance between the radar and its target is calculated by recording the time of transmission and the arrival time of the bounced radio waves. The speed and direction of a moving target is determined by calculating several distances of the target from the bounced waves over some intervals of time.

Radar is always coupled with an antenna. An *antenna* is a device that reads electric power and converts it to radio waves and vice versa. The first antenna was built in 1888 by German physicist Heinrich Hertz. In transmission, a radio transmitter supplies an electric current oscillating at radio frequency to the antenna and the antenna radiates the energy from the current as electromagnetic waves. In reception, an antenna intercepts some of the power from an electromagnetic wave for the receiver to amplify.

An antenna can be arranged to transmit and receive radio waves in omnidirectional and directional positions. An omnidirectional antenna transmits and receives radio waves in equal directions equally as the waves are propagated in a circular formation. A directional antenna focuses on some angle in its transmission and reception to cover an area called a sector.

Today, antennas are a widely used consumer product. At home, antennas are essential items for televisions, radios, mobile cellular telephones, remote controls, and Wi-Fi devices. An antenna may be hidden, as in most of today's cellular phones, or visible, as in television parabolic dishes and cellular base station towers.

9.2.1 Transmission Models

Transmission from an antenna can be made in two ways: omnidirectional and directional. *Omnidirectional transmission* is a transmission in the form of concentric circles that propagate outward in its range with the antenna as its center. Directional transmission is based on an angle that focuses on a sector as its search space. Both omnidirectional and directional transmissions are equally important and are applied in detecting targets within a region.

In an omnidirectional antenna, transmission is measured at time t where a ring at discrete interval Δt is called a *corona*. Figure 9.1 shows a corona formed at Δt between the radius of transmission r at time t and R at time $t + \Delta t$. The transmission consumes a small amount of energy at a low transmission range and increases nonlinearly with a larger range. Omnidirectional transmission is useful especially in determining the distance of the target from its antenna.

One or more objects can be detected inside a corona by performing two consecutive transmissions, first with the range of r at time $t = t_1$ followed by a higher transmission at range R at time $t = t_2$. The corona is formed with $\Delta R = R - r$. Obviously, if the transmissions at $t = t_1$ and $t = t_2$ detect the sets of targets U and V, respectively, then it can be concluded that the corona has a set of targets from $U \cap V$. Assuming the transmission at $t = t_1$ detects targets $U = \{u_1, u_2, u_3\}$ while at $t = t_2$ has $V = \{u_1, u_2, u_3, u_4, u_5\}$, then the corona has $U \cap V = \{u_4, u_5\}$. This gives the approximate locations of the targets from their center.

Directional transmission is an angular transmission from the antenna that forms a *sector* of a circle rather than a full circle, as illustrated in Figure 9.2. The mechanism in terms of power saving is similar to omnidirectional where energy consumption is higher as the range of transmission increases. If the area of coverage is smaller, the directional transmission consumes less power because it transmits according to the angle $\theta = \beta - \alpha$, where

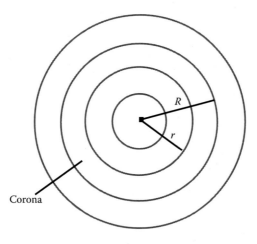

FIGURE 9.1
Omnidirectional transmission.

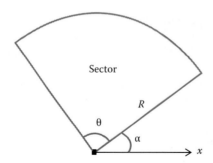

FIGURE 9.2
Directional transmission.

α is the starting transmission at time t and β is the ending transmission at time $t + \Delta t$. The ideal value of θ is normally less than 180° in order for the antenna to focus on the suspected presence of other sensors. When $\theta = 360°$, the directional transmission becomes omnidirectional.

Directional transmission allows an antenna to focus on a sectorial search. With an adjustable range, the antenna can choose the range of sectorial coverage. Of course, power consumption increases as the range increases because more energy is needed for the coverage.

Figure 9.3 shows a target u in a sector covered by the transmission from a directional antenna with the center at C. The sector is indicated by angle θ and radius R from the center C whose coordinates are (c_x, c_y). A sectorial sweep can determine the angle φ of u from the x-axis in the counterclockwise direction. At time $t = t_1$, the antenna transmits with the sectorial angle of θ and range R with angle α from the x-axis in the counterclockwise direction. At $t = t_2$ another transmission is made with a small increase in α but with the same transmission range and sectorial angle of θ. If u is present in both sweeps then its angle φ can be approximated to be equal to α.

Trigonometric laws apply in detecting targets within a sector. With the distance from the center r_u determined through the omnidirectional transmission and at angle φ determined from the directional sweep, the information can then be used to determine the coordinates of u. The coordinates are given by

$$u_x = r_u \cos \varphi, \tag{9.1}$$

$$u_y = r_u \sin \varphi. \tag{9.2}$$

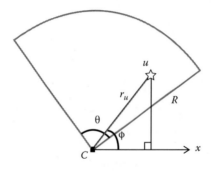

FIGURE 9.3
Detecting target u in a sector.

The reverse can also be applied. Assuming u has the Cartesian coordinates given by (u_x, u_y), then

$$\varphi = \tan^{-1} \frac{u_y - c_y}{u_x - c_x},$$ (9.3)

$$r_u = \sqrt{(u_x - c_x)^2 + (u_y - c_y)^2}.$$ (9.4)

9.2.2. *Code9A*: Centralized Target Detection

Code9A shows the simulation on an antenna for detecting targets in a region. The antenna is assumed to be placed in the middle of a region and is equipped with the capability to perform omnidirectional as well as directional transmissions of signals. The centralized system allows transmissions to be made so that the targets can be detected to produce information such as their distance from the center, their opening angle φ from the x-axis in the counterclockwise direction, and their Cartesian (x, y) coordinates.

Figure 9.4 shows an output from *Code9A*. The output consists of four edit boxes for input, four buttons for activating processes, a drawing area for displaying the antenna and its targets, and a list view window for displaying the results from the antenna sweeps. *Code9A* illustrates the use of both omnidirectional and directional antennas for detecting targets within their range.

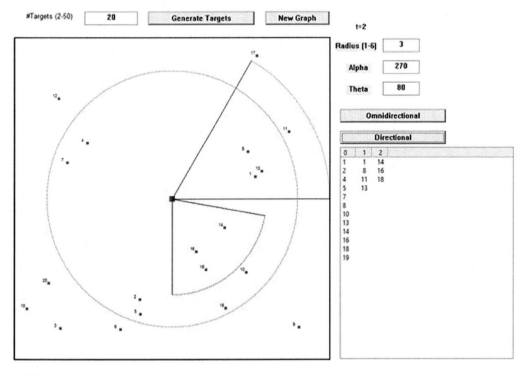

FIGURE 9.4
Output from *Code9A* for the centralized detection system.

The initial output in the drawing area of Figure 9.4 has a red square in the middle (shown on Windows) to represent the center of the antenna. Both transmission and reception of bounced signals are assumed to be based on this point. Input is first entered in an edit box as the number of sensors that need to be detected. The input becomes active once the *GenerateTarget* button is pressed, which produces sensors as small dark boxes in the drawing area.

The simulation is performed starting at time $t = 0$, and this value increases by one by pressing the *Omnidirectional* or *Directional* buttons. The *Omnidirectional* button activates the antenna for a full circle sweep at the indicated radius. The *Directional* button sweeps the sector, which starts at angle α from the x-axis for angular coverage of θ at the indicated radius. In Figure 9.4, the drawing area displays output at time $t = 2$ on directional antenna of radius 3 units in the sector, which starts at $\alpha = 270°$ and sweeps for $\theta = 80°$. Three targets are covered in the sector; namely, u_{14}, u_{16}, and u_{18}. The targets are listed in the list view window together with earlier sweeps at $t = 0$ (omnidirectional with radius of 4 units) and $t = 1$ (directional with $\alpha = 0°$, $\theta = 80°$ and a radius of 4 units).

Code9A has *Code9A.h* and *Code9A.cpp* as the source files. A single class called *CCode9A* is used in the application. Table 9.1 lists the variables and objects used in this project. The targets in this application are represented as the array u, which is an object in the structure *TARGET*. The target u is displayed in the drawing area with the Cartesian home coordinates declared as *CHome* and the corresponding Windows coordinates as *WHome*. The Cartesian coordinates are declared through the structure *PT*. *COrigin* is the location of the antenna in Cartesian coordinates while *WOrigin* is the corresponding Windows coordinates. The number of targets is m out of which z is detected. The targets are numbered in order to differentiate one from another. The detected target number at time t is denoted as $Q[i][t]$.

Input consists of the edit boxes *eAlpha* and *eTheta* for the starting and sectorial angles, respectively. For antenna coverage *eW* is the input for the number of targets while *eRadius* is the range of coverage in the directional antenna. The output is the drawing area bounded by *home* and *end*. Output is also produced in the list view window through *table*.

TABLE 9.1

Important Variables and Objects in *Code9A*

Variable	Type	Description
alpha, beta, theta	double	Starting, ending, and the sectorial angle in the sweep, respectively
bCompute, bTarget, bNGraph	CButton	Buttons for *Compute*, *Generate Targets*, and *New Graph*, respectively
bOmni, bDir	CButton	Buttons for omnidirectional and directional transmissions, respectively
CHome, CEnd	PT	Windows top-left and bottom-right points, respectively, of the drawing area
COrigin, WOrigin	PT, CPoint	The Cartesian and Windows coordinates of the antenna, respectively
eW, eRadius	CEdit	Edit boxes for the number of targets and range of transmission, respectively
eAlpha, eTheta	CEdit	Edit boxes for the starting angle and sectorial angle, respectively
home, end	CPoint	Cartesian top-left and bottom-right points, respectively, of the drawing area
m, z	int	Total number of targets and number of targets detected, respectively
Q[i][t]	int	Target number i at timeslot t
radius	double	Radius of transmission of the antenna
t	int	Timeslot for the simulation
table	CListCtrl	List view window for displaying $Q[i][t]$
u[i]	TARGET	Target i

Figure 9.5 illustrates the organization of *Code9A* showing all the functions. The constructor *CCode9A()* creates the main window, an input edit box for the number of targets, and two buttons, *bNGraph* and *bCompute*. The function calls *Initialize()* to generate random coordinates for the positions of the targets in the drawing area. The targets are drawn using *DrawNodes()*. They are updated and displayed in the drawing area using *OnPaint()*.

The button *bNGraph* calls *OnNewGraph()* to refresh the graph with new coordinates for the targets. The function calls *Initialize()* for the new random coordinates and updates the

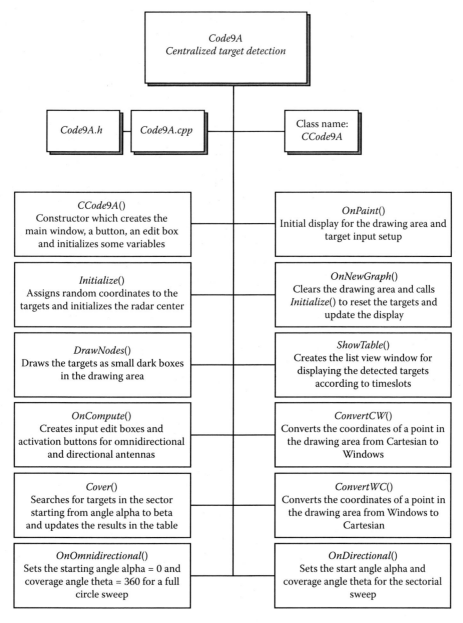

FIGURE 9.5
Organization of *Code9A*.

display through *Invalidate()*. The other button *bCompute* calls *OnCompute()* to read the number of targets from the input box. *OnCompute()* creates edit boxes for the radius, starting angle, and sectorial angle as inputs.

OnCompute() also creates *bOmni* for the omnidirectional and *bDir* for the directional buttons. These two buttons are important as they activate the two modes of transmission. *bOmni* calls *OmniDirectional()* to perform the search for targets according to the omnidirectional transmission, while *bDir* does the directional transmission by calling *OnDirectonal()*.

In the sectorial search, *OnDirectional()* reads *alpha* and *theta* values before calling *Cover()* to determine the angle φ and distance from the center r_u based on the target's coordinates. These two values are evaluated from Equations 9.3 and 9.4, respectively. Relative to the center, the sector may point toward the first, second, third, or fourth quadrants. These relative directions can be determined by comparing the coordinates of the target with its center. The following codes inside *Cover()* verify the direction:

```
tau=atan(u[i].CHome.y-c.y)/(u[i].CHome.x-c.x);
if (u[i].CHome.x>=c.x && u[i].CHome.y>=c.y)
     phi=fabs(tau);
if (u[i].CHome.x<=c.x && u[i].CHome.y>=c.y)
     phi=PI-fabs(tau);
if (u[i].CHome.x<=c.x && u[i].CHome.y<=c.y)
     phi=PI+fabs(tau);
if (u[i].CHome.x>=c.x && u[i].CHome.y<=c.y)
     phi=2*PI-fabs(tau);
tmp=sqrt(pow(u[i].CHome.x-c.x,2.0)+pow(u[i].CHome.y-c.y,2.0));
if ((phi>=alpha && phi<=beta) && (tmp<=radius))
{
     z++;
     Q[z][t]=i;
}
```

The targets are detected if they belong to the circle or sector in the search operations. A detected target is listed as *Q[i][t]* where *i* is its identification number and *t* is the timeslot of the detection. The information is displayed in the list view window through *ShowTable()*.

The full source codes for *Code9A.h* and *Code9A.cpp* are given below:

```
//Code9A.h
#include <afxwin.h>
#include <fstream>
#include <afxcmn.h>
#define IDC_COMPUTE 500
#define IDC_NEWGRAPH 501
#define IDC_OMNI 502
#define IDC_DIR 503
#define M 50//max #targets
using namespace std;

typedef struct       //Cartesian points
{
     double x,y;
} PT;
```

```cpp
typedef struct
{
     PT CHome;
     CPoint WHome;
} TARGET;
class CCode9A : public CFrameWnd
{
private:
     TARGET u[M+1];
     PT COrigin;
     CPoint WOrigin;
     double radius,alpha,theta;
     int Q[M+1][M+1];
     int idc,m,t,z;    //m=#targets, z=#targets detected, t=time
     CListCtrl table;
     CEdit eW,eRadius,eAlpha,eTheta;
     CStatic sRadius,sAlpha,sTheta,sTime;
     CButton bCompute,bTarget,bNGraph;
     CButton bOmni,bDir;
     CPoint home,end; //Windows home, end
     CFont fArial6,fArial8,fArial10;
     PT CHome,CEnd;    //CHome,CEnd=Cartesian home,end
public:
     CCode9A();
     ~CCode9A()              {}
     void Initialize();
     CPoint ConvertCW(PT);
     PT ConvertWC(CPoint);
     void DrawNodes(),Cover(bool),ShowTable();
     afx_msg void OnCompute();
     afx_msg void OnNewGraph();
     afx_msg void OnOmnidirectional();
     afx_msg void OnDirectional();
     afx_msg void OnPaint();
     DECLARE_MESSAGE_MAP();
};

class CMyWinApp : public CWinApp
{
public:
     virtual BOOL InitInstance();
};
CMyWinApp MyApplication;

BOOL CMyWinApp::InitInstance()
{
     m_pMainWnd=new CCode9A;
     m_pMainWnd->ShowWindow(m_nCmdShow);
     return TRUE;
}

//Code9A.cpp
#include "Code9A.h"
#define PI 3.142
```

```
BEGIN_MESSAGE_MAP(CCode9A,CFrameWnd)
    ON_WM_PAINT()
    ON_BN_CLICKED(IDC_COMPUTE,OnCompute)
    ON_BN_CLICKED(IDC_NEWGRAPH,OnNewGraph)
    ON_BN_CLICKED(IDC_OMNI,OnOmnidirectional)
    ON_BN_CLICKED(IDC_DIR,OnDirectional)
END_MESSAGE_MAP()

CCode9A::CCode9A()
{
    idc=400;
    Create(NULL,L"Target tracking",WS_OVERLAPPEDWINDOW,CR
    ect(0,0,1000,800));
    home=CPoint(10,70); end=CPoint(600,660);
    CHome.x=-5; CHome.y=5; CEnd.x=5; CEnd.y=-5;
    eW.Create(WS_CHILD | WS_VISIBLE | WS_BORDER | SS_CENTER,
        CRect(CPoint(home.x+130,home.y-50),CSize(100,25)),this,idc++);
    bCompute.Create(L"Generate Targets",WS_CHILD | WS_VISIBLE |
    BS_DEFPUSHBUTTON,
        CRect(CPoint(home.x+250,home.y-50),CSize(200,25)),this,
        IDC_COMPUTE);
    bNGraph.Create(L"New Graph",WS_CHILD | WS_VISIBLE |
    BS_DEFPUSHBUTTON,
        CRect(CPoint(home.x+470,home.y-50),CSize(120,25)),this,
        IDC_NEWGRAPH);
    fArial6.CreatePointFont(60,L"arial");
    fArial8.CreatePointFont(80,L"arial");
    fArial10.CreatePointFont(100,L"arial");
    Initialize();
}

void CCode9A::OnNewGraph()
{
    CClientDC dc(this);
    CRect rct;
    CBrush pErase(RGB(255,255,255));
    rct=CRect(home.x,home.y,end.x,end.y);
    dc.FillRect(rct,&pErase);
    Initialize(); Invalidate();
}

void CCode9A::Cover(bool flag)
{
    CClientDC dc(this);
    double beta,phi,tau,tmp;
    int i;
    CString s;
    CPen pRed(PS_SOLID,1,RGB(200,0,0));
    PT a,b,c;
    CPoint A,B,C;
    eRadius.GetWindowTextW(s); radius=_ttof(s);
    radius=((radius<=0 || radius>6)?4:radius);
    c=COrigin; C=ConvertCW(c);
    a.x=c.x+radius*cos(alpha);
```

```
        a.y=c.y+radius*sin(alpha);
        A=ConvertCW(a);
        beta=alpha;
        while (beta<=alpha+theta)
        {
            b.x=c.x+radius*cos(beta);
            b.y=c.y+radius*sin(beta);
            B=ConvertCW(b); dc.SetPixel(B.x,B.y,RGB(200,0,0));
            beta +=0.01;
        }
        if (flag)
        {
            dc.MoveTo(C); dc.LineTo(A);
            dc.MoveTo(B); dc.LineTo(C);
        }
        s.Format(L"t=%d",t); dc.TextOutW(end.x+50,home.y-30,s);

        //determine if u[i] inside the sector
        z=0;
        for (i=1;i<=m;i++)
        {
            tau=atan(u[i].CHome.y-c.y)/(u[i].CHome.x-c.x);
            if (u[i].CHome.x>=c.x && u[i].CHome.y>=c.y)
                phi=fabs(tau);
            if (u[i].CHome.x<=c.x && u[i].CHome.y>=c.y)
                phi=PI-fabs(tau);
            if (u[i].CHome.x<=c.x && u[i].CHome.y<=c.y)
                phi=PI+fabs(tau);
            if (u[i].CHome.x>=c.x && u[i].CHome.y<=c.y)
                phi=2*PI-fabs(tau);
            tmp=sqrt(pow(u[i].CHome.x-c.x,2.0)+pow(u[i].CHome.y-c.y,2.0));
            if ((phi >=alpha && phi<=beta) && (tmp<=radius))
            {
                z++;
                Q[z][t]=i;
            }
        }
        ShowTable();
}


void CCode9A::ShowTable()
{
        CString s;
        int i,j;
        table.DestroyWindow();
        table.Create(WS_VISIBLE | WS_CHILD | WS_BORDER | LVS_REPORT
                    | LVS_NOSORTHEADER,CRect(end.x+20,home.y+200,
                    end.x+350,end.y),this,idc++);
        for (j=0;j<=t;j++)  //j=column
        {
            s.Format(L"%d",j);
            table.InsertColumn(j,s,LVCFMT_CENTER,30);
        }
```

```
        for (i=1;i<=m;i++)   //i=row
        {
            if (Q[i][0]>=1 && Q[i][0]<=m)
            {
                s.Format(L"%d",Q[i][0]); table.InsertItem(i-1,s,0);
            }
            for (j=1;j<=t;j++)
                if (Q[i][j]>=1 && Q[i][j]<=m)
                {
                    s.Format(L"%d",Q[i][j]); table.SetItemText(i-1,j,s);
                }
        }
}


void CCode9A::OnOmnidirectional()
{
        alpha=0; theta=360;
        alpha *= PI/180; theta *= PI/180;
        Cover(0);
        t++;
}


void CCode9A::OnDirectional()
{
        CString s;
        eAlpha.GetWindowTextW(s); alpha=_ttof(s);
        eTheta.GetWindowTextW(s); theta=_ttof(s);
        alpha *= PI/180; theta *= PI/180;
        Cover(1);
        t++;
}


void CCode9A::OnPaint()
{
        CPaintDC dc(this);
        CRect rct;
        CPen pBlack(PS_SOLID,2,RGB(0,0,0));
        dc.SelectObject(fArial10);
        dc.TextOutW(home.x+20,home.y-50,L"#Targets (2-50)");
        rct=CRect(home.x-2,home.y-2,end.x+2,end.y+2);
        dc.SelectObject(pBlack);
        dc.Rectangle(rct);
}


void CCode9A::OnCompute()
{
        CString s;
        bOmni.DestroyWindow(); bDir.DestroyWindow();
        eRadius.DestroyWindow(); eAlpha.DestroyWindow(); eTheta.
        DestroyWindow();
        sRadius.DestroyWindow(); sAlpha.DestroyWindow(); sTheta.
        DestroyWindow();
        eW.GetWindowText(s); m=_ttoi(s);
        m=((m<=0 && m>M)?10:m);
```

```
    Initialize(); DrawNodes();
    sRadius.Create(L"Radius (1-6)",WS_CHILD | WS_VISIBLE | SS_CENTER,
        CRect(CPoint(end.x+10,home.y+5),CSize(80,20)),this,idc++);
    eRadius.Create(WS_CHILD | WS_VISIBLE | WS_BORDER | SS_CENTER,
        CRect(CPoint(end.x+100,home.y),CSize(70,25)),this,idc++);
    sAlpha.Create(L"Alpha",WS_CHILD | WS_VISIBLE | SS_CENTER,
        CRect(CPoint(end.x+30,home.y+45),CSize(50,20)),this,idc++);
    eAlpha.Create(WS_CHILD | WS_VISIBLE | WS_BORDER | SS_CENTER,
        CRect(CPoint(end.x+100,home.y+40),CSize(70,25)),this,idc++);
    sTheta.Create(L"Theta",WS_CHILD | WS_VISIBLE | SS_CENTER,
        CRect(CPoint(end.x+30,home.y+85),CSize(50,20)),this,idc++);
    eTheta.Create(WS_CHILD | WS_VISIBLE | WS_BORDER | SS_CENTER,
        CRect(CPoint(end.x+100,home.y+80),CSize(70,25)),this,idc++);
    bOmni.Create(L"Omnidirectional",WS_CHILD | WS_VISIBLE |
    BS_DEFPUSHBUTTON,
        CRect(CPoint(end.x+20,home.y+130),CSize(200,25)),this,IDC_OMNI);
    bDir.Create(L"Directional",WS_CHILD | WS_VISIBLE | BS_DEFPUSHBUTTON,
        CRect(CPoint(end.x+20,home.y+170),CSize(200,25)),this,IDC_DIR);

}

void CCode9A::DrawNodes()
{
    CClientDC dc(this);
    CString s;
    CBrush bGray(RGB(100,100,100));
    CBrush bRed(RGB(200,0,0));
    CBrush bWhite(RGB(255,255,255));
    CRect rct;
    rct=CRect(home.x,home.y,end.x,end.y);
    dc.FillRect(rct,&bWhite);
    rct=CRect(WOrigin.x-5,WOrigin.y-5,WOrigin.x+5,WOrigin.y+5);
    dc.FillRect(rct,&bRed);
    dc.SelectObject(fArial6);
    for (int i=1;i<=m;i++)
    {
        rct=CRect(u[i].WHome.x,u[i].WHome.y,u[i].WHome.x+5,u[i].
        WHome.y+5);
        dc.FillRect(rct,&bGray);
        s.Format(L"%d",i); dc.TextOutW(u[i].WHome.x-9,u[i].
        WHome.y-8,s);
    }
    dc.SetTextColor(RGB(0,0,0)); dc.SelectObject(fArial8);
}

void CCode9A::Initialize()
{
    t=0;
    srand(time(0));
    for (int i=1;i<=M;i++)
    {
        u[i].WHome.x=home.x+10+rand()%(end.x-home.x-30);
        u[i].WHome.y=home.y+10+rand()%(end.y-home.y-30);
        u[i].CHome=ConvertWC(u[i].WHome);
    }
```

```
        COrigin.x=0.0; COrigin.y=0.0;
        WOrigin=ConvertCW(COrigin);
        bTarget.DestroyWindow();
}

CPoint CCode9A::ConvertCW(PT p)          //converts C to W
{
        CPoint q;
        double m1,c1,m2,c2;
        m1=(double)(end.x-home.x)/(CEnd.x-CHome.x);
        c1=(double)(home.x-CHome.x*m1);
        m2=(double)(end.y-home.y)/(CEnd.y-CHome.y);
        c2=(double)(home.y-CHome.y*m2);
        q.x=p.x*m1+c1; q.y=p.y*m2+c2;
        return q;
}

PT CCode9A::ConvertWC(CPoint q)          //converts C to W
{
        PT p;
        double m1,c1,m2,c2;
        m1=(CEnd.x-CHome.x)/(double)(end.x-home.x);
        c1=CHome.x-(double)home.x*m1;
        m2=(CEnd.y-CHome.y)/(double)(end.y-home.y);
        c2=CHome.y-(double)home.y*m2;
        p.x=q.x*m1+c1; p.y=q.y*m2+c2;
        return p;
}
```

9.3 Wireless Sensor Networks

A *sensor* is a transducer that collects information about its surroundings in the form of electrical or optical signals according to its purpose. A sensor becomes activated when it receives a signal in the form of an electromagnetic wave or other waves. There are also sensors for detecting the temperature, infrared, ultraviolet, and other radiation rays that cannot be seen with the naked eye. Some common sensors are summarized below.

A *thermal sensor* becomes active when the temperature at its surroundings reaches certain values and it responds immediately by sending a message to its controller as an alert. An *acoustic wave sensor* is an electronic device called a transducer which converts an input electrical signal that represents a physical event into a mechanical wave, and then bounces back the electrical signal to the sensor.

A *gas sensor* detects the presence of certain gases and the concentration of gas in its vicinity. Different gases are identified through their voltage readings based on the electric field at which the sensor is ionized. A *chemical sensor* is an electronic device that transforms chemical information into an analytical signal. The chemical information includes the composition, concentration, chemical activity, and the presence of a particular element or ion. Some examples of chemical sensors are carbon monoxide detectors, glucose detectors, and nanosensors.

A *photoelectric sensor* detects the presence of an object and its distance using a light transmitter in the form of infrared and photoelectric receivers. The sensor is activated when the light beam emitted from the transmitter is blocked from reaching its receiver. The device is used in smoke detectors and for purposes such as remote sensing and building intrusion control.

An *ultrasonic sensor* is a device that transmits ultrasonic waves from its sensor head and receives the waves reflected from an object. The sensor is usually used for liquid level detection in processing plants. The emitted ultrasonic waves have the capability to penetrate metal or nonmetallic objects as well as clear or opaque liquid and gases.

A *wireless sensor network* is a network consisting of wireless sensors that are capable of forming a network by themselves. The sensors, more commonly called *motes*, can be one of the types discussed above with the main purpose of detecting changes in their surroundings. Figure 9.6 shows a typical wireless sensor architecture that consists of a central processing unit, a small memory module, a transceiver, a battery pack, and a sensing unit. A wireless sensor has a compact design and small built-up space for housing only essential items related to its operation. The device is powered by a small battery that can last for just a few hours. Because of this limitation, the sensor maintains its lifetime by becoming inactive (sleeping) most of the time and becomes active only if it is needed. For communication, a sensor has a small transmitter and receiver, collectively called a transceiver, which allows it to communicate with other sensors. A typical sensor has a transmission and receiving range of about 5 to 30 meters, which is sufficient to cover a small workplace. Having a large number of sensors of this kind in a geographical region can lead to the formation of a self-organized wireless sensor network. The network covers the region effectively this way as its sensors collectively work in a group.

At any given time a wireless sensor can only do one task: transmit a message, receive a message, sleep, and idle. A wireless sensor can save energy by controlling its range of transmission according the actual requirement during its operation. The adjustable range is an important feature of a sensor that allows it to communicate with other sensors. A high range of transmission makes it possible for the sensor to reach many other sensors but it will have heavy energy consumption, while a lower-range transmission will save energy but the area of coverage will be smaller. Therefore, a common strategy for a sensor is to perform a preliminary survey on the network by communicating with other sensors

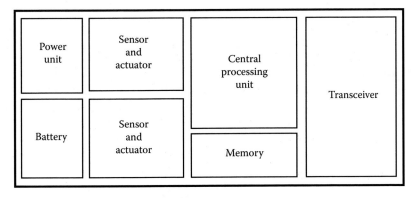

FIGURE 9.6
Typical wireless sensor architecture.

at a low transmission range. High transmissions are applied only when the need for communication with many sensors arises.

Wireless sensor networks have many useful applications in everyday life. In a traffic model called VANET, wireless sensors are used to monitor the traffic flow on streets of a city. The underlying concept is that a motion sensor placed on a street can pick up the speed of the passing cars. Cars traveling at high speeds on a street suggest the street is not busy, whereas slow-moving vehicles indicate the street is congested. With sensors strategically placed at several critical sections of the city the information received by the control center can then be used to draw up the overall traffic condition of the city.

Wireless sensors are also deployed in fire rescue operation. For example, in a massive forest fire an aircraft is used to deploy hundreds of heat-sensitive sensors over the region. The sensors return information about their coverage regions and the information is used by the control center to identify the hot spot areas of the whole region. This helps in preparing the search and rescue operation for the critical areas.

9.3.1 Problems in Wireless Sensor Networks

A wireless sensor network can be modeled as a graph. Each sensor in the network is a node in the graph where the communicating range of the sensors serves as the edges of the graph. Two sensors that are within the communicating range of each other can be modeled as two adjacent nodes in the graph. Hence, a wireless sensor network inherits most of the common properties of a graph. For example, the shortest path algorithm is always applied in transmitting a message from one sensor to another in the network. In broadcasting a message from one sensor to all other sensors the path follows the minimum spanning tree of the graph.

A number of problems are associated with wireless sensor networks. The problems range from the setup of a wireless network to the security issues in the handling of information. In localization, efficient algorithms are applied to locate sensors in an area. The sensors are originally distributed unevenly in an area with no known infrastructure to locate them. The sensors do not possess ids that can identify them and also do not have fixed Internet protocol addresses with regard to their locations. However, with the presence of a transmitter and a receiver in each sensor, communication between the sensors can be established. This helps in the formation of the network as messages sent by individual sensors are used to define the connectivity between them. Some interesting discussions on techniques and algorithms for localization can be found in [4].

Another interesting problem in wireless sensor networks is maximizing the network lifetime. An average sensor can last for only a few hours if it is left operating continuously. Therefore, various strategies have been applied to prolong the lifetime of the network so that it can maintain its presence for a few days instead of a few hours. One common strategy is to alternate the sleep and active cycle in each sensor so that the sensor becomes active only when it is needed. This becomes a scheduling problem where some sensors are active and others are inactive at any given time, similar to having a group of soldiers patrolling a very important area within a hostile region. With the sleep-active cycle the lifetime of each sensor increases and this contributes toward the overall increase in the lifetime of the network.

The network lifetime can also be increased by applying optimal algorithms for energy-efficient routing and data gathering. For example, in transmitting data from one node to another the route should follow Dijkstra's shortest path algorithm. Several graph methods should limit unnecessary and redundant operations that contribute to saving energy for the sensors. Energy can also be saved by controlling the range of transmission in the transmitter of the sensors. A smaller range of transmission uses less power and this may be sufficient for some operations of the sensors. Also, applying the directional transmission method rather than omnidirectional transmission may be practical because it consumes less power. Some interesting articles on the power-saving techniques for wireless sensor networks can be found in [3,5,6]. In [3,5], detection of targets by the sensors is performed through cover sets. A heuristical algorithm using learning automata was introduced to produce the coverage of the targets that maximizes the network lifetime. In [6], a power control routing algorithm based on a genetic algorithm was proposed in routing between the sensors to maximize the network lifetime.

Message transmission between sensors in a wireless sensor network is another common issue. When a sensor transmits a message the transmission is received by all its neighboring sensors. This causes flooding; therefore, the message should be filtered so that only the intended sensors will receive it. Various strategies and efficient message transmission algorithms have been proposed as the solutions to this problem. A good discussion on the problem and the strategies for its solution can be found in [7].

9.3.2 Target Coverage Problem and Its Graph Model

The *target coverage problem* is the problem of grouping sensors into two or more cover sets for scheduling the detection of the targets in a geographical region. Each cover set has at least one sensor for covering the targets. For efficient coverage, two or more cover sets are needed where the members of a cover set may overlap with other sets.

The target coverage problem is defined as follows: Given a set of wireless sensors that have the capabilities to transmit and receive messages within a given range, each transmission from the sensor's antenna consumes power from its battery that is proportional to the range of transmission and angular coverage. High-power consumption results from a high range of transmission and large angular coverage. How can the sensors cover all the targets in the region and control their transmissions to save power in such a way that will maximize the network lifetime? The main idea here is to schedule the sensors in locating the targets on the coverage area in minimum time, which contributes in prolonging the lifetime of network.

Figure 9.7 shows the formation of two cover sets $S_1 = \{s_1, s_2, s_3\}$ and $S_2 = \{s_3, s_4, s_5\}$ from sensors s_1, s_2, s_3, s_4, and s_5 for covering three targets r_1, r_2, and r_3. With this arrangement, S_1 covers r_1 and r_2, while S_2 covers r_3. A graph is formed from this relationship, and the

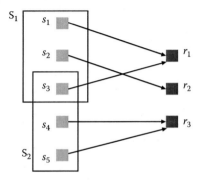

FIGURE 9.7
Two cover sets for detecting three targets.

solution to the target coverage is obtained by applying some common properties of the graph such as matching, mapping, and routing. For example, in broadcasting a message from one sensor to another, the shortest path solution using Dijkstra's algorithm can be applied. If the message needs to be sent to all sensors then the path from the sender is the minimum spanning tree of the graph.

Several targets with known locations are deployed in a two-dimensional Euclidean plane that must be continuously monitored. A number of sensors are scattered randomly and uniformly within the field to cover the targets. A target is covered if it lies within the sensing range of at least one sensor. Note that at least one sensor covers each target and there may be overlapped targets simultaneously covered by adjacent sensors. All deployed sensors are homogeneous in terms of sensing range and initial energy. In this study, the sensors can adopt either an active or passive state. Active sensors monitor the targets while others go into sleep mode to save their energy.

Suppose m targets with known locations and n directional sensors with adjustable sensing ranges are deployed randomly in the vicinity of the targets to monitor them. Initially, sensors have a certain battery lifetime. Each sensor can be in either active or inactive mode. Each active sensor can monitor targets located in one of its nonoverlapping directions at any given time. When a sensor is in active mode, its power consumption is dependent on its sensing range. In contrast, inactive sensors do not monitor any target and their power consumption is negligible.

The technique of scheduling sensor nodes' activity has the advantage of redundancy in sensor deployment. It organizes sensors into a number of cover sets, each of which can monitor all the targets. Additionally, it determines the amount of time each cover set can be activated. Afterward, the cover sets are activated successively for a predetermined duration. When a cover set is active, the sensors belonging to other cover sets are in inactive mode. This significantly increases the network lifetime because inactive sensors consume considerably less energy than the active ones, and if the battery of a sensor frequently oscillates between active and inactive modes, it lasts for a longer duration.

9.3.3 *Code9B*: Directional Sensors for Distributed Target Detection

The detection of targets by sensors is based on the concept that the waves produced by the targets are captured by the sensors. A transmission from a sensor is wave propagation according to certain wavelength. When the wave hits the target it bounces back to the sensor. With the receipt of the bounced wave, a sensor gets the information that the target is in its transmission range.

Code9B is a target detection model using wireless sensors with directional antenna. The program is a distributed model for detecting targets where n sensors are deployed to detect m targets. Figure 9.6 is the sample output from the program displaying a set of 10 sensors in the drawing area distributed at random locations for detecting six targets. The input consists of edit boxes for the number of sensors and the number of targets. Each directional sensor is beamed in the counterclockwise direction with the starting angle α and ending angle β measured from the x-axis.

The simulation in *Code9B* is performed to test the network on target detection with a constraint on its energy consumption. Each sensor in the network is assumed to have an initial power value of 10 units at the start-up and this value decreases by one unit every time the sensor is activated. A sensor is said to be activated when it beams at a sectorial angle of θ in the direction at the starting angle of α from the x-axis. The range of transmission is assumed to a constant for all sensors. At time t a random number of sensors are selected for activation. Those selected have their allocated power value reduced by one unit. When the power decreases to zero value the sensor is considered dead and it ceases its duty altogether.

The targets in this model are scattered randomly. It is assumed that the targets can receive transmission signals from the sensors in the area of transmission and reflect the signals back to the sensors. A signal received by the sensor this way indicates the presence of the target in the area of coverage. A small range of transmission means a small area of coverage, which suggests that the location of the target can easily be known. In contrast, a large area of coverage that arises from a large transmission range requires further fine tuning in order to get the precise location of the target.

Figure 9.8 displays 6 active sensors out of 10 at an instance of time $t = 5$. For simplicity in the model description, the sensors and targets are numbered for identification purposes. Each sensor beams at the indicated starting angle of α and ending angle of β to give the sectorial angle of θ, where $\theta = \beta - \alpha$. Both α and β are assigned as random values from 0 to 360 (degrees), and their assigned values are shown in the static boxes. There are six targets that are marked as red circles.

The targets are considered detected if they belong to a coverage sector. Therefore, from the figure three targets are detected:

u_1 by v_1 and v_9,

u_2 by v_8,

u_5 by v_1 and v_2

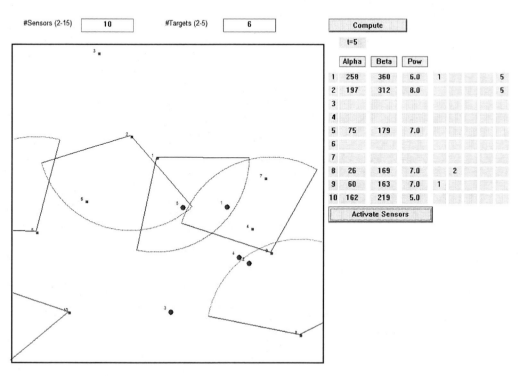

FIGURE 9.8
Output from *Code9B*.

Code9B has a single class called *CCode9B*, and the source files *Code9B.h* and *Code9B.cpp*. The variables for the sensors are grouped into a structure called *SENSOR* and the targets have *TARGET* as its structure. *SENSOR* is represented by the array *v* as its object while *TARGET* has the array *u*.

The sensor v_i for $i = 1, 2,...,n$ has *CHome* and its corresponding *WHome* as its home coordinates. *Power* denotes the power allocated to the sensor that initially set at 10. The state of the sensor at time *t* is represented by the variable *status*: $v_i.status = 0$ denotes that the sensor is sleeping, $v_i.status = 1$ means the sensor is active, while $v_i.status = 2$ is an indicator that the sensor is running out of power or is dead. When the sensor is active it transmits with range *R* for the sectorial angle of *theta* that starts at *alpha* and ends at *beta*. The targets captured by the sensor when it is active is stored as the array *nR*.

Table 9.2 lists important variables and objects in the project. The number of sensors and targets are *n* and *m*, respectively. The values are read from the edit boxes *enS* and *enT*. Two buttons form the main events: *bCompute* for producing the sensor and target nodes and *bSensor* for selecting sensors for activation. Sensors and their targets are distributed according to their random coordinates in the drawing area, which has *CHome* as its top-left point and *CEnd* as its bottom-right point.

A number of static boxes are used to display the current values from the transmissions: *sTime* for the current timeslot, *sv* for the sensor number, *sPow* for the power status of the sensor, *sA* for angle α, *sB* for angle β, and *sR* for the targets that the sensor captured.

Figure 9.9 shows the organization of *Code9B*. The processing starts with the constructor *CCode9B* that creates the main window, the *Compute* button, and the input boxes for the number of sensors and targets. The function also initializes some variables. The initial display from *OnPaint()* in the main window consists of the drawing area for the sensor and target nodes.

The mouse's left click on the *Compute* button triggers *OnCompute()* to ask for the number of sensors and targets that then creates these objects by calling *Initialize()*. *OnCompute()* calls *DrawNode()* to display the nodes in the drawing area. The function also creates the *Activate Sensor* button and the static boxes for displaying the transmission angles from each sensor.

TABLE 9.2

Important Variables and Objects in *Code9B*

Variable	Type	Description
v, u	SENSOR, TARGET	Sensor and target, respectively
bCompute, bSensor	CButton	Buttons for *Compute* and *Active Sensors*, respectively
n, m	int	Number of sensors and targets, respectively
CHome, CEnd	PT	Windows top-left and bottom-right points, respectively, of the drawing area
WHome, WEnd	CPoint	Cartesian top-left and bottom-right points, respectively, of the drawing area
sv, sPOW	CStatic	Display boxes for sensor identification number and its current power
sA, sB, sR	CStatic	Display boxes for starting and ending angles and the targets detected by the sensor
sTime	CStatic	Display box for timeslot *t*
enS, enT	CEdit	Input boxes for the number of sensors and targets, respectively

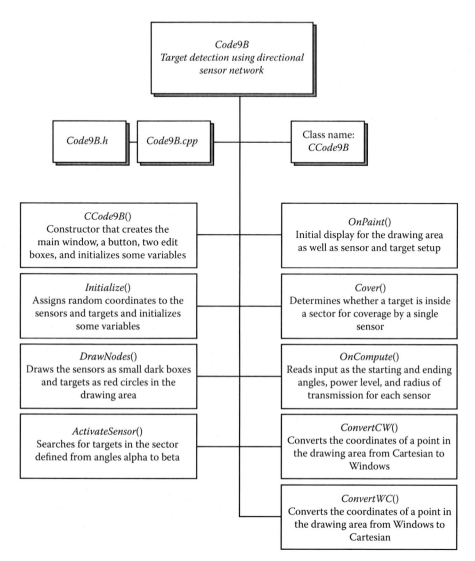

FIGURE 9.9
Organization of *Code9B*.

The *Activate Sensor* button is an event that is handled by the function *ActivateSensor*(). The function is the main engine in this application as it activates a random number of sensors to perform the directional transmissions for detecting the targets. Each selected sensor v_i is assigned with an *alpha* value as the starting angle for the sectorial sweep of *theta*. A target can be verified to be in the coverage sector by comparing its coordinates relative to the sensor's coordinates. The relative position can be in the first, second, third, or fourth angular quadrants. The exact quadrant of the target's position is determined through *Cover*().

An activated sensor at time t has its power level reduced by one unit. Every detected target u_i by the sensor at time t updates the value of $Q[i][t]$ and this matrix value is displayed through *ShowTable*().

The full listing of the codes are given below:

```
//Code9A.h
#include <afxwin.h>
#include <fstream>
#define IDC_COMPUTE 500
#define IDC_ANGLE 501
#define M 15        //max #sensors
#define N 5         //max #targets
using namespace std;

typedef struct                     //Cartesian points
{
    double x,y;
} PT;

typedef struct
{
    PT CHome;
    CPoint WHome;
    double alpha,beta;   //alpha=start, beta=WEnd
    double R;                       //radius
    double Power;
    int status;                     //0=sleep, 1=active, 2=dead
    int nR[N+1];
} SENSOR;

typedef struct
{
    PT CHome;
    CPoint WHome;
} TARGET;

class CCode9B: public CFrameWnd
{
private:
    SENSOR v[M+1];
    TARGET u[N+1];
    int idc,m,n,t;
    CEdit enS,enT;
    CStatic sv[M+1],sPow[M+1],sA[M+1],sB[M+1],sR[M+1][N+1],sTime;
    CButton bCompute,bSensor;
    CPoint WHome,WEnd;            //Windows WHome, WEnd
    CFont fArial6,fArial8,fArial10;
    PT CHome,CEnd;                //CHome,CEnd=Cartesian WHome,WEnd
                                 public:
    CCode9B();
    ~CCode9B()              {}
    void Initialize();
    CPoint ConvertCW(PT);
    PT ConvertWC(CPoint);
    void Cover(CPoint,CPoint,double);
    void DrawNodes();
```

```
        afx_msg void OnCompute();
        afx_msg void OnPaint();
        afx_msg void ActivateSensor();
        DECLARE_MESSAGE_MAP();
};

class CMyWinApp: public CWinApp
{
public:
        virtual BOOL InitInstance();
};
CMyWinApp MyApplication;

BOOL CMyWinApp::InitInstance()
{
        m_pMainWnd=new CCode9B;
        m_pMainWnd->ShowWindow(m_nCmdShow);
        return TRUE;
}

//Code9B.cpp
#include "Code9B.h"
#define PI 3.142

BEGIN_MESSAGE_MAP(CCode9B,CFrameWnd)
        ON_WM_PAINT()
        ON_BN_CLICKED(IDC_COMPUTE,OnCompute)
        ON_BN_CLICKED(IDC_ANGLE,ActivateSensor)
END_MESSAGE_MAP()

CCode9B::CCode9B()
{
        idc=400;
        Create(NULL,L"WSN Target Coverage",WS_OVERLAPPEDWINDOW,CR
        ect(0,0,1000,800));
        WHome=CPoint(10,70); WEnd=CPoint(600,660);
        CHome.x=-5; CHome.y=5; CEnd.x=5; CEnd.y=-5;
        enS.Create(WS_CHILD | WS_VISIBLE | WS_BORDER | SS_CENTER,
            CRect(CPoint(WHome.x+130,WHome.y-50),CSize(100,25)),this,idc++);
        enT.Create(WS_CHILD | WS_VISIBLE | WS_BORDER | SS_CENTER,
            CRect(CPoint(WHome.x+400,WHome.y-50),CSize(100,25)),this,idc++);
        bCompute.Create(L"Compute",WS_CHILD | WS_VISIBLE | BS_DEFPUSHBUTTON,
            CRect(CPoint(WHome.x+600,WHome.y-50),CSize(150,25)),this,
            IDC_COMPUTE);
        fArial6.CreatePointFont(60,L"arial");
        fArial8.CreatePointFont(80,L"arial");
        fArial10.CreatePointFont(100,L"arial");
}

void CCode9B::OnPaint()
{
        CPaintDC dc(this);
        CRect rct;
```

```
        CPen pBlack(PS_SOLID,2,RGB(0,0,0));
        dc.SelectObject(fArial10);
        dc.TextOutW(WHome.x+20,WHome.y-50,L"#Sensors (2-15)");
        dc.TextOutW(WHome.x+290,WHome.y-50,L"#Targets (2-5)");
        rct=CRect(WHome.x-2,WHome.y-2,WEnd.x+2,WEnd.y+2);
        dc.SelectObject(pBlack);
        dc.Rectangle(rct);
}


void CCode9B::OnCompute()
{
        CString s;
        int i,j;
        enS.GetWindowText(s); m=_ttoi(s);
        enT.GetWindowText(s); n=_ttoi(s);
        m=((m<=M)?m:M); n=((n<=N)?n:N);
        Initialize();
        sTime.Create(L"t=0",WS_CHILD | WS_VISIBLE | SS_CENTER,
            CRect(CPoint(WEnd.x+30,WHome.y-15),CSize(50,20)),this,idc++);
        sA[0].Create(L"Alpha",WS_CHILD | WS_VISIBLE | WS_BORDER | SS_CENTER,
            CRect(CPoint(WEnd.x+30,WHome.y+20),CSize(50,20)),this,idc++);
        sB[0].Create(L"Beta",WS_CHILD | WS_VISIBLE | WS_BORDER | SS_CENTER,
            CRect(CPoint(WEnd.x+90,WHome.y+20),CSize(50,20)),this,idc++);
        sPow[0].Create(L"Pow",WS_CHILD | WS_VISIBLE | WS_BORDER | SS_CENTER,
            CRect(CPoint(WEnd.x+150,WHome.y+20),CSize(50,20)),this,idc++);
        bSensor.Create(L"Activate Sensors",WS_CHILD | WS_VISIBLE |
        BS_DEFPUSHBUTTON,
            CRect(CPoint(WEnd.x+10,WHome.y+50+25*m),CSize(200,30)),this,
            IDC_ANGLE);
        for (i=1;i<=m;i++)
        {
            s.Format(L"%d",i);
            sv[i].Create(s,WS_CHILD | WS_VISIBLE | SS_CENTER,
                CRect(CPoint(WEnd.x+10,WHome.y+50+25*(i-1)),CSize(18,20)),
                this,idc++);
            sA[i].Create(L"",WS_CHILD | WS_VISIBLE | SS_CENTER,
                    CRect(CPoint(WEnd.x+30,WHome.y+50+25*(i-1)),
                    CSize(50,20)),this,idc++);
            sB[i].Create(L"",WS_CHILD | WS_VISIBLE | SS_CENTER,
                CRect(CPoint(WEnd.x+90,WHome.y+50+25*(i-1)),CSize(50,20)),
                this,idc++);
            s.Format(L"%.1lf",v[i].Power);
            sPow[i].Create(s,WS_CHILD | WS_VISIBLE | SS_CENTER,
                CRect(CPoint(WEnd.x+150,WHome.y+50+25*(i-1)),CSize(50,20)),
                this,idc++);
            for (j=1;j<=5;j++)
                sR[i][j].Create(L"",WS_CHILD | WS_VISIBLE | SS_CENTER,
                CRect(CPoint(WEnd.x+210+30*(j-1),
                WHome.y+50+25*(i-1)),CSize(25,20)),this,idc++);
        }
        DrawNodes();
}
```

```
void CCode9B::DrawNodes()
{
    CClientDC dc(this);
    CString s;
    int i;
    CBrush bGray(RGB(100,100,100));
    CBrush bWhite(RGB(255,255,255));
    CBrush bRed(RGB(200,0,0));
    CRect rct;
    rct=CRect(WHome.x,WHome.y,WEnd.x,WEnd.y);
    dc.FillRect(rct,&bWhite);
    dc.SelectObject(fArial6);
    for (i=1;i<=m;i++)
    {
        rct=CRect(v[i].WHome.x,v[i].WHome.y,v[i].WHome.x+5,v[i].
        WHome.y+5);
        dc.FillRect(rct,&bGray);
        s.Format(L"%d",i); dc.TextOutW(v[i].WHome.x-9,v[i].WHome.y-8,s);
    }
    dc.SelectObject(bRed);
    dc.SetTextColor(RGB(200,0,0)); dc.SelectObject(fArial6);
    for (i=1;i<=n;i++)
    {
        rct=CRect(u[i].WHome.x,u[i].WHome.y,u[i].WHome.x+10,u[i].
        WHome.y+10);
        s.Format(L"%d",i); dc.TextOutW(u[i].WHome.x-8,u[i].WHome.y-8,s);
        dc.Ellipse(rct);
    }
    dc.SetTextColor(RGB(0,0,0)); dc.SelectObject(fArial8);
}

void CCode9B::ActivateSensor()
{
    CClientDC dc(this);
    CRect rct;
    CString s;
    int i,j,k;
    double h,tmp,phi,theta;      //h is increment
    PT a,b,c;
    CPoint A,B,C;                //C=center, A=begin, B=WEnd
    DrawNodes(); t++;
    s.Format(L"t=%d",t); sTime.SetWindowTextW(s);
    for (i=1;i<=m;i++)
    {
        for (j=1;j<=n;j++)
            sR[i][j].SetWindowTextW(L"");
        if (v[i].status!=2)
            v[i].status=rand()%2;
        if (v[i].status==1)
        {
            v[i].Power-;
            if (v[i].Power<=0)
```

```
{
    v[i].Power=0; v[i].status=2;
}
v[i].alpha=(double)(1+rand()%360);
theta=(double)(1+rand()%180);
if (theta==0.0)
    theta=30.0;
v[i].beta=v[i].alpha+theta;
if (v[i].beta>360)
    v[i].beta=360;
s.Format(L"%.0lf",v[i].alpha); sA[i].SetWindowText(s);
s.Format(L"%.0lf",v[i].beta); sB[i].SetWindowText(s);
s.Format(((v[i].Power>0)?L"%.1lf":L"dead"),v[i].Power);
sPow[i].SetWindowTextW(s);
v[i].alpha *= PI/180; v[i].beta *= PI/180;
h=(v[i].beta-v[i].alpha)/200;
c=v[i].CHome; C=ConvertCW(c);
a.x=c.x+v[i].R*cos(v[i].alpha);
a.y=c.y+v[i].R*sin(v[i].alpha);
A=ConvertCW(a);
Cover(A,C,v[i].alpha);
rct=CRect(WHome.x+3,WHome.y+3,WEnd.x-3,WEnd.y-3);
tmp=v[i].alpha;
while (tmp<=v[i].beta)
{
    b.x=c.x+v[i].R*cos(tmp);
    b.y=c.y+v[i].R*sin(tmp);
    B=ConvertCW(b);
    if (rct.PtInRect(B))
        dc.SetPixel(B.x,B.y,RGB(0,0,200));
    tmp += h;
        }
Cover(B,C,v[i].beta);
for (j=1;j<=n;j++)
{
    phi=atan((u[j].CHome.y-v[i].CHome.y)
        /(u[j].CHome.x-v[i].CHome.x));
    if (u[j].CHome.x>v[i].CHome.x && u[j].CHome.y>v[i].
    CHome.y)
        phi=phi;
    if (u[j].CHome.x<=v[i].CHome.x && u[j].CHome.y>=
    v[i].CHome.y)
        phi=PI-fabs(phi);
    if (u[j].CHome.x<=v[i].CHome.x && u[j].CHome.y<=
    v[i].CHome.y)
        phi=PI+fabs(phi);
    if (u[j].CHome.x>=v[i].CHome.x && u[j].CHome.y<=
    v[i].CHome.y)
        phi=2*PI-fabs(phi);
    tmp=sqrt(pow(u[j].CHome.x-v[i].CHome.x,2.0)
        +pow(u[j].CHome.y-v[i].CHome.y,2.0));
    if ((phi >=v[i].alpha && phi<=v[i].beta) && (tmp<=
    v[i].R))
```

```
                    {
                        k=++v[i].nR[0];    //nR[0] is #targets
                        v[i].nR[k]=j;      //nR[k] is target k
                        s.Format(L"%d",j);
                        sR[i][j].SetWindowTextW(s);
                    }
                }
            }
        if (v[i].status==0 || v[i].status==2)
        {
            s.Format(L"",v[i].alpha); sA[i].SetWindowText(s);
            s.Format(L"",v[i].beta); sB[i].SetWindowText(s);
            s.Format((v[i].status==2)?L"dead":L"");
            sPow[i].SetWindowTextW(s);
        }
    }
}


void CCode9B::Cover(CPoint D,CPoint C,double theta)
{
    CClientDC dc(this);
    PT d,c;
    CRect rct;
    rct=CRect(WHome.x,WHome.y,WEnd.x,WEnd.y);
    c=ConvertWC(C); d=ConvertWC(D);
    if (rct.PtInRect(D))
    {
        dc.MoveTo(D); dc.LineTo(C);
    }
    else
    {
        if (D.x>WEnd.x)           //right of box
        {
            d=ConvertWC(WEnd);
            d.y=c.y+(d.x-c.x)*tan(theta);
        }
        if (D.y<WHome.y)          //higher than box
        {
            d=ConvertWC(WHome);
            d.x=c.x+(d.y-c.y)/tan(theta);
        }
        if (D.x<WHome.x)          //left of box
        {
            d=ConvertWC(WHome);
            d.y=c.y+(d.x-c.x)*tan(theta);
        }
        if (D.y>WEnd.y)
        {
            d=ConvertWC(WEnd);
            d.x=c.x+(d.y-c.y)/tan(theta);
        }
        D=ConvertCW(d);
        dc.MoveTo(D); dc.LineTo(C);
    }
}
```

```
void CCode9B::Initialize()
{
    int i,j;
    t=0;
    srand(time(0));
    for (i=1;i<=M;i++)
    {
        sv[i].DestroyWindow(); sPow[i].DestroyWindow();
        sA[i].DestroyWindow(); sB[i].DestroyWindow();
        for (j=1;j<=N;j++)
            sR[i][j].DestroyWindow();
        v[i].WHome.x=WHome.x+10+rand()%(WEnd.x-WHome.x-30);
        v[i].WHome.y=WHome.y+10+rand()%(WEnd.y-WHome.y-30);
        v[i].CHome=ConvertWC(v[i].WHome);
        v[i].Power=10.0; v[i].R=3.0;
        v[i].nR[0]=0;
    }
    for (i=1;i<=N;i++)
    {
        u[i].WHome.x=WHome.x+10+rand()%(WEnd.x-WHome.x-30);
        u[i].WHome.y=WHome.y+10+rand()%(WEnd.y-WHome.y-30);
        u[i].CHome=ConvertWC(u[i].WHome);
    }
    sPow[0].DestroyWindow(); bSensor.DestroyWindow();
    sTime.DestroyWindow(); sA[0].DestroyWindow(); sB[0].DestroyWindow();
}


CPoint CCode9B::ConvertCW(PT p)              //converts C to W
{
    CPoint q;
    double m1,c1,m2,c2;
    m1=(double)(WEnd.x-WHome.x)/(CEnd.x-CHome.x);
    c1=(double)(WHome.x-CHome.x*m1);
    m2=(double)(WEnd.y-WHome.y)/(CEnd.y-CHome.y);
    c2=(double)(WHome.y-CHome.y*m2);
    q.x=p.x*m1+c1; q.y=p.y*m2+c2;
    return q;
}


PT CCode9B::ConvertWC(CPoint q)              //converts C to W
{
    PT p;
    double m1,c1,m2,c2;
    m1=(CEnd.x-CHome.x)/(double)(WEnd.x-WHome.x);
    c1=CHome.x-(double)WHome.x*m1;
    m2=(CEnd.y-CHome.y)/(double)(WEnd.y-WHome.y);
    c2=CHome.y-(double)WHome.y*m2;
    p.x=q.x*m1+c1; p.y=q.y*m2+c2;
    return p;
}
```

10

Network Routing Application

10.1 Network Routing Problem

Routing is a common problem about finding a feasible or optimal path for connecting two or more points in the network. Here, the network refers to any system with full infrastructure support such as transportation lines in roads and railways, circuits in electronic boards, and arteries in the human body. The main objective of routing is to provide a path that minimizes the communication or transportation costs between points. The costs in this context may refer to operational factors such as time, speed, distance, and monetary value. Hence, routing is inherently an optimization problem that has its roots in graph theory.

Routing is implemented in common networks such as printed circuit boards, telephone networks (circuit switching), social networks, electronic data networks, and transportation networks (resource distribution). In a printed circuit board, for example, routing between two points is derived from the necessity to connect the two points in order to minimize the congestion in the circuit. Space is very limited in the circuit, and at the same time the circuit needs to accommodate many pins, vias, transistors, diodes, and other electronic components. Therefore, a good routing technique that leads to low congestion makes it possible to achieve this objective.

10.2 Routing in a Reconfigurable Mesh Network

A *reconfigurable mesh network* is a massively parallel rectangular network that is capable of changing its connectivity according to the program requirements. A rectangular reconfigurable mesh network is arranged in a $m \times n$ grid with m rows and n columns of nodes. The network is supported by fast algorithms for bus configuration, communication, and constant-time computation with low complexity.

The reconfigurable mesh network is implemented as a parallel computation model that captures the salient features for solving problems in areas such as graph theory and digitized images. The conceptual model is machine-independent, which means that it can easily be adopted to any parallel computational model for a variety of high-performance computational problems. In the PRAM model [1], an array of processors is interconnected by a reconfigurable bus system for producing patterns of computation according to the problem requirements. Among the problems suitable for adoption using this arrangement are low-level arithmetic operations including number sorting and permutation whose results are produced in low complexity [2].

An architecture called a field-programmable gate array (FPGA) allows a designer to configure an integrated circuit after manufacturing using reconfigurable mesh technology and is discussed in [3]. The on-the-chip work relates FPGA to many core systems that have capabilities for multitasking and multithreading. Further extension of the work is the optical reconfigurable field programmable gate array (OFPGA) [4], which speeds up communication between processors at the speed of light in the reconfigurable architecture.

Figure 10.1 shows a 4 × 4 rectangular network (left) and toroid (right). The nodes are the shaded squares labeled from 1 to 16 while the circles indicate their ports. The rectangular mesh network has two sets of nodes: exterior nodes along the left, right, bottom, and top boundaries, and interior nodes. Each interior node in the network has a degree of four while the exterior nodes have degrees of two (at the corners) and three (other than the corners). A special type of rectangular mesh called a *toroid* has nodes that all have a degree of four. This is a good example of a regular graph. In this case, the nodes in the left boundary have their west port connected with the east ports of the nodes in the right boundary. Similarly, the north ports of the top nodes are connected with the south ports of the bottom nodes.

Each node in a reconfigurable mesh network has four ports labeled as north, south, east, and west, strategically located at the boundaries to allow fast communication with the node's neighbors. Figure 10.2 shows all the possible paths between the north (*n*), south (*s*), east (*e*), and west (*w*) ports in a node. As a rule, a path can be drawn internally between any two ports in the node. Two paths can be drawn simultaneously as long as they don't overlap with each other, as shown in the last three nodes. In the last node, two paths appear to be crossing each other at a point. However, one of the paths is raised above the other at the indicated point of intersection to avoid the crossing.

An interesting feature about the nodes in a rectangular mesh network is that the ports can be reconfigured dynamically to allow communication among the nodes. Communication between two nodes in a rectangular mesh network is performed by activating the corresponding ports. For example, node 10 in Figure 10.1 (left) can communicate with its right neighbor, node 14, by setting its east port connected to its neighbor's west port. Similarly, an interior node of the network can communicate with its four neighbors in the same manner. Nodes 10 and 11 communicate with each other through the south port (10) and north (11), respectively.

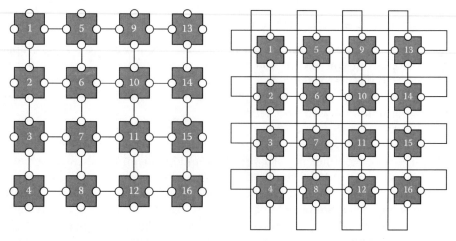

FIGURE 10.1
Rectangular mesh (left) and toroid (right).

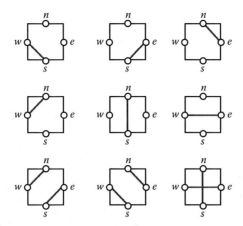

FIGURE 10.2
Communication ports in the processor and their possible connections.

10.2.1 *Code10A*: Bus Drawing in a Mesh Network

Code10A is a manually interactive approach for drawing buses for a reconfigurable mesh network. A *bus* is defined as a path that is successfully constructed between two nodes in the network. The project is a simple form of computer-aided design (CAD) that allows a user to design a network on the computer interactively with the aid of a mouse. A more sophisticated CAD software such as AutoCAD has many more features and is commonly used by professionals for drawing buildings, bridges, vehicles, and mechanical objects.

Figure 10.3 is an output from *Code10A* showing a rectangular mesh network of size 8×8 having two buses between two pairs of nodes. The nodes are labeled as $v_{i,j}$ where $i = 1, 2,\ldots,8$ is the column number and $j = 1, 2,\ldots,8$ is the row number. It can be seen that any two nodes $v_{i,j}$ and $v_{k,l}$ for $k, l = 1, 2,\ldots,8$ can communicate with each other if

$$i = k \text{ and } |l - j| = 1$$

or

$$j = l \text{ and } |i - k| = 1.$$

Code10A allows the construction of several buses that are displayed as a set of unique colors. A bus can start from any boundary node and ends at another boundary node simply by left-clicking the mouse at the respective nodes. The path of the bus is drawn by left-clicking the mouse at the starting node, continuing by left-clicking its neighboring node, and so forth until another boundary node is reached. Referring to Figure 10.3, there are three buses in the network:

$$v_{5,8} \to v_{5,7} \to v_{5,6} \to v_{5,5} \to v_{6,5} \to v_{7,5} \to v_{8,5}$$

$$v_{2,1} \to v_{2,2} \to v_{2,3} \to v_{3,3} \to v_{3,4} \to v_{4,4} \to v_{5,4} \to v_{6,4} \to v_{6,5} \to v_{6,6} \to v_{6,7} \to v_{6,8},$$

$$v_{1,4} \to v_{2,4} \to v_{3,4} \to v_{3,5} \to v_{3,6} \to v_{4,6} \to v_{5,6} \to v_{6,6} \to v_{7,6} \to v_{7,7} \to v_{7,8}$$

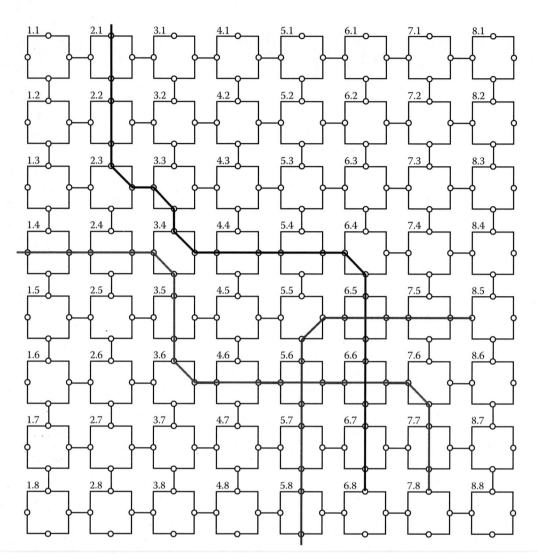

FIGURE 10.3
Output from *Code10A* showing three buses.

A bus can be drawn between any two nodes as long as it obeys the rule for avoiding path collision, which is that the path of a bus should not overlap partially or completely with another path. However, two buses can cross at one or more points when one path is raised above the other at the crossing point to avoid collision. In this case, the paths of the three buses in the figure do not overlap although they appear to be crossing at $v_{5,4}$, $v_{5,4}$, and $v_{5,4}$.

Code10A has two files: *Code10A.h* and *Code10A.cpp*. A single class called *CCode10A* represents the application class in this project. In the header file *Code10A.h*, the rectangular mesh of size $M \times N$ is defined with $M = 8$ and $N = 8$.

Table 10.1 lists bus variables used in the project. Two structures called *NODE* and *CurrentPosition* are created to manage the objects and variables. *NODE* represents the nodes in the rectangular mesh while *CurrentPosition* is the current bus being drawn through clicking the mouse. Tables 10.2 and 10.3 summarize the members of these structures. In *NODE*, each processor is located at *home* as the rectangular object *rct*, and it has the ports *w*, *e*, *n*, and *s*.

TABLE 10.1

Important Objects and Variables in *Code10A*

Variable	Type	Description
BusColor	*int*	Color of the bus as defined by *pBus*
pBus	*CPen*	Color of the pen object

TABLE 10.2

Important Objects and Variables in *NODE*

Variable	Type	Description
home	*CPoint*	Top left-hand coordinates of the node
w, e, n, s	*CPoint*	West, east, north. and south ports of the node
rcl	*CRect*	The rectangular object of the node

TABLE 10.3

Important Objects and Variables in *CurrentPosition*

Variable	Type	Description
status	*bool*	Current mouse position, 0 = inactive, 1 = active
dir	*int*	Direction of bus, 1 = east, 2 = west, 3 = north, 4 = south
jointPort	*int*	
cI, cJ, pI, pJ	*int*	Current (*c*) and previous position (*p*) with 1 = east, 2 = west, 3 = north, 4 = south for *I* column and *J* row
joint, A, B	*CPoint*	

CurrentPosition sets the current status of a bus through the object *cPos*. The bus has a *status* value of 1 to denote its drawing as active and 0 if it is no longer active. The direction of the bus is *dir*, which takes one of the following values: 1 is east, 2 is west, 3 is north, and 4 is south. The same values are used for the port called *jointPort*, which is used in the node for accommodating the bus. The path of the bus has *cI* and *cJ* to denote the current row and column numbers of the node, respectively. The previous node of the bus is denoted by *pI* and *pJ* for the row and column numbers, respectively.

Figure 10.4 is the organization of *Code10A* showing all the functions. The constructor *CCode10A()* creates the main window and initializes the home coordinates of all the nodes and their ports in the network. The ports are labeled as east (*e*), west (*w*), north (*n*), and south (*s*). The function also defines the colors of the buses through the object *cPen*. The current position of the drawing is represented as the Boolean variable *cPos.status* with an initial value of 0 denoting the position is inactive. This position becomes 1 when a boundary node has been clicked to denote the beginning of a bus.

OnPaint() provides the initial display and constantly updates the output of the network. The current position of a bus is represented by *cPos*. The initial display is recognized as *cPos.status* = 0 with a rectangular grid network of size 8 × 8. Each node is a square and is labeled as *i, j* where *i* is the column number and *j* is the row number. The ports in each node are drawn as small circles and displayed through *DrawPort()*. There is no bus in the initial display.

A bus starts when a boundary node is left-clicked and ends when another boundary node is reached. The event is recognized as *ON_WM_LBUTTONDOWN* and it is handled

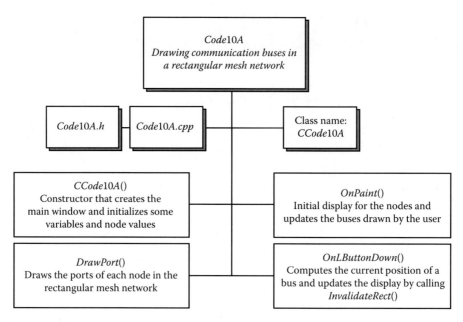

FIGURE 10.4
Organization of *Code*10A.

by *OnLButtonDown*(). The function immediately updates the drawing by calling *OnPaint*()
through *InvalidateRect*().

A mouse click at a boundary node updates the value as *cPos.status* = 1. A bus starts at this
node and continues when a neighboring node is clicked. In updating the bus, the current
node position *cPos* (*cI* for column number and *cJ* for row number) is compared to its previous position (*pI* for column number and *pJ* for row number). The bus is updated by its line
draw using a unique color passing through the respective ports.

The full listing of the codes in *Code*10A.*h* and *Code*10A.*cpp* are given below:

```
//Code10A.h
#include <afxwin.h>
#define M 8
#define N 8
#define rSIZE 40              //node size
#define pSIZE 6                  //port size
#define rSPACING 20              //node spacing

class CCode10A: public CFrameWnd
{
private:
    typedef struct
    {
        CRect rct;
        CPoint home;
        CPoint w,e,n,s;
    } NODE;
    NODE Node[M+1][N+1];
    typedef struct
    {
```

```
            bool status;              //b=begin, a=active, e=end
            int dir;                  //1=east, 2=west, 3=north, 4=south
            int jointPort;              //1=east, 2=west, 3=north, 4=south
            int cI,cJ,pI,pJ;          //c=current, p=previous
            CPoint joint,A,B;
      } CurrentPosition;
      CurrentPosition cPos;
      CPen pBus[10];
      int BusColor;
public:
      CCode10A();
      ~CCode10A()                           {}
      afx_msg void OnPaint();
      afx_msg void OnLButtonDown(UINT nFlags,CPoint pt);
      void DrawPort(int,int,CPoint);
      DECLARE_MESSAGE_MAP();
};

class CMyWinApp: public CWinApp
{
public:
  virtual BOOL InitInstance();
};
CMyWinApp MyApplication;

BOOL CMyWinApp::InitInstance(void)
{
    m_pMainWnd=new CCode10A;
    m_pMainWnd->ShowWindow(m_nCmdShow);
    return TRUE;
}


//Code10A: Reconfigurable Mesh
#include "Code10A.h"

BEGIN_MESSAGE_MAP(CCode10A,CFrameWnd)
    ON_WM_PAINT()
    ON_WM_LBUTTONDOWN()
END_MESSAGE_MAP()

CCode10A::CCode10A()
{
    int i,j;
    int pColor[]={RGB(0,0,200),RGB(255,0,0),RGB(0,150,0),
        RGB(255,50,255),RGB(0,0,150),RGB(200,200,0),RGB(0,200,200),
        RGB(55,0,255),RGB(0,150,0),RGB(0,100,255)};
    Create(NULL,L"Reconfigurable Mesh Model",WS_OVERLAPPEDWINDOW,
        CRect(0,0,750,550));
    cPos.status=0;        //b=begin
    BusColor=0;
    Node[1][1].home=CPoint(30,20);
    Node[1][1].rct=CRect(Node[1][1].home,Node[1][1].home+CPoint(50,50));
    for (j=1;j<=M;j++)
```

```
                for (i=1;i<=N;i++)
                {
                        Node[i][j].home=Node[1][1].home
                                +CPoint((rSIZE+rSPACING)*(i-1),
                                        (rSIZE+rSPACING)*(j-1));
                        Node[i][j].rct=CRect(Node[i][j].home,
                                Node[i][j].home+CPoint(rSIZE,rSIZE));
                        Node[i][j].w=Node[i][j].home+CPoint(0,rSIZE/2);
                        Node[i][j].e=Node[i][j].home+CPoint(rSIZE,rSIZE/2);
                        Node[i][j].n=Node[i][j].home+CPoint(rSIZE/2,0);
                        Node[i][j].s=Node[i][j].home+CPoint(rSIZE/2,rSIZE);
                }
        for (i=0;i<=9;i++)
            pBus[i].CreatePen(PS_SOLID,2,pColor[i]);
}

void CCode10A::OnPaint()
{
    int i,j;
    CString s;
    CRect cPort;
    CPoint a,b;
    CPaintDC dc(this);
    CFont fArial;
    fArial.CreatePointFont (60,L"Arial");

    //draw the PEs and their ports
    dc.SelectObject(fArial);
    for (j=1;j<=M;j++)
        for (i=1;i<=N;i++)
        {
            dc.Rectangle(&Node[i][j].rct);
            s.Format(L"%d,%d",i,j);
            dc.TextOutW(Node[i][j].home.x,Node[i][j].home.y-10,s);
            DrawPort(i,j,Node[i][j].w);
            DrawPort(i,j,Node[i][j].e);
            DrawPort(i,j,Node[i][j].n);
            DrawPort(i,j,Node[i][j].s);
        }

    //draw the grid lines
    for (j=1;j<=M;j++)
        for (i=1;i<=N;i++)
        {
            //horizontal grids
            if (i!=N)
            {
                a=Node[i][j].e+CSize(pSIZE/2,0);
                b=Node[i+1][j].w-CSize(pSIZE/2,0);
            }
            //vertical grids
            if (j!=M)
            {
                a=Node[i][j].s+CSize(0,pSIZE/2);
                b=Node[i][j+1].n-CSize(0,pSIZE/2);
```

```
                    }
                    dc.MoveTo(a); dc.LineTo(b);
              }

        //draw the buses whenever a PE is clicked
        if (cPos.status)    //if active
        {
                dc.SelectObject (&pBus[BusColor]);
                dc.MoveTo(cPos.joint);
                dc.LineTo(cPos.A); dc.LineTo(cPos.B);
                if (cPos.dir==1)     //eastbound
                   cPos.jointPort=2;
                if (cPos.dir==2)     //westbound
                   cPos.jointPort=1;
                if (cPos.dir==4)     //southbound
                   cPos.jointPort=3;
                if (cPos.dir==3)     //northbound
                   cPos.jointPort=4;
        }
        if (!cPos.status)                //first time clicked
        {
            dc.SelectObject(&pBus[BusColor]);
            if (cPos.dir==5)     //terminate eastward
            {
                dc.MoveTo(cPos.joint);
                dc.LineTo(cPos.A); dc.LineTo(cPos.B);
            }
            if ((cPos.cI==1 || cPos.cI==N) || (cPos.cJ==1 || cPos.cJ==M))
            {
                if (cPos.cI==1)
                   cPos.jointPort=2;
                if (cPos.cI==N)
                   cPos.jointPort=1;
                if (cPos.cJ==1)
                   cPos.jointPort=3;
                if (cPos.cJ==M)
                   cPos.jointPort=4;
                dc.MoveTo(cPos.A); dc.LineTo(cPos.B);
                cPos.status=1;
            }
        }
    }
}

void CCode10A::DrawPort(int i,int j,CPoint Q)
{
    CClientDC dc(this);
    CRect rct;
    rct=CRect(Q-CSize(pSIZE/2,pSIZE/2),Q+CSize(pSIZE/2,pSIZE/2));
    dc.Ellipse(&rct);
}
```

```
void CCode10A::OnLButtonDown (UINT nFlags,CPoint pt)
{
    int i,j;
    CPoint OutUpLeft,OutDownRight;
    CPoint InUpLeft,InDownRight;
    CRect OutsideRect,InsideRect;
    cPos.pI=cPos.cI; cPos.pJ=cPos.cJ;
    cPos.cI=1+(pt.x-30)/(rSIZE+rSPACING);
    cPos.cJ=1+(pt.y-30)/(rSIZE+rSPACING);
    if (Node[cPos.cI][cPos.cJ].rct.PtInRect(pt))
    {
        if (cPos.status)
        {
            if (cPos.cI-cPos.pI==1 && cPos.cJ==cPos.pJ)
            {
                cPos.dir=1;                    //eastbound
                cPos.joint=cPos.B;
                cPos.A=Node[cPos.pI][cPos.pJ].e;
                cPos.B=Node[cPos.cI][cPos.cJ].w;
                if (cPos.jointPort==3)
                {
                    InUpLeft=cPos.joint;
                    InDownRight=cPos.A;
                }
                if (cPos.jointPort==2)
                {
                    InUpLeft=cPos.joint+CPoint(0,-1);
                    InDownRight=cPos.A+CPoint(0,1);
                }
                if (cPos.jointPort==4)
                {
                    InUpLeft=cPos.joint+CPoint(0,-rSIZE/2);
                    InDownRight=cPos.A+CPoint(0,rSIZE/2);
                }
                OutUpLeft=cPos.A+CPoint(0,-rSIZE/2);
                OutDownRight=cPos.B+CPoint(0,rSIZE/2);
            }
            if (cPos.cI-cPos.pI==-1 && cPos.cJ==cPos.pJ)
            {
                cPos.dir=2;                    //westbound
                cPos.joint=cPos.B;
                cPos.A=Node[cPos.pI][cPos.pJ].w;
                cPos.B=Node[cPos.cI][cPos.cJ].e;
                if (cPos.jointPort==3)
                {
                    InUpLeft=cPos.joint+CPoint(-rSIZE/2,0);
                    InDownRight=cPos.A+CPoint(rSIZE/2,0);
                }
                if (cPos.jointPort==1)
                {
                    InUpLeft=cPos.A+CPoint(0,-1);
                    InDownRight=cPos.joint+CPoint(0,1);
                }
                if (cPos.jointPort==4)
```

```
            {
                InUpLeft=cPos.A;
                InDownRight=cPos.joint;
            }
            OutUpLeft=cPos.B+CPoint(0,-rSIZE/2);
            OutDownRight=cPos.A+CPoint(0,rSIZE/2);
        }
        if (cPos.cI==cPos.pI && cPos.cJ-cPos.pJ==1)
        {
            cPos.dir=4;                    //southbound
            cPos.joint=cPos.B;
            cPos.A=Node[cPos.pI][cPos.pJ].s;
            cPos.B=Node[cPos.cI][cPos.cJ].n;
            if (cPos.jointPort==2)
            {
                InUpLeft=cPos.joint;
                InDownRight=cPos.A;
            }
            if (cPos.jointPort==1)
            {
                InUpLeft=cPos.A+CPoint(0,-rSIZE/2);
                InDownRight=cPos.joint+CPoint(0,rSIZE/2);
            }
            if (cPos.jointPort==3)
            {
                InUpLeft=cPos.joint+CPoint(-1,0);
                InDownRight=cPos.A+CPoint(1,0);
            }
            OutUpLeft=cPos.A+CPoint(-rSIZE/2,0);
            OutDownRight=cPos.B+CPoint(rSIZE/2,0);
        }
        if (cPos.cI==cPos.pI && cPos.cJ-cPos.pJ==-1)
        {
            cPos.dir=3;                    //northbound
            cPos.joint=cPos.B;
            cPos.A=Node[cPos.pI][cPos.pJ].n;
            cPos.B=Node[cPos.cI][cPos.cJ].s;
            if (cPos.jointPort==1)
            {
                InUpLeft=cPos.A;
                InDownRight=cPos.A+CPoint(rSIZE/2,rSIZE/2);
            }
            if (cPos.jointPort==4)
            {
                InUpLeft=cPos.A+CPoint(-1,0);
                InDownRight=cPos.A+CPoint(2,rSIZE);
            }
            if (cPos.jointPort==2)
            {
                InUpLeft=cPos.A+CPoint(-rSIZE/2,0);
                InDownRight=cPos.A+CPoint(0,rSIZE/2);
            }
            OutUpLeft=cPos.B+CPoint(-rSIZE/2,0);
            OutDownRight=cPos.A+CPoint(rSIZE/2,0);
        }
```

```
                    InsideRect=CRect(InUpLeft,InDownRight);
                    InvalidateRect(InsideRect);
                    OutsideRect=CRect(OutUpLeft,OutDownRight);
                    InvalidateRect(OutsideRect);

                    int SumSquare=(cPos.cI-cPos.pI)*(cPos.cI-cPos.pI)
                        +(cPos.cJ-cPos.pJ)*(cPos.cJ-cPos.pJ);
                    if (SumSquare!=1)
                    {
                        cPos.status=0;
                        BusColor++;
                        if (BusColor==10)
                                    BusColor=0;
                    }
            }
            if (!cPos.status)
            {
                cPos.pI=cPos.cI; i=cPos.pI;
                cPos.pJ=cPos.cJ; j=cPos.pJ;
                if (cPos.cI==1)  //starting from west
                {
                    cPos.B=Node[i][j].w;
                    cPos.A=cPos.B+CPoint(-10,0);
                    OutUpLeft=Node[i][j].home+CPoint(-10,0);
                    OutDownRight=Node[cPos.cI][cPos.cJ].
                        home+CPoint(0,rSIZE);
                }
                if (cPos.cI==N)  //starting from east
                {
                    cPos.B=Node[i][j].e;
                    cPos.A=cPos.B+CPoint(10,0);
                    OutUpLeft=Node[i][j].home+CPoint(rSIZE,0);
                    OutDownRight=Node[i][j].home+CPoint(rSIZE+10,rSIZE);
                }
                if (cPos.cJ==1)  //starting from north
                {
                    cPos.B=Node[i][j].n;
                    cPos.A=cPos.B+CPoint(0,-10);
                    OutUpLeft=Node[i][j].home+CPoint(0,-10);
                    OutDownRight=Node[i][j].home+CPoint(rSIZE,0);
                }
                if (cPos.cJ==M)  //starting from south
                {
                    cPos.B=Node[i][j].s;
                    cPos.A=cPos.B+CPoint(0,10);
                    OutUpLeft=Node[i][j].home+CPoint(0,rSIZE);
                    OutDownRight=Node[i][j].home+CPoint(rSIZE,rSIZE+10);
                }
                cPos.joint=cPos.B;
                OutsideRect=CRect(OutUpLeft,OutDownRight);
                InvalidateRect(OutsideRect);
            }
        }
    }
}
```

10.3 Single-Row Routing

Single-row routing is a combinatorial optimization problem that can be classified as NP-complete. In general, the problem can be defined as drawing the paths (called nets) of pairs of pins that are arranged in a single axis in such a way that the nets will not cross. The problem has primary applications in the design of the multilayer printed circuit boards for routing pins and vias through several layers in the circuit. A typical PCB has hundreds to millions of electronic components that require proper planning in its design. The design for these components consists of two main problems: the placement of the components and routing between the components. Obviously, the placement of the components is similar to the placement of the nodes of a graph. In optimization, the facility layout problem discusses techniques for getting an optimal network for placing the components strategically, which involves graph modeling. Routing between the components involves common graph theoretical problems such as the Hamiltonian path, minimum spanning tree, and shortest path problems.

The primary objective in single-row routing is to produce a realization that minimizes congestion in the network. The single-row routing problem is formulated as follows: Given a set of m evenly-spaced pins (terminals or vias), p_i, for $i = 1, 2, \ldots, m$, arranged horizontally from left to right in a single row called a *single-row axis*, the problem is to draw paths connecting the pairs of pins so as to minimize the congestion in the network. Each path joining a pair of pins is called a *net*, with net k denoted as n_k. The paths cannot cross each other and they must be drawn from left to right with reverse directions not allowed.

Figure 10.5 is an example of a single-row network having $m = 10$ pins. The figure shows the realization of the single-row network from the initial requirement of five pin-pairings, as follows:

$$n_1 = (p_3, p_6),$$

$$n_2 = (p_4, p_{10}),$$

$$n_3 = (p_1, p_7),$$

$$n_4 = (p_5, p_8)$$

and

$$n_5 = (p_2, p_9).$$

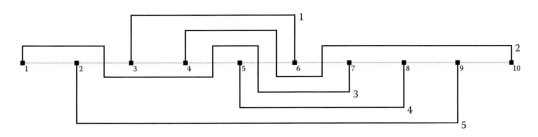

FIGURE 10.5
Realization of the single-row routing with 10 pins.

The pins are arranged according to the list given by

$$L = \{n_1, n_2, n_3, n_4, n_5\}.$$

Each pair of pins forms a net. There are $m/2 = 5$ noncrossing nets in the realization. Each net n_k is formed from the horizontal intervals, (b_k, e_k) for $k = 1, 2, 3, 4, 5$ in the node axis, where b_k and e_k are the beginning (left) and end (right) pins of the intervals, respectively.

The realization is produced from the ordered list L given by

$$L = \{n_1, n_2, n_3, n_4, n_5\}.$$

The list L is the arrangement of the nets according to the priority order from left to right. A different list such as $L = \{n_4, n_2, n_1, n_3, n_5\}$ will produce a different realization with a higher or lower congestion.

A crossing on the node axis, as shown through a line between nodes (p_2, p_3) in Figure 10.5, is called a *dogleg* or interstreet crossing. There are four other doglegs in this example: one each in (p_4, p_5) and (p_6, p_7), and two in (p_5, p_6). A high number of doglegs within a single pin interval means there are many net crossings in that interval. As the nets become too close to each other, the heat that they generate accumulates and may cause problems to the network. Therefore, a high number of doglegs within a single pin interval is to be avoided and it becomes part of the objectives in single-row routing.

The area above the node axis is called the *upper street*, while that below is the *lower street*. Congestion in a street refers to the number of horizontal tracks in that street. Congestion is said to be high if the value is large and low if the value is small. Congestion in the network, denoted as Q, is defined as the maximum of the congestion in the upper and lower streets of the network. It can be shown from Figure 10.5 that congestion in the upper street, or $C_u = 3$ while congestion in the lower street is $Q_l = 4$. That gives the overall congestion as

$$Q = \max(Q_u, Q_l) = 4.$$

Hence, the main objective of single-row routing is to achieve minimum congestion. Congestion depends very much on the order of the nets represented as the list L. The order of the nets $L = \{n_k\}$ is defined as the priority position from of the nets in the list. For example, for the network with five nets in Figure 10.5 the order in the list is $L = \{n_1, n_2, n_3, n_4, n_5\}$, which places the nets according to the priority order $n_1 \rightarrow n_2 \rightarrow n_3 \rightarrow n_4 \rightarrow n_5$. This produces a network with a congestion value of 4. Obviously, the congestion will be different with another list such as $L = \{n_4, n_2, n_1, n_3, n_5\}$. Hence, the single-row routing problem is inherently a permutation problem and the solution to the problem can be obtained through a massive search by running through all the permutations of the combinations.

The order of the nets in the list L is associated with the energy function, or cost function, of the network E. The energy of the network is computed from the energy scheme that is constructed by placing the nets according to their priority order from top to bottom. Figure 10.6 shows the energy scheme for the list $L = \{n_1, n_2, n_3, n_4, n_5\}$ that produces the realization in Figure 10.5. The scheme starts by placing the nets $n_k = (b_k, e_k)$ according to their orders from top to bottom. Each net is represented as a horizontal interval from b_k to e_k. Next, thin lines called *reference lines* are drawn continuously in the order from the first pin to the second, third, and so forth until the last pin. The reference lines cross some of the horizontal lines, and the points of crossings are the doglegs. Each dogleg creates a new *segment* for

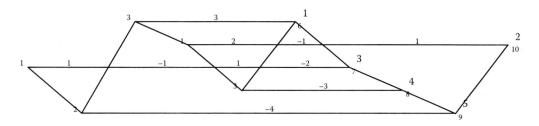

FIGURE 10.6
Energy scheme for the realization from Figure 10.5.

the interval. As a general rule, a net with r doglegs has $r + 1$ segments. A net without any dogleg has only one segment that is the default interval.

It is obvious that each net has now been partitioned into one or more segments. It can be seen from Figure 10.6 that $n_3 = (p_1, p_7)$ has three doglegs and four segments. The doglegs $d_{n_3,i}$ for $i = 1, 2, 3$ are located in the subintervals (p_2, p_3), (p_4, p_5), and (p_5, p_6). The segments are

$$S = \{s_{3,1}, s_{3,2}, s_{3,3}, s_{3,4}\}$$
$$= \{(p_1, d_{3,1}), (d_{3,1}, d_{3,2}), (d_{3,2}, d_{3,3}), (d_{3,3}, p_7)\}.$$

There are also two doglegs and two crossings in (p_5, p_6).

Each segment $s_{k,j}$ in net n_k is associated with an energy value that is determined from its height in the upper or lower street. The *height* $h_{k,j}$ of segment j in net k is the number of levels the segment lies below or above the node axis. The energy of the segment is computed by referring to the pins in its subinterval. If the pins are below the subinterval, then the energy has a positive value; otherwise, the energy is negative. For $s_{3,1} = (p_1, d_{3,1})$ in Figure 10.6, the only pin is p_2, which lies below it. Hence, the energy of $s_{3,1}$ is positive and since the subinterval is one level higher than the pin the energy of the segment is +1, or $e_1 = +1$. In contrast, the segment $s_{3,2} = (d_{3,1}, d_{3,2})$ has $e_2 = -1$ since the interval is one level below its contained pins p_3 and p_4. Similarly, $s_{3,3} = (d_{3,2}, d_{3,3})$ has $e_3 = +1$. The last segment $s_{3,4} = (d_{3,1}, p_7)$ is two levels below p_6, which gives $e_4 = -2$.

If the segment has more than one covered pin, then its energy level is determined from the maximum of the levels from the pins. The top-most net $n_1 = (p_3, p_6)$ does not have any doglegs. Hence, the net has only one segment. The interval covers p_4 and p_5. It can be seen that the segment is one level up from p_4 and three levels up from p_5. Therefore, $e_1 = +3$, which is the maximum of the two.

We obtain the overall energies of the nets as follows:

n_1: $\{e_1 = +1\}$
n_2: $\{e_1 = +2, e_2 = -1, e_3 = +1\}$
n_3: $\{e_1 = +1, e_2 = -1, e_3 = +1, e_4 = -2\}$
n_4: $\{e_1 = -3\}$
n_5: $\{e_1 = -4\}$

The energy scheme provides a preview of the realization of the whole network. The realization is obtained by drawing the nets according to their segment heights, where the positive value denotes the upper street and the negative is the lower street. A segment with a +3 energy value is drawn three levels up in the upper street while another segment with a –2 energy value is two levels down in the lower street.

The main objective in single-row routing is to minimize the congestion of the network. This objective is achieved by finding the overall energy of the network, given by

$$E = \sum_{k=1}^{m} \sum_{j=1}^{m_k} |h_{k,j}|.$$ (10.1)

In this equation, $h_{k,j}$ is the height of segment j in net k, while m is the number of nets in the problem and m_k is the number of segments in net k. The realization produced in Figure 10.5 with $L = \{n_1, n_2, n_3, n_4, n_5\}$ has an energy of $E = 19$, which suggests it may not be optimal. A better solution can be obtained by reforming the list with different orderings of nets.

Due to its importance, the single-row routing problem has been much discussed in the literature. The problem was first proposed in [5] with a divide-and-conquer approach to deal with the complicated wiring problem in the very large scale integrated (VLSI) circuit design. The method begins with a systematic decomposition of the general multilayer routing problem into a number of independent single-layer and single-row routing problems. This approach defines single-row routing problems for every horizontal and vertical line of points in the original problem. The solutions of these subproblems are then combined to contribute toward the overall solution to the original problem.

The single-row routing problem has been shown to be NP-complete with a large number of interacting degrees of freedom [6]. Most solutions to the problem have been expressed in the form of heuristic algorithms based on graph theory (as in Deogun and Sherwani [7]), exhaustive search (as in Tarng et al. [8]), and greedy algorithms (as in Du and Liu [9]). In Salleh and Zomaya [10], a simulated annealing model was proposed that produces the optimal solution for the congestion Q. The technique was further improved in [11] in a model called *enhanced simulated annealing technique for single-row routing* (ESSR), which considers both the street congestion (Q) and the number of doglegs (D). The model is based on an energy function E as a collective set representing both Q and D. Since the two parameters are allowed to vary freely during the annealing steps, the energy may, in some cases, produce an optimum solution in one while ignoring the other.

In terms of graph theory, the single-row routing problem can be modeled as mapping the nets. As proposed by Deogun and Sherwani [7], each net is made up of an interval from the left and right pins. Figure 10.7 shows the *overlap graph* of the network in Figure 10.5.

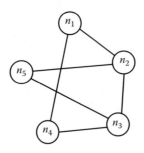

FIGURE 10.7
Overlap graph of the network from Figure 10.5.

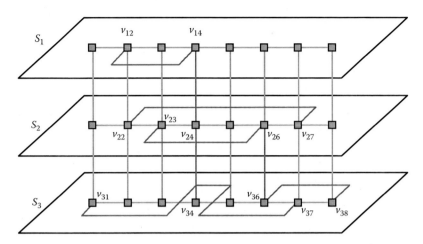

FIGURE 10.8
Rectangular mesh network formed from three layers of single-row networks.

Each interval (net) in the network makes up a node of the graph where an edge between two nodes means the two intervals overlap. The overlap graph describes the relationship of the intervals where an edge between n_1 and n_4 means the two intervals overlap. There is no edge between nodes n_1 and n_5 as the former is contained inside the latter with no overlap.

Traditionally, single-row routing is one of the techniques employed for designing the routes between the electronic components of a PCB. Figure 10.8 shows a typical application involving three layers of PCB where each layer has its own single-row network. Obviously, the arrangement of m pins each in n layers forms a rectangular mesh of size $n \times m$ network. The pins in each layer are arranged in a single axis so that they can be connected to any other pin in that layer through the single-row routing technique. The pins are also configured as nodes in a rectangular mesh network to enable them to communicate with pins in other layers. In this case, the horizontal pins are nodes in the same layer and they can communicate with nodes from other layers through the mesh routes.

10.3.1 *Code*10*B*: Realizing Single-Row Routing

The single-row routing problem is illustrated in *Code*10*B*. Figure 10.9 shows the output in the form of a realization from a set of 20 pins for 10 pairs of nets together with its energy scheme diagram. The nets are ordered 1 to 10 according to priority from top to bottom as shown in the energy scheme diagram. The visual output also displays the congestion value Q, the number of doglegs D, and the total energy of the network E. The results from the realization are $Q = 7$, $D = 32$, and $E = 129$.

The output from *Code*10*B* consists of three regions: a pin-pairing selection at the bottom, an energy scheme diagram in the middle, and the realization of the network on top. The pairs of pins are selected by left-clicking the mouse on the boxes at the pin-pairing area. Each selected pair is assigned with its own unique color. The realization and the energy scheme diagram are computed and displayed once the selection of the pairs of pins has been completed. The energy scheme diagram displays the priority order of the nets and the height of each segment in the nets.

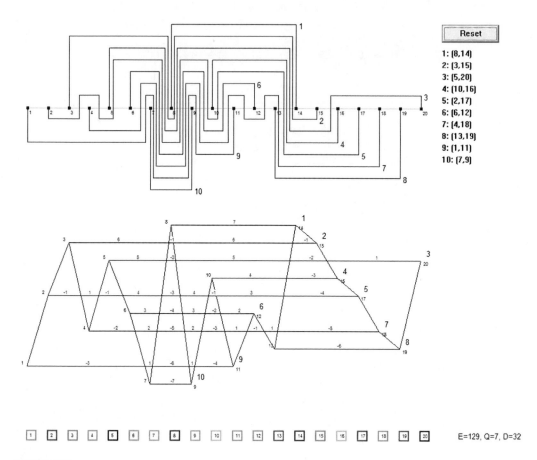

FIGURE 10.9
Output from *Code10B* showing the results of 20-pin single-row routing.

In designing the visual solution for the single-row routing problem, a good understanding of the variables and parameters of the problem is required. There must be an input area that collects the information about the pin-pairs. The processing of the input data is triggered after the last pair of pins has been selected. The output follows with the graphical displays of the realization and the energy scheme diagram. The following algorithm summarizes the solution:

Given:
    A set of m pins paired in two for $m/2$ nets. The nets are ordered in $L = \{n_k\}$ for $k = 1, 2, \ldots, m/2$.
Initialization:
    Set the total energy $E = 0$.
    Define the drawing area consisting of the realization and energy scheme sections using Windows and assign the corresponding Cartesian coordinates.
    Define the pin-pairing area.
    Assign coordinates to the pins in the single-row axis.
    Display the initial display of the pin-pairs using Windows coordinates and assign the corresponding Cartesian coordinates.

Process:

Allow the user to choose nets from the pairs through clicks of the mouse.

Draw the energy scheme diagram using horizontal lines for representing the net intervals according to their order.

Draw reference lines in the energy scheme diagram for connecting p_i successively for $i = 1, 2,\ldots, m$.

Locate the doglegs $d_{k,j}$ in the energy scheme diagram from the intersection of the reference line with the horizontal lines.

Partition each net n_k into segments $s_{k,j}$ based on the doglegs $d_{k,j}$.

Compute the Cartesian coordinates of the left and right points of each segment $s_{k,j}$.

Compute the energy of each segment $s_{k,j}$.

Compute the total energy E, congestion Q and total number of doglegs D.

Output:

Visual realization of the single-row network and the results E, Q, and D.

*Code*10B has *Code*10B.h and *Code*10B.cpp as its source files. The application class is *CCode*10B. The constant m in the header file represents the total number of pins that are preassigned as 20. Some important variables and objects in the class are listed in Table 10.4. Basically, the drawing consists of the realization and energy sections. The *CPoint* objects *home* and *end* are the Windows corner points of the rectangular region for the node axis in the realization section. The corresponding top-left sections for the pin click and energy scheme sections are *cHome* and *eEnd*, respectively. Both sections share the same subinterval width *wInt* as the node axis in order to maintain the same horizontal positions for the pins. The number of nets is *nNets*, which is constantly updated when the pairs of pins are selected.

Input in this application is provided through the left clicks of the mouse on the pins which updates the value of *nNets*. This variable is initially set to 0. A control variable called *bFlag* updates this value: *bFlag* = 0 is the initial value, *bFlag* = 1 means that the first pin has been selected, and *bFlag* = 2 denotes that the second pin in the pair has been selected, which then increases *nNets* value by one. With *nNets* = *m*, the pins have all been assigned to the nets, and this triggers the next process, which is calculation of the energies of the segments.

There are three structures called *PIN, NET,* and *SEGMENT* that represent the pins, nets, and segments of the nets, respectively. *PIN* defines the array p as its object with members of the structure summarized in Table 10.5.

TABLE 10.4

Important Objects/Variables in *Code*10B

Variable	Type	Description
home, end	CPoint	Left and right points of the single-row axis
cHome	CPoint	Left point of the selection area
eHome	CPoint	Left point of the energy scheme diagram
nNets	int	Number of selected nets
wInt	int	Width of the intervals in the single-row axis
bFlag	int	Left button click status with *bFlag* = 0 indicates no action, *bFlag* = 1 indicates selection on the left pin, and *bFlag* = 2 is the selection of its pair
Q, D, E	int	Q = congestion, D = number of doglegs, E = total energy
Reset	CButton	Reset button that refreshes the network to the starting position

TABLE 10.5

Structural Members of *PIN*

Variable	Type	Description
home	CPoint	Top left-hand point of *p*
cHome	CPoint	Left point of *p* in the selection area
eHome	CPoint	Left point of *p* in the energy scheme diagram
rct	CRect	Rectangular object of *p*
net	int	Net number where the pin belongs
DL	int	Dogleg number after the stated pin number

From the above structure pin *i* is represented as *p*[*i*]. Its home coordinates on the single-row axis are *p*[*i*].*home.x* and *p*[*i*].*home.y*. The pin is represented in the pin-pairing area as *p*[*i*].*cHome* and in the energy scheme area as *p*[*i*].*eHome*. Pin *i* is drawn in Windows as the rectangular object *p*[*i*].*rct*.

The structure *NET* for nets is represented as the array object *q*. Several variables and objects from this structure are listed and summarized in Table 10.6. A net is represented as an interval bounded on the left by *b* and on the right by *e*. The segments of the nets are represented by the array *Seg* with *nSegs* as their total number. The dogleg number *DL* is related to the segments. The position of the net arranged from top to bottom is *order*, which is the priority order of the net. Finally, every net has *E*, which is the sum of the energies from the segments.

From the above declarations, net n_k is represented as *q*[*k*]. The net's left and right pins are *q*[*k*].*b* and *q*[*k*].*e*, while its priority order is *q*[*k*].*order*. The number of segments in the net is *q*[*k*].*nSeg* with segment *j* represented by *q*[*k*].*Seg*[*j*].

NET also has *SEGMENT* as its member, which implies that a net can be partitioned into one or more segments. *NET* refers to *SEGMENT* for describing the properties of each segment in it. The members of *SEGMENT* are described briefly in Table 10.7.

From the Table 10.7, we see that *q*[*k*].*Seg*[*j*].*b* is the pin number of the left pin that bounds the segment. The real left and right coordinates of segment *j* in net *k* are *q*[*k*].*Seg*[*j*].*sb* and

TABLE 10.6

Important Objects and Variables in *NET*

Variable	Type	Description
b, e	int	Beginning and ending pins of the net
E	int	Energy
order	int	Priority order number of the net
DL	int	Dogleg number in the net
nSeg	int	Number of segments of the net
Seg	SEGMENT	Net partition into segments

TABLE 10.7

Important Objects and Variables in *SEGMENT*

Variable	Type	Description
b, e	int	Left and right pins that bound each segment
sb, se	CPoint	True left and right points of each segment in the net
E	int	Energy of each segment

$q[k].Seg[j].se$, respectively. The energy of the segment of the net is $q[k].Seg[j].E$. From Figure 10.6, the following values are obtained:

$q[k].nSeg = 3$		
$q[2].Seg[1].b = 4$	$q[2].Seg[1].e = 6$	$q[2].Seg[1].E = +2$
$q[2].Seg[1].b = 5$	$q[2].Seg[2].e = 7$	$q[2].Seg[3].E = -1$
$q[2].Seg[3].b = 6$	$q[2].Seg[3].e = 10$	$q[2].Seg[3].E = +1$

Functions in *Code10B* are created according to specific jobs. Figure 10.10 is the organization of *Code10B* showing all the functions.

Code10B starts with the constructor *CCode10B()*, which creates the main window and defines the initial values of the left and right points of the single-row axis (*home* and *end*). The function also defines the values of the left points in the selection (*cHome*) and energy scheme (*eHome*) areas. With these values, the width of each interval *wInt* of the single-row axis is computed.

The constructor calls *Initialize()* to initialize the values of the pins and several other variables. With these values, *OnPaint()* displays the boxes representing the pins in the selection area. This allows the selection of pins as pairs to form the nets through the left clicks of the mouse in *OnLButtonDown()*. A variable called *bFlag* changes values to indicate the control from the mouse click: *bFlag* = 0 means no activity, *bFlag* = 1 selects the left pin, while *bFlag* = 2 selects the right pin that completes the pairing. The number of selected nets is updated as *nNets*, which corresponds to the number of pairs of pins.

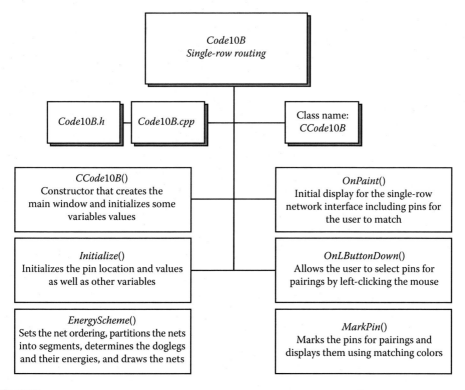

FIGURE 10.10
Organization of *Code10B*.

The equality *nNets* = *m*/2 indicates that all the nets formed from the pins have been selected. This triggers *EnergyScheme*(), which computes the energies of the segments in every net. The energy scheme starts by placing the nets according to their priority orders from top to bottom. The nets are drawn as horizontal lines in the intervals from their left to right pins. Reference lines are then drawn continuously starting from the first pin, to the second, third, and so forth until the last pin. In running from left to right, the reference lines cross the horizontal lines at several points, which are referred to as doglegs. In this case, the doglegs are said to have partitioned the nets into segments. The following codes in *EnergyScheme*() create the doglegs and segments:

```
for (k=1;k<=nNets;k++)
{
    q[k].nSeg=0;
    q[k].DL[0]=0; r=1;
    q[k].DL[r]=q[k].e;
    q[k].Seg[r].b=q[k].b;
    for (j=q[k].b+1;j<q[k].e;j++)
        if ((q[k].order<q[p[j].net].order && q[k].order>q[p[j+1].
        net].order)
                || (q[k].order>q[p[j].net].order && q[k].
                    order<q[p[j+1].net].order))
            {
                q[k].DL[r]=j; q[k].DL[0]++;
                w=++p[j].DL[0]; p[j].DL[w]=k;
                q[k].Seg[r].e=j+1;
                q[k].nSeg++;
                r++;
                q[k].Seg[r].b=j;
            }
    q[k].Seg[r].e=q[k].e; q[k].nSeg++;
    D +=q[k].DL[0];
}
```

Next, the energy of each segment is computed according to the maximum of the heights from the pins bounded in the segment. The energy has a positive value if the net lies above the bounded pins and a negative value if the net is below. The absolute value sum of the energies gives the total energy *E*. At the same time, the congestion *Q* and the total number of doglegs are also computed. The following codes show how these activities are performed:

```
for (k=1;k<=nNets;k++)
    for (i=1;i<=q[k].nSeg;i++)
    {
        max=0;
        for (j=q[k].Seg[i].b+1;j<q[k].Seg[i].e;j++)
        {
            r=1;
            for (w=1;w<=nNets;w++)
                if (w!=k || w!=p[j].net)
                    if (p[j].home.x>p[q[w].b].home.x
                            && p[j].home.x<p[q[w].e].home.x)
                        if ((q[w].order>q[k].order
```

```
                                    && q[w].order<q[p[j].net].
                                       order)
                                 || (q[w].order<q[k].order
                                    && q[w].order>q[p[j].net].
                                       order))
                          r++;
              max=((max<r)?r:max);
        }
        q[k].Seg[i].E=((q[k].order<q[p[q[k].Seg[i].b+1].net].
              order)?max:-max);
        q[k].Seg[i].sb.y=p[q[k].Seg[i].b].home.y-20*q[k].
              Seg[i].E;
        q[k].Seg[i].se.y=q[k].Seg[i].sb.y;
        E += abs(q[k].Seg[i].E);
        if (Q<abs(q[k].Seg[i].E))
              Q=abs(q[k].Seg[i].E);
    }
```

The doglegs are quite tricky to draw. In a single-pin interval, one or more doglegs may exist. Therefore, the pin interval will have to be subdivided into several subintervals based on the number of doglegs to allow each dogleg to cross the axis comfortably. For r doglegs inside (p_i, p_{i+1}) it is necessary to divide this interval into $r + 1$ subintervals so that the tracks from the nets will not cross each other. For a dogleg $d = (d.x, d.y)$ between $p_i = (p_i.x, p_i.y)$ and $p_{i+1} = (p_{i+1}.x, p_{i+1}.y)$ the x-coordinate of the dogleg, d, x, is obtained by comparing their gradients that have the same value, as follows:

$$\frac{p_i.y - p_{i+1}.y}{p_{i+1}.x - p_i.x} = \frac{p_i.y - d.y}{d.x - p_i.x}.$$

Since d lies on the single-row axis,

$$d.y = home.y.$$

This gives

$$d.x = p_i.x + (p_i.y - d.y)\frac{p_{i+1}.x - p_i.x}{p_{i+1}.y - p_i.y}$$

$$= p_i.x + (p_i.y - home.y)\frac{p_{i+1}.x - p_i.x}{p_{i+1}.y - p_i.y}.$$

We can now redraw the segments using the true coordinates of the doglegs. For a net having r doglegs there are $r + 1$ segments. The segments are

$$(p_b, d_1), (d_1, d_2), (d_2, d_3), \ldots, (d_{r-1}, d_r), (d_r, p_e).$$

The energies of the segments discussed earlier will now apply to the segments based on their true x- and y-coordinates. With these properly assigned coordinates, the tracks of each net can now be drawn from the left to right pins passing through doglegs, if any, to produce the final realization. The following codes show how this is done:

```
for (k=1;k<=nNets;k++)
    for (i=1;i<=q[k].nSeg;i++)
    {
        if (i==1)
        {
            q[k].Seg[i].sb.x=p[q[k].b].home.x;
            q[k].Seg[i].se.x=p[q[k].DL[i]].home.x;
            if (q[k].DL[i]<q[k].Seg[i].e)
            {
                tmp=(double)(p[q[k].DL[i]].home.x+(home.y-q[k].
                    Seg[i].se.y)
                    *(p[q[k].DL[i]+1].home.x-p[q[k].DL[i]].home.x)
                    /(q[k].Seg[i+1].sb.y-q[k].Seg[i].se.y));
                q[k].Seg[i].se.x=(int)tmp;
            }
        }
        if (i>1 && i<q[k].nSeg)
        {
            q[k].Seg[i].sb.x=q[k].Seg[i-1].se.x;
            q[k].Seg[i].se.x=p[q[k].DL[i]].home.x;
            if (q[k].DL[i]<q[k].Seg[i].e)
            {
                tmp=(double)(p[q[k].DL[i]].home.x+(home.y-q[k].
                    Seg[i].se.y)
                    *(p[q[k].DL[i]+1].home.x-p[q[k].DL[i]].home.x)
                    /(q[k].Seg[i+1].sb.y-q[k].Seg[i].se.y));
                q[k].Seg[i].se.x=(int)tmp;
            }
        }
        if (i==q[k].nSeg)
        {
            if (i>1)
            {
                q[k].Seg[i].sb.x=q[k].Seg[i-1].se.x;
                q[k].Seg[i].se.x=p[q[k].e].home.x;
            }
            dc.SelectObject(fHelvetica);
            s.Format(L"%d",k);
            dc.TextOutW(q[k].Seg[i].se.x+5,q[k].Seg[i].se.y-5,s);
        }
        dc.SelectObject(fArial); dc.SetTextColor(RGB(200,0,0));
        s.Format(L"%d",q[k].Seg[i].E);
        dc.TextOutW((q[k].Seg[i].sb.x+q[k].Seg[i].se.x)/2,
            eHome.y-10+30*(k-1),s);
        dc.SetTextColor(RGB(0,0,0));
        dc.MoveTo(q[k].Seg[i].sb.x,home.y);
        dc.LineTo(q[k].Seg[i].sb); dc.LineTo(q[k].Seg[i].se);
        dc.LineTo(q[k].Seg[i].se.x,home.y);
    }
```

The full listing of the codes are given below:

```
//Code10B.h
#include <afxwin.h>
#define m 20      //#pins
#define IDC_RESET 301
using namespace std;

class CCode10B: public CFrameWnd
{
private:
    CPoint home,end; //axis home,end
    CPoint cHome;      //click home
    CPoint eHome;      //energy diagram home
    int nNets;         //#selected nets
    int wInt;          //width of intervals in axis
    int bFlag;         //button flag
    int Q,D,E;         //Q=congestion, E=energy, D=doglegs
    CButton Reset;
    CFont fArial,fHelvetica;

    typedef struct
    {
        int b,e;
        int E;
        CPoint sb,se;
    } SEGMENT;

    typedef struct
    {
        CPoint home; //pin home at axis
        CPoint eHome;//energy home
        CPoint cHome;//click home
        CRect rct;     //rectangular object
        int net;       //net where p belongs
        int DL[m+1]; //dogleg
    } PIN;
    PIN p[m+1];

    typedef struct
    {
        int b,e;
        int E[m+1];
        int order;
        int DL[m+1]; //doglegs
        SEGMENT Seg[m+1];
        int nSeg;
    } NET;
    NET q[m+1];

public:
    CCode10B();
    ~CCode10B()                {}
    afx_msg void OnPaint();
```

```
      afx_msg void OnLButtonDown(UINT,CPoint);
      afx_msg void OnReset();
      void Initialize();
      void MarkPin(int);
      void EnergyScheme();
      DECLARE_MESSAGE_MAP()
};

class CMyWinApp: public CWinApp
{
public:
      virtual BOOL InitInstance();
};
CMyWinApp MyApplication;

BOOL CMyWinApp::InitInstance()
{
      m_pMainWnd=new CCode10B;
      m_pMainWnd->ShowWindow(m_nCmdShow);
      return TRUE;
}

//Code10B.cpp
#include "Code10B.h"

BEGIN_MESSAGE_MAP(CCode10B,CFrameWnd)
      ON_WM_PAINT()
      ON_WM_LBUTTONDOWN()
      ON_BN_CLICKED (IDC_RESET,OnReset)
END_MESSAGE_MAP()

CCode10B::CCode10B()
{
      home=CPoint(30,200); end=CPoint(750,200);
      eHome=CPoint(home.x,400);
      cHome=CPoint(home.x,750);
      Create(NULL,L"Code10B: Single-row routing",
          WS_OVERLAPPEDWINDOW,CRect(0,0,1000,850));
      fArial.CreatePointFont(60,L"Arial");
      fHelvetica.CreatePointFont(100,L"Helvetica");
      Reset.Create(L"Reset",WS_CHILD | WS_VISIBLE | BS_DEFPUSHBUTTON,
          CRect(CPoint(end.x,60),CSize(100,30)),this,IDC_RESET);
      wInt=(end.x-home.x)/m;
      Initialize();
}

void CCode10B::OnPaint()
{
      CPaintDC dc(this);
      CString s;
      CRect rct;
      CBrush bRct(RGB(0,0,200));
      CPen pGray(PS_SOLID,1,RGB(200,200,200));
      CPen pDark(PS_SOLID,1,RGB(0,0,0));
```

```
        dc.SelectObject(pGray);
        for (int k=1;k<=m;k++)
        {
            dc.MoveTo(p[1].home.x,p[1].home.y+2);
            dc.LineTo(p[m].home.x,p[m].home.y+2);
        }
        s.Format(L"Click on the nodes from left to right for pairings");
        dc.TextOutW(p[1].cHome.x,p[1].cHome.y+30,s);
        dc.SelectObject(pDark); dc.SelectObject(fArial);
        for (int k=1;k<=m;k++)
        {
            dc.FillRect(p[k].rct,&bRct);
            s.Format(L"%d",k);
            dc.TextOutW(p[k].home.x+3,p[k].home.y+5,s);
            rct=CRect(CPoint(p[k].cHome),CSize(17,17));
            dc.Rectangle(rct);
            dc.TextOutW(p[k].cHome.x+3,p[k].cHome.y+5,s);
        }
}


void CCode10B::OnReset()
{
    CClientDC dc(this);
    CBrush bClear(RGB(255,255,255));
    CRect rct;
    GetClientRect(rct);
    dc.FillRect(rct,&bClear);
    Initialize(); Invalidate();
}


void CCode10B::Initialize()
{

    for (int i=1;i<=m;i++)
    {
        p[i].home=CPoint(30+wInt*(i-1),home.y);
        p[i].rct=CRect(CPoint(p[i].home.x-2,p[i].home.y),CSize(5,5));
        p[i].cHome=CPoint(p[i].home.x,cHome.y);
        p[i].DL[0]=0;           //#doglegs at pin i
    }
    bFlag=0; nNets=0; Q=0; E=0; D=0;
}


void CCode10B::OnLButtonDown(UINT nFlags,CPoint pt)
{
    CClientDC dc(this);
    CString s;
    CRect rct;
    for (int k=1;k<=m;k++)
    {
        rct=CRect(CPoint(p[k].cHome),CSize(17,17));
        if (rct.PtInRect(pt))
        {
            bFlag++;
```

```
                if (bFlag==1)
                {
                    nNets++;
                    q[nNets].b=k;
                    MarkPin(k);
                }
                if (bFlag==2)
                {
                    q[nNets].e=k;
                    s.Format(L"%d: (%d,%d)",nNets,q[nNets].b,q[nNets].e);
                    dc.TextOutW(end.x,100+20*(nNets-1),s);
                    MarkPin(k);
                    if (nNets==m/2)
                        EnergyScheme();
                    bFlag=0;
                }
            }
        }
}

void CCode10B::EnergyScheme()
{
    CClientDC dc(this);
    CString s;
    CPen pBlue(PS_SOLID,1,RGB(0,0,200));
    CPen pDark(PS_SOLID,1,RGB(0,0,0));
    int i,j,k;
    int w,r,max;
    double tmp;
    for (k=1;k<=nNets;k++)
    {
        q[k].order=k;    //order from top to bottom
        p[q[k].b].net=k; p[q[k].e].net=k; //begin,end belongs to net
        p[q[k].b].eHome=CPoint(p[q[k].b].home.x,eHome.y+30*(k-1));
        p[q[k].e].eHome=CPoint(p[q[k].e].home.x,eHome.y+30*(k-1));
        dc.SelectObject(pDark);
        dc.MoveTo(p[q[k].b].eHome); dc.LineTo(p[q[k].e].eHome);
        dc.SelectObject(fArial);
        s.Format(L"%d",q[k].b); dc.TextOutW(p[q[k].b].home.x-10,
                eHome.y-10+30*(k-1),s);
        s.Format(L"%d",q[k].e); dc.TextOutW(p[q[k].e].home.x+5,eHome.
                y+30*(k-1),s);
        dc.SelectObject(fHelvetica);
        s.Format(L"%d",k);
        dc.TextOutW(p[q[k].e].eHome.x+10,p[q[k].b].eHome.y-20,s);
    }
    dc.SelectObject(pBlue);
    dc.MoveTo(p[1].eHome);
    for (i=1;i<m;i++)
        dc.LineTo(p[i+1].eHome);
```

```
//Determine the doglegs of every net, partition every net into
        segments
for (k=1;k<=nNets;k++)
{
    q[k].nSeg=0;
    q[k].DL[0]=0; r=1;
    q[k].DL[r]=q[k].e;
    q[k].Seg[r].b=q[k].b;
    for (j=q[k].b+1;j<q[k].e;j++)
        if ((q[k].order<q[p[j].net].order && q[k].order>q[p[j+1].
            net].order)
                || (q[k].order>q[p[j].net].order && q[k].order
                    <q[p[j+1].net].order))
            {
                q[k].DL[r]=j; q[k].DL[0]++;
                w=++p[j].DL[0]; p[j].DL[w]=k;
                q[k].Seg[r].e=j+1;
                q[k].nSeg++;
                r++;
                q[k].Seg[r].b=j;
            }
    q[k].Seg[r].e=q[k].e; q[k].nSeg++;
    D +=q[k].DL[0];
}
//determine the energy of segment i of net k
for (k=1;k<=nNets;k++)
    for (i=1;i<=q[k].nSeg;i++)
    {
        max=0;
        for (j=q[k].Seg[i].b+1;j<q[k].Seg[i].e;j++)
        {
            r=1;
            for (w=1;w<=nNets;w++)
                if (w!=k || w!=p[j].net)
                    if (p[j].home.x>p[q[w].b].home.x &&
                        p[j].home.x<p[q[w].e].home.x)
                        if ((q[w].order>q[k].order &&
                            q[w].order<q[p[j].net].order)
                                || (q[w].order<q[k].order &&
                                q[w].order>q[p[j].net].order))
                                r++;
            max=((max<r)?r:max);
        }
        q[k].Seg[i].E=((q[k].order<q[p[q[k].Seg[i].b+1].net].
            order)?max:-max);
        q[k].Seg[i].sb.y=p[q[k].Seg[i].b].home.y-20*q[k].Seg[i].E;
        q[k].Seg[i].se.y=q[k].Seg[i].sb.y;
        E+=abs(q[k].Seg[i].E);
        if (Q<abs(q[k].Seg[i].E))
            Q=abs(q[k].Seg[i].E);
    }
//compute the dogleg coordinates, draw the nets
for (k=1;k<=nNets;k++)
    for (i=1;i<=q[k].nSeg;i++)
        {
```

```
if (i==1)
{
    q[k].Seg[i].sb.x=p[q[k].b].home.x;
    q[k].Seg[i].se.x=p[q[k].DL[i]].home.x;
    if (q[k].DL[i]<q[k].Seg[i].e)
    {
        tmp=(double)(p[q[k].DL[i]].home.x+(home.y-
            q[k].Seg[i].se.y)
            *(p[q[k].DL[i]+1].home.x-p[q[k].DL[i]].
              home.x)
            /(q[k].Seg[i+1].sb.y-q[k].Seg[i].
              se.y));
        q[k].Seg[i].se.x=(int)tmp;
    }
}
if (i>1 && i<q[k].nSeg)
{
    q[k].Seg[i].sb.x=q[k].Seg[i-1].se.x;
    q[k].Seg[i].se.x=p[q[k].DL[i]].home.x;
    if (q[k].DL[i]<q[k].Seg[i].e)
    {
        tmp=(double)(p[q[k].DL[i]].home.x+(home.y-
            q[k].Seg[i].se.y)
            *(p[q[k].DL[i]+1].home.x-p[q[k].DL[i]].
              home.x)
            /(q[k].Seg[i+1].sb.y-q[k].Seg[i].
              se.y));
        q[k].Seg[i].se.x=(int)tmp;
    }
}
if (i==q[k].nSeg)
{
    if (i>1)
    {
        q[k].Seg[i].sb.x=q[k].Seg[i-1].se.x;
        q[k].Seg[i].se.x=p[q[k].e].home.x;
    }
    dc.SelectObject(fHelvetica);
    s.Format(L"%d",k);
    dc.TextOutW(q[k].Seg[i].se.x+5,q[k].Seg[i].
        se.y-5,s);
}
dc.SelectObject(fArial); dc.SetTextColor
    (RGB(200,0,0));
s.Format(L"%d",q[k].Seg[i].E);
dc.TextOutW((q[k].Seg[i].sb.x+q[k].Seg[i].se.x)/2,
    eHome.y-10+30*(k-1),s);
dc.SetTextColor(RGB(0,0,0));
dc.MoveTo(q[k].Seg[i].sb.x,home.y);
dc.LineTo(q[k].Seg[i].sb); dc.LineTo(q[k].Seg[i].se);
dc.LineTo(q[k].Seg[i].se.x,home.y);
}
```

```
        dc.SelectObject(fHelvetica);
        s.Format(L"E=%d, Q=%d, D=%d",E,Q,D);
        dc.TextOutW(end.x+30,cHome.y,s);
}

void CCode10B::MarkPin(int k)
{
        CClientDC dc(this);
        CString s;
        CRect rct;
        CPen pNode[m+1];
        int Color[]={0,RGB(200,0,0),RGB(0,200,0),RGB(0,0,200),
                    RGB(200,200,0),RGB(200,0,200),
                    RGB(0,200,200),RGB(50,200,50),RGB(200,50,50),
                    RGB(50,200,50),RGB(100,200,50)};
        for (int i=1;i<=m/2;i++)
            pNode[i].CreatePen(PS_SOLID,2,Color[i]);
        rct=CRect(CPoint(p[k].cHome),CSize(17,17));
        dc.SelectObject(pNode[nNets]);
        dc.Rectangle(rct);
        dc.SelectObject(fArial);
        s.Format(L"%d",k);
        dc.TextOutW(p[k].cHome.x+5,p[k].cHome.y+3,s);
}
```

References

Chapter 1

1. J. L. Gross and J. Yellen (2005), *Graph Theory and Its Applications*, Second Edition, CRC Press, Boca Raton, FL.
2. J. L. Gross and J. Yellen (eds.) (2003), *Handbook of Graph Theory*, CRC Press, Boca Raton, FL.
3. F. Harary (1972), *Graph Theory*, Addison-Wesley Longman, Reading, MA.

Chapter 2

1. S. Salleh, A. Y. Zomaya, S. Olariu and B. Sanugi (2005), *Numerical Simulations and Case Studies Using Visual C++.Net*, Wiley-Interscience, Hoboken, NJ.
2. S. Salleh, A. Y. Zomaya and S. Abu Bakar (2008), *Computing for Numerical Methods Using Visual C++*, Wiley-Interscience, Hoboken, NJ.
3. R. M. Jones (2000), *Introduction to MFC Programming with Visual C++*, Prentice-Hall, Upper Saddle River, NJ.

Chapter 3

1. D. E. Van de Bout and T. K. Miller (1990), Graph partitioning using neural network, *IEEE Trans. Neural Netw.*, 1(2): 192–202.
2. F. Harary (1994), *Graph Theory*, Westview Press, Boulder, CO (paperback).
3. J. L. Gross and J. Yellen (2005), *Graph Theory and its Applications*, Second Edition, CRC Press, Boca Raton, FL.
4. Z. Zhou, C. Li, C. Huang and R. Xu (2014), An exact algorithm with learning for the graph coloring problem, *Comput. Oper. Res.*, 51: 282–301.
5. I. Blochliger and N. Zufferey (2008), A graph coloring heuristic using partial solutions and a reactive tabu scheme, *Comput. Oper. Res.*, 35(3): 960–975.
6. S. M. Douiri and S. Elbernoussi (2015), Solving the graph coloring problem via hybrid genetic algorithm, *J. King Saud University—Engineering Sciences*, 27(1): 114–118.
7. D. D. Liu (1992), T-colorings of graphs, *Discrete Math.*, 101(1): 203–212.
8. A. Graf (1999), Distance graphs and T-coloring problem, *Discrete Math.*, 196(1): 153–166.
9. B. Aspvall and J. R. Gilbert (1983), Graph coloring using eigenvalue decomposition, Technical Report #83-545, Department of Computer Science, Cornell University.
10. R. L. Burden and J. D. Faires (2004), *Numerical Analysis*, Eighth Edition, Brooks Cole Publishing, Belmont, CA.
11. S. Salleh, A. Y. Zomaya and S. Abu Bakar (2008), *Computing for Numerical Methods Using Visual C++*, Wiley-Interscience, Hoboken, NJ.

Chapter 4

1. W. Jin, S. Chen and H. Jiang (2013), Finding the k shortest paths in a time-schedule network with constraints on arcs, *Comput. Oper. Res.*, 40(12): 2975–2982.
2. S. Hess, M. Quddus, N. Rieser-Schussler and A. Daly (2015), Developing advanced route choice models for heavy goods vehicles using GPS data, *Transp. Res. E-Log.*, 77: 29–44.
3. L. Hsu and C. Lin (2009), *Graph Theory and Interconnection Networks*, CRC Press, Boca Raton, FL.
4. D. T. Lee, C. D. Yang and C. K. Wong (1996), Rectilinear paths among rectilinear obstacles, *Discrete Appl. Math.*, 70(3): 185–215.

Chapter 5

1. W. Kocay and D. L. Kreher (2004), *Graphs, Algorithms, and Optimization*, CRC Press, Boca Raton, FL.
2. S. Pemmaraju and S. Skiena (2003), Minimum spanning tree, in *Computational Discrete Mathematics: Combinatorics and Graph Theory in Mathematica*, Cambridge University Press, Cambridge, England, pp. 335–336.
3. S. Skiena (1990), Minimum spanning tree, in *Implementing Discrete Mathematics: Combinatorics and Graph Theory with Mathematica*, Addison-Wesley, Redwood City, CA, pp. 232–236.
4. M. Frodigh, P. Johansson and P. Larsson (2000), Wireless ad hoc networking—The art of networking without a network, *Ericsson Rev.* (English edition), 77(4): 248–263.

Chapter 6

1. J. D. Eblen, C. A. Phillips, G. L. Rogers and M. A. Langston (2012), The maximum clique enumeration problem: Algorithms, applications, and implementations, *BMC Bioinformatics*, 13(Suppl. 1) Suppl. 10: S5.
2. V. Boginski, S. Butenko and P. M. Pardalos (2003), On structural properties of the market graph, *Innov. Financ. Econ. Networks*, 29–45.
3. J. Pattillo, N. Youssef and S. Butenko (2012), Clique relaxation models in social network analysis, in *Handbook of Optimization in Complex Networks*, M. T. Thai and P. M. Pardalos (eds.), Springer, New York, pp. 143–162.
4. S. R. Bulo and M. Pelillo (2010), A new spectral bound on the clique number of graphs, in *Structural, Syntactic and Statistical Pattern Recognition, Lecture Notes in Computer Science*, E. Hancock, R. Wilson, T. Windeatt, I. Ulusoy and F. Escolano (eds.), Berlin, Heidelberg: Springer Berlin Heidelberg, pp. 680–689.
5. R. A. Rossi, D. F. Gleich, A. H. Gebremedhin and Md. Mostofa Ali Patwary (2013), *Parallel Maximum Clique Algorithms with Applications to Network Analysis and Storage*, Cornell University Library, arXiv preprint arXiv:1302.6256.
6. J. Ugander, B. Karrer, L. Backstrom and C. Marlow (2011), *The Anatomy of the Facebook Social Graph*, Cornell University Library, arXiv preprint arXiv:1111.4503.
7. S. E. Schaeffer (2007), Graph clustering, *Comput. Sci. Rev.*, 1(1): 27–64.
8. M. Oliveira and J. Gama (2012), An overview of social network analysis. *Wiley Interdisciplinary Reviews: Data Mining and Knowledge Discovery*, 2(2): 99–115. doi:10.1002/widm.1048.

Chapter 7

1. B. Delaunay (1934), Sur la sphere vide, *Bull. Acad. Sci. USSR VII Class. Sci. Mat. Nat.*, *Classe des sciences mathématiques et na*, 6: 793–800.
2. R. A. Jarvis (1973), On the identification of the convex hull of a finite set of points in the plane, *Inf. Process. Lett.*, 2: 18–21.
3. R. L. Graham (1972), An efficient algorithm for determining the convex hull of a finite planar set, *Inf. Process. Lett.*, 1: 132–133.
4. M. J. Fadili, M. Melkemi and A. El Moataz (2004), Non-convex onion-peeling using a shape hull algorithm, *Pattern Recognit. Lett.*, 25(14): 1577–1585.
5. A. Bowyer (1981), Computing Dirichlet tessellations, *Comput. J.*, 24(2): 162–166.

Chapter 8

1. A. S. Jain and S. Meeran (1999), Deterministic job-shop scheduling: Past, present and future, *Eur. J. Op. Res.*, 113(2): 390–434.
2. A. Allahverdi, C. T. Ng, T. C. E. Cheng and M. Y. Kovalyov (2008), A survey of scheduling problems with setup times or costs, *Eur. J. Op. Res.*, 187(3): 985–1032.
3. S. DasBit and S. Mitra (2003), Challenges of computing in mobile cellular environment, *Comput. Commun.*, 26(18): 2090–2105.
4. R. Hwang, M. Gen and H. Katayama (2008), A comparison of multiprocessor task scheduling algorithms with communication costs, *Comput. Op. Res.*, 35(3): 976–993.
5. M. Drozdowski (1996), Scheduling multiprocessor tasks, *Eur. J. Op. Res.*, 94(2): 215–230.
6. H. El-Rewini, T. G. Lewis and H. H. Ali (1994), *Task Scheduling in Parallel and Distributed Systems*, Prentice-Hall, Englewood Cliffs, NJ.
7. A. Agarwal, S. Colak, V. S. Jacob and H. Pirkul (2006), Heuristics and augmented neural networks for task scheduling with non-identical machines, *Eur. J. Op. Res.*, 175(1): 296–317.
8. F. A. Omara and M. M. Arafa (2010), Genetic algorithms for task scheduling problem, *J. Parallel Distrib. Comput.*, 70(1): 13–22.
9. S. Salleh and A. Y. Zomaya (1998), Multiprocessor scheduling using mean-field annealing, *Future Gener. Comput. Syst.*, 14(5): 393–408.
10. S. Salleh, A. Y. Zomaya, S. Olariu and B. Sanugi (2005), *Numerical Simulations and Case Studies Using Visual C++.Net*, Wiley-Interscience, Hoboken, NJ.

Chapter 9

1. I. Maherin and Q. Liang (2014), Radar sensor network for target detection using Chernoff information and relative entropy, *Phys. Com.*, 13(3): 244–252.
2. M. A. Guvensen and A. G. Yavuz (2011), On coverage issues in directional sensor networks, *Ad Hoc Netw.*, 9(7): 1238–1255.
3. H. Mohamadi, S. Salleh and M. N. Razali (2014), Heuristic methods to maximize network lifetime in directional sensor networks, *J. Netw. Comput. Appl.*, 46: 26–35.
4. L. Cheng, C. Wu, Y. Zhang, H. Wu, M. Li and C. Maple (2012), A survey of localization in wireless sensor network, *Int. J. Distrib. Sens.*, N. 1–12.

5. H. Mohamadi, A. S. Ismail and S. Salleh (2014), Solving target coverage problem using cover sets in wireless sensor networks based on learning automata, *Wireless Pers. Commun.*, 75: 447–463.

6. Y. Chen, X. Hu, H. Yang and L. Ge (2013), Power control routing algorithm for maximizing lifetime in wireless sensor networks, *Lecture Notes in Electrical Engineering: Advances in Mechanical and Electronic Engineering*, 178: 129–136.

7. K. Akkaya and M. Younis (2005), A survey on routing protocols for wireless sensor networks, *Ad Hoc Netw.*, 3: 325–349.

Chapter 10

1. R. Miller (1993), Parallel computations on reconfigurable meshes, *IEEE Trans. Comput.*, 42(6): 678–692.

2. S. Olariu, M. C. Pinotti and S. Zheng (1999), How to sort N items using a sorting network of fixed I/O size, *IEEE Trans. Parallel Distrib. Syst.*, 10(5): 487–499.

3. H. Giefers and M. Platzner (2014), An FPGA-based reconfigurable mesh many-core, *IEEE Trans. Comput.*, 63(12): 2919–2932.

4. T. Fujimori and M. Watanabe (2015), Parallel-operation-oriented optically reconfigurable gate array, *Lect. Notes Comput. Sci.*, 9017: 3–14.

5. H. C. So (1974), Some theoretical results on the outing of multilayer printed wiring board, *Proc. IEEE Symp. Circuits Systems*, pp. 296–303.

6. R. Raghavan and S. Sahni (1983), Single row routing, *IEEE Trans. Comput.*, 32(3): 209–220.

7. J. S. Deogun and N. A. Sherwani, A decomposition scheme for single-row routing problems, Technical Report Series #68, Department of Computer Science, University of Nebraska, 1988.

8. T. T. Tarng, M. M. Sadowska and E. S. Kuh (1984), An efficient single-row algorithm, *IEEE Trans. Comput. Aid. D.*, 3(3): 178–183.

9. D. H. Du and L. H. Liu (1987), Heuristic algorithms for single row routing, *IEEE Trans. Comput.*, 36(3): 312–319.

10. S. Salleh and A. Y. Zomaya (1999), *Scheduling for Parallel Computing Systems: Fuzzy and Annealing Techniques*, Kluwer Academic Publishers, Boston.

11. S. Salleh, B. Sanugi, H. Jamaluddin, S. Olariu and A. Y. Zomaya (2002), Enhanced simulated annealing technique for the single-row routing problem, *J. Supercomput.*, 21(3): 285–302.

Index